Data Acquisition and Processing in Cultural Heritage

Data Acquisition and Processing in Cultural Heritage

Special Issue Editors

Gabriele Bitelli
Fulvio Rinaudo
Diego Gonzalez-Aguilera
Pierre Grussenmeyer

MDPI • Basel • Beijing • Wuhan • Barcelona • Belgrade

Special Issue Editors

Gabriele Bitelli
University of Bologna
Italy

Fulvio Rinaudo
Politecnico di Torino—DAD
Italy

Diego Gonzalez-Aguilera
University of Salamanca
Spain

Pierre Grussenmeyer
Photogrammetry and
Geomatics Group
France

Editorial Office
MDPI
St. Alban-Anlage 66
4052 Basel, Switzerland

This is a reprint of articles from the Special Issue published online in the open access journal *ISPRS International Journal of Geo-Information* (ISSN 2220-9964) from 2018 to 2019 (available at: https://www.mdpi.com/journal/ijgi/special_issues/Cultural_Heritage_IJGI).

For citation purposes, cite each article independently as indicated on the article page online and as indicated below:

LastName, A.A.; LastName, B.B.; LastName, C.C. Article Title. *Journal Name* **Year**, *Article Number*, Page Range.

ISBN 978-3-03921-740-3 (Pbk)
ISBN 978-3-03921-741-0 (PDF)

Contents

About the Special Issue Editors . . . . **vii**

Preface to "Data Acquisition and Processing in Cultural Heritage" . . . . **ix**

Caterina Balletti and Martina Ballarin
An Application of Integrated 3D Technologies for Replicas in Cultural Heritage
Reprinted from: *ISPRS Int. J. Geo-Inf.* **2019**, *8*, 285, doi:10.3390/ijgi8060285 . . . . **1**

Miaole Hou, Su Yang, Yungang Hu, Yuhua Wu, Lili Jiang, Sizhong Zhao and Putong Wei
Novel Method for Virtual Restoration of Cultural Relics with Complex Geometric Structure
Based on Multiscale Spatial Geometry
Reprinted from: *ISPRS Int. J. Geo-Inf.* **2018**, *7*, 353, doi:10.3390/ijgi7090353 . . . . **30**

Andrea Piemonte, Gabriella Caroti, Isabel Martínez-Espejo Zaragoza, Filippo Fantini and Luca Cipriani
A Methodology for Planar Representation of Frescoed Oval Domes: Formulation and Testing
on Pisa Cathedral
Reprinted from: *ISPRS Int. J. Geo-Inf.* **2018**, *7*, 318, doi:10.3390/ijgi7080318 . . . . **50**

Filiberto Chiabrando, Giulia Sammartano, Antonia Spanò and Alessandra Spreafico
Hybrid 3D Models: When Geomatics Innovations Meet Extensive Built Heritage Complexes
Reprinted from: *ISPRS Int. J. Geo-Inf.* **2019**, *8*, 124, doi:10.3390/ijgi8030124 . . . . **71**

Arnadi Murtiyoso, Pierre Grussenmeyer, Deni Suwardhi and Rabby Awalludin
Multi-Scale and Multi-Sensor 3D Documentation of Heritage Complexes in Urban Areas
Reprinted from: *ISPRS Int. J. Geo-Inf.* **2018**, *7*, 483, doi:10.3390/ijgi7120483 . . . . **101**

Bashar Alsadik, Nagham Amer Abdulateef and Yousif Husain Khalaf
Out of Plumb Assessment for Cylindrical-Like Minaret Structures Using Geometric
Primitives Fitting
Reprinted from: *ISPRS Int. J. Geo-Inf.* **2019**, *8*, 64, doi:10.3390/ijgi8020064 . . . . **121**

Young Hoon Jo and Seonghyuk Hong
Three-Dimensional Digital Documentation of Cultural Heritage Site Based on the Convergence
of Terrestrial Laser Scanning and Unmanned Aerial Vehicle Photogrammetry
Reprinted from: *ISPRS Int. J. Geo-Inf.* **2019**, *8*, 53, doi:10.3390/ijgi8020053 . . . . **142**

Grazia Tucci, Erica Isabella Parisi, Giulio Castelli, Alessandro Errico, Manuela Corongiu, Giovanna Sona, Enea Viviani, Elena Bresci and Federico Preti
Multi-Sensor UAV Application for Thermal Analysis on a Dry-Stone Terraced Vineyard in Rural
Tuscany Landscape
Reprinted from: *ISPRS Int. J. Geo-Inf.* **2019**, *8*, 87, doi:10.3390/ijgi8020087 . . . . **156**

Pablo Rodríguez-Gonzálvez, Ángel Guerra Campo, Ángel L. Muñoz-Nieto, Luis J. Sánchez-Aparicio and Diego González-Aguilera
Diachronic Reconstruction and Visualization of Lost Cultural Heritage Sites
Reprinted from: *ISPRS Int. J. Geo-Inf.* **2019**, *8*, 61, doi:10.3390/ijgi8020061 . . . . **177**

Jihn-Fa Jan
Application of Open-Source Software in Community Heritage Resources Management
Reprinted from: *ISPRS Int. J. Geo-Inf.* **2018**, *7*, 426, doi:10.3390/ijgi7110426 . . . . **202**

Piotr Siekański, Jakub Michoński, Eryk Bunsch and Robert Sitnik
CATCHA: Real-Time Camera Tracking Method for Augmented Reality Applications in Cultural
Heritage Interiors
Reprinted from: *ISPRS Int. J. Geo-Inf.* **2018**, *7*, 479, doi:10.3390/ijgi7120479 . . . . **221**

Chris Panou, Lemonia Ragia, Despoina Dimelli and Katerina Mania
An Architecture for Mobile Outdoors Augmented Reality for Cultural Heritage
Reprinted from: *ISPRS Int. J. Geo-Inf.* **2018**, *7*, 463, doi:10.3390/ijgi7120463 . . . . **240**

About the Special Issue Editors

Gabriele Bitelli is Full Professor of Geomatics at Alma Mater Studiorum—University of Bologna, Italy, at the Department of Civil, Chemical, Environmental, and Materials Engineering. His research interests are related to the applications of integrated geospatial techniques for cultural heritage surveying and management, topographic and photogrammetric monitoring, remote sensing for environmental analysis and change detection, deformation control for structures and territory, 3D modeling, thermal mapping, and digital approaches for historical cartography. He has been involved in a number of EU and national research projects, and he is author of over 300 scientific publications in international and national journals or conference proceedings.

Fulvio Rinaudo is a Full Professor of Geomatics in the Department of Architecture and Design (Politecnico di Torino).

Professor Rinaudo graduated in Civil Engineering, and received his Ph.D. in Geodetic and Cartographic Sciences in 1990. He is an author of more than 290 scientific papers, and his research activity is mainly oriented toward the applications of photogrammetry and scanner systems, GIS, and HBIM design with particular attention to cultural heritage assets.

Since 2005 he has been an active member of CIPA-HD (ICOMOS Int. Sc. Comm. for Cultural Heritage Documentation), and since 2008 he has been chair of the "Data Acquisition and Processing in Cultural Heritage" Working Group of the International Photogrammetry Society (ISPRS). He teaches Architecture Metric Survey as well as GIS & Modelling of Cultural Heritage at the Polytechnic University of Turin and at the School of Specialization in Architectural Heritage and Landscape; Heritage Documentation at the Catholic University of Louvain (Belgium); and Land Surveying at Turin Tashkent Polytechnic University (Uzbekistan). He is member of the Teaching Committee of the International Ph.D. School "Technicians for Culture" at the University of Turin.

Diego Gonzalez-Aguilera (born 9 April 1976, Ávila, Spain). He is a Full Professor at the University of Salamanca. He was Director of the Cartographic and Land Engineering Department from 2012 to 2019. He holds a PhD in Photogrammetry and Computer Vision from the University of Salamanca (2005), an M.Sc. degree in Geodesy and Cartography Engineering (2001), and a B.Sc. degree in Surveying Engineering (1999). He is the Director of the TIDOP Research Unit (http://tidop.usal.es), which was created in 2005, and is also the cofounder and CTO of the ITOS3D start-up (http://itos3d.com).

Dr. Gonzalez-Aguilera has been a Visiting Professor at the Computer Vision and Robotics Institute (INRIA) in Grenoble (France) (2005), Visiting Professor at the University of Massachusetts (Boston, USA) (2014), and at Dublin City University (DCU) (2017).

Dr. Gonzalez-Aguilera is Editor-in-Chief of the MDPI journal Drones (https://www.mdpi.com/journal/drones), Co-chair of the International Society of Photogrammetry and Remote Sensing, Commission II-WG II/8, and Director of the Master's degree in Geotechnologies Applied to Engineering and Architecture at the University of Salamanca. He has authored more than 150 impact journal JCR articles and is a co-inventor of 11 patents and 23 intellectual properties (software). Based on his research, he has received eight international awards from the more important institutions in remote sensing (ISPRS/ASPRS/CIPA-ICOMOS).

Dr. Gonzalez-Aguilera works on projects in the fields of Computer Vision, Photogrammetry, Remote Sensing, Laser Scanning, 3D Modelling, and Machine Learning, leading more than 15 research projects per year.

Pierre Grussenmeyer is a Full Professor at INSA Strasbourg (Graduate School of Science and Technology), France. He received an Engineering degree (Master's) in Geodetic Surveying and Topography in 1984 from ENSAIS, currently INSA Strasbourg, and a Ph.D. degree in photogrammetry in 1994 from the University of Strasbourg (in relationship with IGN, Paris). He founded the Photogrammetry and Geomatics Group at INSA Strasbourg (France) in 1996 and has led this group since then. His group is affiliated with the ICUBE UMR 7357 Laboratory (The Engineering science, computer science and imaging laboratory) of the University of Strasbourg. He has been VP for Research at INSA Strasbourg since 2019. His current research interests include close-range photogrammetry (civil engineering and metrology), architectural photogrammetry and laser scanning (cultural heritage documentation, Heritage-BIM), mobile recording systems, integration, and accuracy of data in 3D city and building models (BIM).

Preface to "Data Acquisition and Processing in Cultural Heritage"

Advances in the knowledge of the tangible components (position, size, shape) and intangible components (identity, habits) of an historic building or site involves fundamental and complex tasks in any project related to the conservation of cultural heritage (CH). In recent years, new geotechnologies have proven their usefulness and added value to the field of cultural heritage (CH) in the tasks of recording, modeling, conserving, and visualizing. In addition, current developments in building information modeling (HBIM), allow integration and simulation of different sources of information, generating a digital twin of any complex CH construction. As a result, experts in the area have increased the number of available sensors and methodologies. However, the quick evolution of geospatial technologies makes it necessary to revise their use, integration, and application in CH. This process is difficult to adopt, due to the new options which are opened for the study, analysis, management, and valorization of CH. Therefore, the aim of the present Special Issue is to cover the latest relevant topics, trends, and best practices in geospatial technologies and processing methodologies for CH sites and scenarios as well as to introduce the new tendencies. This book originates from the Special Issue "Data Acquisition and Processing in Cultural Heritage", focusing primarily on data and sensor integration for CH; documentation/restoration in CH; heritage 3D documentation and modeling of complex CH sites; drone inspections in CH; software development in CH; and augmented reality in CH. It is hoped that this book will provide the advice and guidance required for any CH professional, making the best possible use of these sensors and methods in CH.

Gabriele Bitelli, Fulvio Rinaudo, Diego Gonzalez-Aguilera, Pierre Grussenmeyer
Special Issue Editors

 International Journal of
Geo-Information

Article

An Application of Integrated 3D Technologies for Replicas in Cultural Heritage

Caterina Balletti * and Martina Ballarin

Università Iuav di Venezia, S. Croce 191, 30135 Venezia, Italy; martinab@iuav.it
* Correspondence: balletti@iuav.it

Received: 28 March 2019; Accepted: 15 June 2019; Published: 18 June 2019

Abstract: In recent decades, 3D acquisition by laser scanning or digital photogrammetry has become one of the standard methods of documenting cultural heritage, because it permits one to analyze the shape, geometry, and location of any artefact without necessarily coming into contact with it. The recording of three-dimensional metrical data of an asset allows one to preserve and monitor, but also to understand and explain the history and cultural heritage shared. In essence, it constitutes a digital archive of the state of an artefact, which can be used for various purposes, be remodeled, or kept safely stored. With the introduction of 3D printing, digital data can once again take on material form and become physical objects from the corresponding mathematical models in a relatively short time and often at low cost. This possibility has led to a different consideration of the concept of virtual data, no longer necessarily linked to simple visual fruition. The importance of creating high-resolution physical copies has been reassessed in light of different types of events that increasingly threaten the protection of cultural heritage. The aim of this research is to analyze the critical issues in the production process of the replicas, focusing on potential problems in data acquisition and processing and on the accuracy of the resulting 3D printing. The metric precision of the printed model with 3D technology are fundamental for everything concerning geomatics and must be related to the same characteristics of the digital model obtained through the survey analysis.

Keywords: cultural heritage; replica; photogrammetry; laser-scanning; 3D printing

---

1. Introduction

The use of sensors as well as 3D acquisition and modeling techniques became routine in the industrial production process during the 1990s, and the related technologies have since experienced strong growth. This occurred because of the use of the techniques in "rapid prototyping" (RP) for the generation of a physical part from a digital model created using CAD systems, a practice that offers remarkably reduced times and lower costs in the production process.

Great advances have been made in these techniques in recent years, thanks to the widespread market penetration of low-cost desktop 3D printers. The rapid rise of these technologies extends even to "consumer" applications, admitting new users to enter the world of 3D printing. However, practitioners in cultural heritage (CH) tend to apply slightly different processes in arriving at the physical reproduction. The more typical industrial practice is to produce a replica synthetically, ex-novo, and then evaluate the discrepancies or problems in relation to the initial theoretical model. In CH the more common process is "reverse modeling," which begins from the existing physical object. From this, the practitioner acquires a set of 3D data, used to generate a digital model through a series of steps. The operator can also potentially declare the copy's degree of reliability, or the level of fidelity to the original, through validation of each step in the process.

The current work is organized in two sections. The first provides a brief review of the state of 3D printing, considering the subtractive and additive techniques most commonly used and the different areas in which they are applied. The review devotes particular attention to the 3D technologies used in the field of CH and to the significance and applications of 3D printing in conservation/restoration and museum use. The second part describes two applications drawn from a series of cases dealt with over the course of the past year and selected on the basis of their ability to illustrate the main problems in the technological path. The examples deal with the acquisition of data from two objects of different shapes and sizes, using different methodologies and tools, which then lead to different methods of data processing. The purpose of the case studies is to analyze the criticalities in the path from the surveying to the physical representation of the object, concentrating on potential problems in data acquisition and processing and then in the metric precision of the resulting 3D printing.

The selection of case studies excluded large objects (e.g., architectural scale) a priori, since these would very rarely be printed at a true scale, and any examples that may be known, by literature or web, concern the creation of new objects and not replicas of existing ones. Given that the aim is to analyze the potential precision of the technological process, from point cloud to solid print, i.e., the fidelity of the copy (a scale of 1:1) to the original, the selection instead concentrated on objects that allow for direct, measured verification of the accuracy of the copy. Finally, the research presented here focuses on the 3D printing technique known as selective laser sintering (SLS), which affords high-quality results at relatively low costs. The current paper does not report on other types of printing, which on the basis of years of experience and previous analysis are considered less satisfactory [1].

2. Overview

2.1. 3D Printing Techniques

The so-called "digital manufacturing" techniques are classified into two main groups: subtractive and additive. Additive techniques, also called RP, first originated in the late 1980s, but have recently undergone extraordinary development [2,3].

Various authors have already published extensive research on the use of solid printing techniques in CH contexts [4]. This part of the paper therefore provides only an overview of the different techniques available.

The names of the two macro groups indicate their fundamental differences: "subtractive" techniques involve the removal of material to generate the physical model, applying processes similar to those of sculpture; "additive" techniques involve the deposition or solidification of material within a printing chamber.

The devices used in subtractive techniques are classified as "lathes" or "milling machines," reflecting the traditional instruments from which they derive. Both types of devices are operated "automatically," relying on computer-aided manufacturing (CAM) software that translates the digital file model into a series of commands that can be interpreted by the machine.

The structures inherent to subtractive devices imply that in most cases the instrument can only move along two axes, which gives rise to certain restrictions. This means that they are best suited to production of the so-called "2.5D" models, or to the reproduction of objects without strong three-dimensionality, such as bas-reliefs. Some geometries cannot be used in computer numerical control (CNC) machining (even with 5-axis CNC systems), as such tools cannot access all the surfaces of a component. There are a number of limitations, including tool access and clearances, hold or mount points, as well as the inability to machine square corners due to tool geometry [5].

The additive technique achieves a true "3D printing" of objects, through successive deposition of layers of material. The production of the printed model requires the application of "slicing software" to a prior digital model, which breaks it down into the necessary sequence of sections. The resolution of the details depends both on the thickness of the layer and the dimensions of the movement imposed on the tool: the smaller the movement of the governing motors, the greater the degree of detail.

The additive and subtractive processes are otherwise similar: both require dedicated software capable of translating the digital data (surface model) into a series of machine commands, through creation of a specific numerical "G-Code." The choice of technique depends on different factors, such as the geometry, size, and materials. CNC milling is ideal when manufacturing workpieces need to be extremely robust, precise, and/or heat-resistant and can handle the post-processing required by most 3D printers and the refining and cleaning up of the final part. Additive techniques are more versatile and allow for the reproduction of highly complex, topology-optimized geometries with finer detail, such as hollows and enclosures, which cannot be produced using any of the subtractive technologies. All additive techniques share the above features, but they differ in the way the sections take form and in the variety of materials that can be used, resulting in varying quality and resolution in the final product.

Considering the pros and cons, it is best to think of all of them as complementary technologies, working individually and together to enhance the manufacturing ecosystem in their own ways.

2.2. 3D Technologies for Heritage Purposes

The development of 3D technologies has led to important results in the conservation, enhancement, and use of CH, confirmed by major projects such as Europeana, the European digital library, and "Horizon 2020 Reflective Societies 7-2014: Advanced 3D modeling for accessing and understanding European cultural assets." Countless interdisciplinary projects involving computer graphics experts, industrial engineers, and personnel specialized in the documentation, conservation, and use of CH, have also achieved excellent results.

Solid printing is one of the latest developments within the range of technologies applied to the CH context. However, it is only one step in the extensive process of data acquisition and management: the final moment of the workflow, from recording the shape of the object to its physical reproduction. Because it is part of this process, the technologies of solid printing are closely linked to those for metric data acquisition (point cloud production) and digital data management (virtual modeling). As a result, the literature on digital modeling applications tends to be intermingled with that on solid printing. In fact, the more general term "3D technologies" now covers a wide range of tools, methods, and applications of modeling and printing.

The macro group of 3D technologies includes, first of all, the tools for acquisition of three-dimensional metric data. Scholars, researchers, and curators increasingly turn to the tools of triangulation-based laser scanners, time of flight or phase-based laser scanners, and digital photogrammetry, for recording and analysis of the shape and geometry of monuments, archaeological finds, and artwork.

In addition to recording shape, 3D tools and methods offer a further advantage of positioning the object in space. This characteristic is highly useful in relation to CH, since it allows for analysis of the relations among objects within a context, such as an urban area, archaeological site, or monument, or even within a single object, for example, for purposes of reconstructing portions of a broken statue, or decorative elements on the surface of a vase.

The use of printing processes based on metric surveying enables production of objects that replicate the true shapes. The "point cloud," meaning the digital point model obtained by photogrammetry or laser scanning, can be saved and archived, providing a record of the true state of an object in a given space–time context, and these models can then also be modified or replicated as desired. Researchers often use these three-dimensional digital models and printed replicas for purposes of data collection and analysis, thereby avoiding physical contact with sensitive objects and exhibits. Researchers and technical experts can also share digital models on a global scale, and the public can enjoy museum collections in richer ways, through the experience of physical replicas [6–9].

3D and graphics technologies have also recently been used in the modeling, rendering, and creation of synthetic images and videos for the production of virtual tours, enabling access to the CH of specific sites and museum exhibitions for the broad public. The videogame technologies serving as the basis of these techniques result in fluid and realistic interaction with the models, making them a useful cognitive tool, as described in [10]. The various applications of the tools allow researchers, students, and even those with casual interests to move in a virtual environment and interact with the models.

These same technologies are applied as vehicles for conservation, reconstruction, documentation, research, and promotion within the broad world of CH.

Normally, for a virtual reconstruction to succeed in an application, it must adhere to the form and proportions of the original object. As also stated by Bruno et al. [11], the application of 3D scanning techniques achieves the precise data that are necessary for these purposes [12]. The literature offers many examples of digitization for applications involving paintings (normally in two dimensions) and three-dimensional sculptures, buildings, and small objects, in materials ranging from wood to stone, marble, ceramics, and metal [13–16]. Data acquisition techniques, such as 3D scanning and above all digital photogrammetry [17], enable the creation of photo-realistic models, which in turn let the public to explore and enjoy cultural properties from a scale of tiny objects to massive areas [18–25].

The most recent arrival in the CH "developmental path" is the production of the printed model, which in turn relates to strong growth in other sectors, such as industrial design and production, medicine, and dentistry [26–28]. The growth of 3D printing in CH, while strong, has mostly been limited to applications involving exhibitions and other museum functions. Within this realm, most of the applications discussed in the literature concern the themes of casting and integrative restoration.

The production of scientifically accurate casts allows for the safeguarding of artistic and especially archaeological heritage from unforeseen failures and risks of inadequate conservation. Museum collections typically include many artworks and countless archaeologically recovered objects that cannot be exhibited for reasons of space, but which have great scientific and art-historical value. Many objects are displayed only after decades in storage, and others will never be presented to the public. The passage of time can have serious consequences on their state of preservation and obviously reflects on their functionality. The classic solution for the mitigation of risks, particularly for archaeological materials, is scientific casting, which can substitute the original object for purposes of study and analysis. However, the traditional casting process inevitably raises complexities of direct physical contact with the original objects, which are often fragile. The process therefore requires careful analysis of the object's physical characteristics, preservation status, and painstaking planning of the entire operation, including choices of materials and equipment, in order to avoid irreparable damage. This process is very costly in terms of time, human, and financial resources. Today, applying the methods of three-dimensional documentation, using laser scanners and photogrammetry, heritage scientists can acquire the necessary data and produce extremely accurate digital models. The resulting printed model is then at least equivalent to the traditional cast, but since it is achieved without directly contacting the object, the process ensures the improved protection of the original property [29–39].

Three-dimensional data-acquisition and casting technologies are now frequently applied in the field of conservation/restoration. Bigliardi et al. [40], for example, report on a pioneering research project involving the use of solid printing for restorative reintegration of architectural elements, specifically for the case of an "egg-and-dart" cornice of the Cristoforo Sorte room in the Palazzo Ducale of Mantua. Missing portions of the cornice were restored using RP: a well-preserved part was acquired by photogrammetry and then replicated, using a fused deposition modeling (FDM) printer with a polyactic acid filament (PLA) as the material. The replicated piece was then placed in one of the gaps. The digital model was also used for the production of a negative, or mould, which could then be used for traditional, larger-scale production of the decoration, using a soft lime-and-aggregate mortar. A particularly interesting aspect of this restoration project is the double application of 3D printing: in the production of a cast, a mould, and then the traditional plaster decoration; and in the more innovative one of direct use of the printed model. However, this second method is still subject to uncertainties, primarily because the models, especially when printed using the filament technique, are not considered of sufficient quality for art-historical applications, mainly because of the materials used. The replicas must therefore in most cases be coated with some other material, achieving greater similarity with the original. Secondly, given the recent origin of this type of application, the durability and resistance of the materials are also not fully known. These are important considerations, particularly in the area of conservation/restoration, where reversibility, compatibility, and duration of interventions are essential.

What is also important, however, is that the application of 3D technologies has resulted in the achievement of the objectives, within much less time than would have been possible using solely traditional methods, and has gained the necessary data without directly contacting the original materials.

The case of the Madonna di Pietranico, a terracotta statue damaged by the 2009 earthquake in Abruzzo, Italy, provides another interesting example [41]. The statue was reduced to a multitude of pieces and the application of digitization and 3D graphics techniques made it possible to visualize hypotheses for a reassembly of the pieces, without the need to touch or manipulate them, thereby risking further damage. Although a full decade has now passed since the execution of this project, it remains interesting because of the way it drew on a full range of 3D technologies, and it is still considered an important milestone of "computerized" restoration applications.

2.3. Solid Printing in Museum Exhibitions

As noted, the most rapid and extensive diffusion of 3D technologies, within the overall area of CH, has been in museums. In this sector, the concept of "usability" of CH has always been manifested: a reality that relates well the potentials of 3D technologies for the attraction and engagement of a very wide public, including even those who would not be regular visitors. The most avant-garde institutions now recognize modern technologies as an opportunity to add new "reading" tools to their traditional visitor paths (e.g., the Louvre in Abu Dhabi and the Modigliani project at the Tate Modern in London [42,43]). The aim is to move the traditionally conceived museum exhibit towards a multi-level and multi-sensory experience. Alongside art objects in traditional display cases, modern museums increasingly provide interpretive supports, offering new ways of interaction. Monitors, projectors, and physical replicas are used to add information, but also to transform the way in which the user approaches art itself [44,45].

A further consideration is that, rather than relying only on sight, the museum visitor could also engage in the equally important senses of touch, smell, and hearing. Recent discoveries in the fields of cognitive psychology and neuroscience [46,47] have in fact altered our conception of learning and of multisensory experience itself. We now speak of "situated learning," indicating that our ways of thinking and our learning results are defined by the space in which we move, the environment in which we live, the social context that surrounds us, and our actions. This conception enables progression beyond the traditional paradigm of "education," in which an expert transmits his knowledge to a passive audience [48,49]. The visitors to a museum can engage personally in the cognitive process, creating knowledge themselves. Given that our knowledge of our surrounding reality depends on all the senses at our disposal, the museum, as a vehicle of knowledge, cannot fail to take this aspect into consideration [50,51].

In the wake of these theories, museums are devoting particular attention to "touch." Because it presents risks to their preservation, touching the objects in museums has typically not been allowed. However, we are witnessing new attempts to allow this type of use through the creation of displays that facilitate forms of direct tactile interaction for the visitor, with objects that they would otherwise only be able to observe.

Touch and manipulation of objects, whether physical or virtual, offers the advantages of creating a stronger connection than mere sight, between the individual and the object [52]. In this context, 3D technologies provide a privileged vehicle for engaging users with material culture [53–55]. Haptic interfaces and printed replicas can enable both tactile and kinaesthetic feedback and simulate the physical properties of original materials: the user can go beyond the simple appreciation of shape and color and can now sense the weight, structure, texture, and friction of the object [56,57].

One of the stimuli for the current developments has been the new consideration of an often-ignored public: the blind and visually impaired. The greater sensitivity towards this often-marginalized group has led to the investigation of all the experiential possibilities in the museum offer, beyond the visual one. In some cases, sight has been replaced by verbal descriptions or two-dimensional tactile reproductions. Still, the quality of these interventions has tended to remain inadequate and has not yet reached the potentials of directly sensing "true" objects. Museums have therefore developed many projects intended to exploit the potentials of 3D technologies, including solid printing, for the tactile exploration of works of art by the blind or visually impaired [58–61].

3D printing has also been used where three-dimensionality does not seem at issue, for example, in the documentation and conservation of paintings. "Alchemy," by Jackson Pollock, for example, was scanned at high resolution for creation of a model that is correct metrically (therefore in its geometry) and in color (radiometric value), which is obviously a fundamental feature of painting [62]. In this case, the three-dimensional model had a dual function: (1) study and research and (2) access and dissemination. The model was first used by experts to study, analyze, and interpret the traces and signs in the painted surface, distinctive to Pollock's technique. However, the analyses also served in dissemination: the 3D model was presented in an interactive kiosk, as part of a temporary exhibition at the Peggy Guggenheim Collection in Venice. In this environment, the visitor could navigate within the model and read explanations of specific features. The complexity of the surface and the amount of detail in the work led curators and authors to think that the complexity of the surface could only be fully appreciated through touch. As a result, a 1:1 replica was also printed, without resorting to the radiometric data, so that users could sense the geometry of the painting without being "disturbed" by the color, which was still appreciable from the nearby original.

The intersection of different fields of research, and the means and techniques of communication, has also given rise to new forms of representation of, and access to, CH. The association of printed copies and other media, for example, can overcome the "static" nature of the traditional physical model, which tends to restrict the modeled object to a specific historical moment. Working in this direction, the Photogrammetry Laboratory of the Università Iuav di Venezia, together with the Visualizing Venice research group, collaborated in the installation of an exhibition at Palazzo Ducale. This project featured, among other things, projections on printed three-dimensional models, recounting the stories of the lagoon and its islands [63–65] (Figure 1). With the benefit of 3D printed models and mapped projections, the non-expert audience could "participate" in the city's historical transformations. The combination of these two media indeed supersedes the concept of the 3D model as a frozen representation of a single historical moment. Instead, visitors can witness the continuous flow of history.

One of the new objectives in applied research is to add information to original objects by creating "sensorized" models. The literature reports many cases of production of three-dimensional copies solely for visualization purposes [66,67], but there are also examples of the integration of printed 3D replicas and tactile or pulsating sensors in exhibits. Particularly interesting in this regard is the Virtex Project (VIRTual EXhibition), which involves the integration of a series of sensors and buttons within a printed object, making it interactive: the user, by touching the replica at certain points, activates a video monitor that briefly describes its history [68].

Solid printing can create remarkable added value in the museum context in comparison to the simple mass production and marketing of objects. Indeed, the global market has recently seen the appearance of three-dimensional reproductions of famous paintings (i.e. Van Gogh [69]), which individuals can purchase in place of the more traditional, and cheaper, posters and cards.

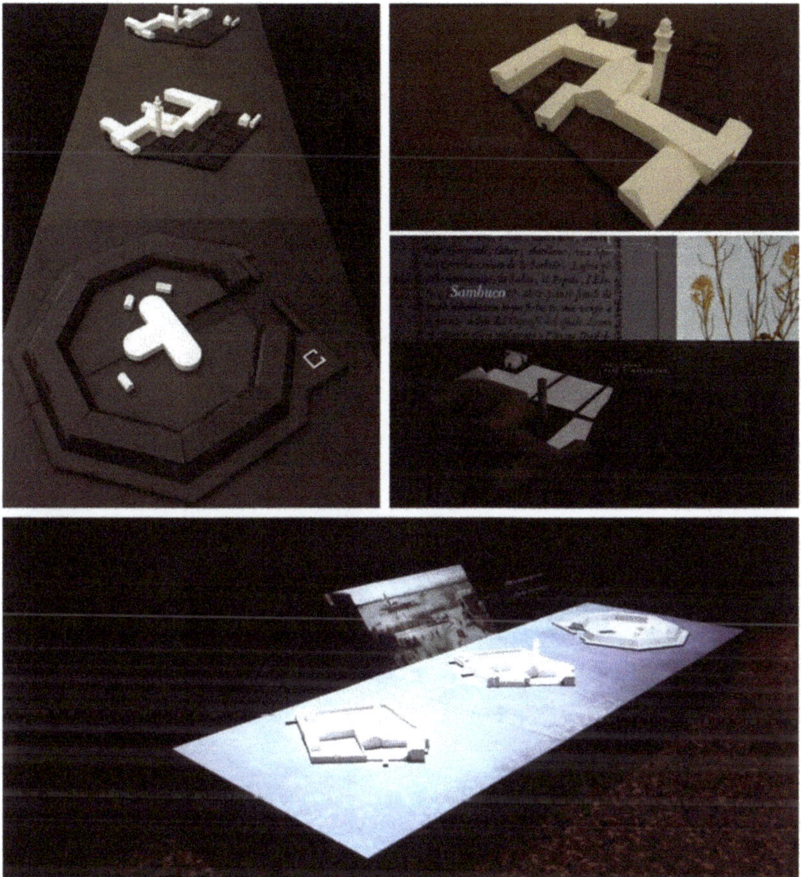

Figure 1. Printed three-dimensional models for a video-mapping installation about the history of the Venetian lagoon and its islands.

2.4. Solid Printing in Geomatics

Given its remarkable development in other fields, interest in solid printing has also extended to the sphere of geomatics, i.e. the gathering, processing, and delivering of spatially referenced geographic information: so much that 3D printing, similarly to paper or digital products, is considered among potential outputs when planning a survey.

However, the path from the surveyed model to its physical representation implies a series of steps that result in simplifications and thus the loss of conformity between the original object and its reproduction.

The typical tool for acquisition is a laser scanner, although several multi-image software products that can create a comparable point cloud from a set of digital photographs have now been developed. The point cloud is a pivot in the acquisition process—a set of transit data requiring further processing, depending on the research needs and objectives. The raw data serve as a digital copy of the real object, a dataset representing three-dimensional form, geometry, measurement, and matter, which can be archived and interpreted in support, for example, of multidisciplinary analyses. As the final product, designers typically choose orthophotos or traditional drawings, such as plans and sections.

The production of the solid print can only occur if the slicing software detects a unique closed surface of the mesh.

The production of the physical model requires machine operations that are themselves characterized by different levels of accuracy. Different printers utilize different methods and materials to translate the virtual model into the physical replica, as suggested by their names: computer numerical control (CNC), sterolithography apparatus (SLA), fused deposition modeling (FDM), and selective laser sintering (SLS). These different choices imply different accuracies and resolutions. The digitally produced replica will be more or less simplified, depending on the selection of printing technique and specific device. Solid printing is limited to a maximum thickness of a few tenths of a millimeter, requiring that the digital model be "sliced." This phase too has implications. The model can be sliced longitudinally, transversely, or diagonally, with different thicknesses of slicing, depending on the needs and characteristics of the printer. A related aspect is that the object must often be tilted on the printing plate, as the extruders deposit the layers, to avoid evident traces on the outer shell of the replica (Figure 2). The thickness of the section also effects the level of detail in the final model.

Figure 2. Different quality of replica due to the slicing orientation of the model on the printing plate. (**left**) Longitudinally with evident traces of layers; (**right**) diagonally, without traces of layers.

Point, digital, and physical models, therefore, are representations with different purposes and consequently different characteristics. Beginning from the point cloud, the overall procedure of attaining a "drawn" product is well established, but the corresponding process for obtaining a solid print has not yet been optimized or canonized. In fields other than geomatics, the final product is often printed from models created from scratch on the computer, consisting of data already ideal for the process: simple, closed surfaces, defined in a number of polygons appropriate to the intended degree of detail and requirements of printer operation. However, these conditions never occur in a geomatics survey, where the data are instead extremely redundant, discontinuous, and affected by noise and require very extensive processing.

3. Case Studies

The aim of the research is to evaluate the precision of 3D printing relative to the field of CH. In this context, tolerances are clearly higher than those for the areas of mechanical industry. However, distinctions must be made between models of different kinds: those produced to provide new and deeper ways of understanding the world, particularly in museum contexts, where tactile exploration of original objects is generally prohibited, versus those intended for scientific aims of study, analysis, and conservation for archaeological objects and artworks.

In the first case, printed models are intended to provide richer enjoyment by the public. The most important aspects are therefore plausibility and "realism," implying greater attention to choices concerning materials, colors, and the weight, texture, and feel of the object.

In the case of an object of important art-historical value, reproduced for scientific purposes, the model must be a perfect copy of the original, particularly in its geometry and form. Greater emphasis must then be placed on the problems of correctly constructing the digital models, and the conformity of the printed version with the original.

The accuracy of the printed model is fundamental to everything concerning geomatics, and has to be related to the same characteristics of the digital model obtained through the surveying process: the precision of the printed product must be evaluated in relation to the precision of the instruments used in the analysis. This level of accuracy must be related to that of the instruments of analysis through which the artefacts are converted into digital format. We cannot forget that the process that leads to the realization of a material copy must go through a numeric model and that in this process there is a progressive loss of definition, both from the qualitative and quantitative points of view. The simplification operations that the digital data must undergo might cause a difference between the geometry of the printed object and that of the original object.

3.1. Bearded Man of Vado all'Arancio

The first case study, as described in [70], is an archaeological finding (Figure 3) of the upper Palaeolithic (11,600 years B.P.) named "Uomo barbuto di Vado all'Arancio" (Bearded Man of Vado all'Arancio). The survey was done within a project carried out in collaboration with the Archaeological Museum of Massa Marittima, where the object is held, and its purpose was the production of a path accessible to blind or partially sighted people through audio-tactile works.

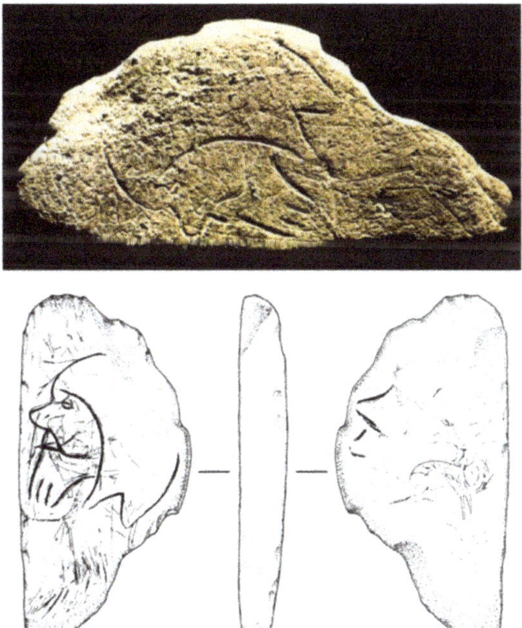

Figure 3. Image and drawings of the limestone slab of the "Uomo barbuto di Vado all'Arancio".

It is an engraved limestone slab found near Massa Marittima (GR) [71]. On the main face, we can still see his nose, his eye, his long moustache, his straight and thin mouth, his beard, and what could be his hair or a headgear. On the opposite side, there are still traces of what could have been another human face, but the drawing looks unfinished. The size of the object is very small: its dimensions are 8.2 × 4.1 × 1.1 cm.

Because of the small size of the object, we decided to use a triangulation-based laser scanner with the projection of a laser light blade (Range 7, Konika Minolta). This instrument guarantees a sub-millimetric precision (up to 40 μm) and can be used with two different lenses: tele and wide-angle.

For this case study, we used a tele lens. As already stated, the object was very small and the engravings on its surface were so light that they could barely be perceived by touch. In order to acquire these small deformations of the surface, we had to obtain the highest resolution.

The object was acquired through 23 scans, taken at an approximate distance of 700–800 mm and underwent the standard processing pipeline, using the software Geomagic Studio [72]:

- cleaning and deleting marginal parts that are mostly affected by noise;
- refining the registration through a global alignment (average distance: 0.251 mm; standard deviation: 0.474 mm);
- merging the scans into a single file, removing the redundant parts;

in the end, filling the small holes that were left.

At the end of data processing, the model was composed by approximately 1 million triangles, so we did not need to decimate it.

The digital models realized were then printed using an online printing service: Sculpteo [73], one of many websites that allow users to upload a 3D model they design, have it printed by a wide range of 3D printers, techniques, and materials, and have it delivered to their home by express delivery in any part of the world.

Using this website, we decided to test different printers, SLS-based techniques, and materials, as we wanted to see which one yielded the best results. We printed four test copies: three with one machine, but using materials with two different colors, and one with another printer that allows for the creation of fully colored replicas.

The two white copies (Figure 4) were printed with an EOS Formiga P395 using a plastic material (nylon). The first copy, on the left, is Nylon PA 12, created from a fine polyamide powder and available in different colors, and allowed us to obtain a good resolution for the smallest details (100 μm).

Figure 4. (**left**) Replica in Nylon PA 12; (**right**) replica in glass-filled nylon.

The second copy, on the right, was printed with Nylon 3200 Glass-Filled (glass-filled nylon), which is made of a mix of polyamide powder and glass beads. The grey copy (Figure 5) is made of alumide, a mix of polyamide powder and fine aluminium particles, which gives the final product a shiny look. There is only one resolution available for this material as the smallest layer has a thickness of 150 μm.

Figure 5. Replica in alumide.

The fourth replica was realized with a ZPrinter 650s by 3D Systems that uses a fine powder, similar to sandstone, which is painted during the printing process (Figure 6).

Figure 6. Colored replica.

As the final aim was to also have blind people be able to access the object inside a museum exhibition, and the surface shows engravings that are only visible to the human eye and cannot be felt by touch, we decided to create a somewhat "augmented" printed model.

The digital.stl file was imported into Mudbox, software by Autodesk for digital sculpting and painting. Here, we used texturization as a guideline to increase the depth of the model, where we could see the signs of engraving. This replica was not used for further analyses, as its metrical accuracy was compromised by this process (Figure 7).

We decided to analyze all the replicas produced, as well as those created by the same printer, comparing the survey data of the original finding with those of the printed copies, in order to check the reliability of the test itself and the behaviors of the different materials. The surveying process of the printed replicas followed the same pipeline used for the original object. The instrument used was the same triangulation-based laser scanner, Range 7, with the tele lens. In order to guarantee the same precision on the acquisition and to control the thermal expansion, the instrument was recalibrated.

The main difference in the process was the number of scans and the use of a rotating stage to reduce the acquisition time and perform an automatic first rough alignment. This system uses an auto focus but has no significant effect on the final model. In fact, during the acquisition of the original object, we saw that both methods guaranteed the recording of suitable data in terms of both the depth of field and the number of scans acquired.

Figure 7. Comparison between the colored 3D replica and the "augmented" one.

The acquisition and processing phases were kept unchanged for all the replicas. For every dataset, each scan was cleaned and realigned. The accuracy of the alignment of the single scans for each one of the replicas' datasets was high: an average distance of approximately 0.03 mm with a standard deviation of 0.04 mm was obtained. Only the colored scan showed an average distance of 0.07 mm with a standard deviation of 0.10 mm.

As with the original finding, due to its geometry and the high number of scans, it was not necessary to perform a long post-processing phase. Only a few small holes were filled, and there was no need to decimate the meshes. The digital data of the printed objects acquired were then compared to the one that was used for 3D printing.

The module used for the analysis of the two datasets is contained within the alignment menu of Geomagic Studio (Geomagic), which uses the ICP (iterative closest point) algorithm employed to minimize the difference between clouds of points and for the orientation of the scans.

The two models obtained by the acquisition of the monochromatic printings show comparable results (standard deviation: ± 0.036 mm for the gray nylon one and ± 0.046 mm for the white nylon one). Apart from some small differences between the two comparisons, probably caused by global alignments, the data are concordant.

In the operating result obtained, as represented in Scheme 1, the metric comparison was not between the original and its printed replica, but between the scanning of the printed replica and the original numerical model, a model from which hard copies are produced. Therefore, the precision of the numerical model of the original is irrelevant in the evaluation of the accuracy of the 3D printer. From the test conducted, the precision of the mesh alignment of the printed copy is the same as the difference between the digital model of the original and the digital model of the copy. It can therefore be stated that the printing process followed did not result in any loss of precision and that the expected theoretical precision linked to perception is linked only to the first phase of acquisition and to the alignment of the scans of the original.

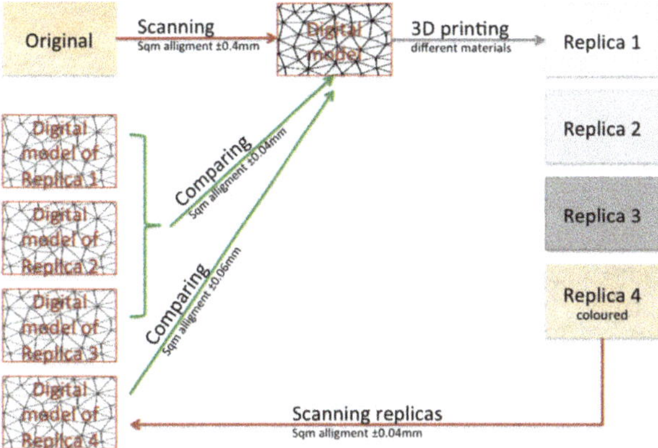

Scheme 1. A schematic summary of the workflow to compare replicas to the digital model of the finding.

On the other hand, the one printed with the multicolor printer shows an average distance higher than that of the other ones (standard deviation: ± 0.065 mm for the colored one). In order to be certain of the results, we performed another acquisition of the colored printed object. In this way, we were able to discard possible causes of imprecision, such as the variation of the instrument temperature and a non-correct alignment of the single scans.

Figure 8 shows, in the upper part, the results of the comparison between the original and the data relating to the first laser scanning acquisition of the colored replica, whereas in the lower part, the original data are always compared with the second copy acquisition. Although with small differences, due to mesh registration, this second test confirmed the overall results of the first one. The printed model appears generally larger than the original one, probably due to the printing process itself. In fact, the printer first creates a uniform layer of powder and then colors it with another head.

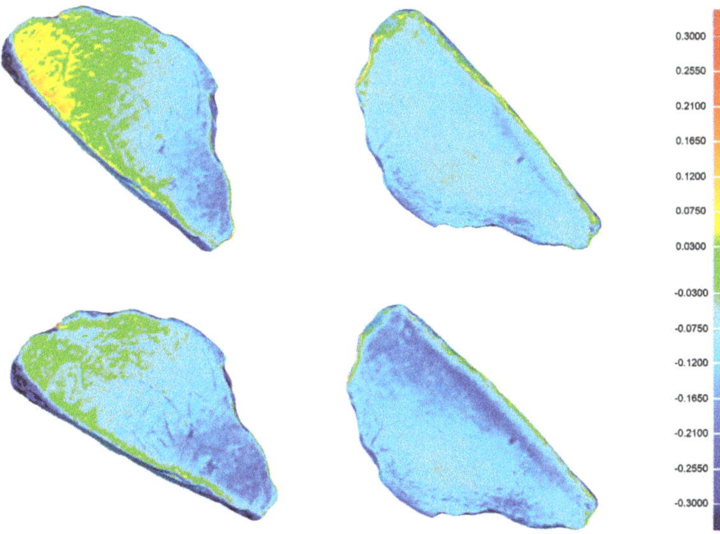

Figure 8. Results of the comparison between the original data and its colored replica. (**above**) the first test (the original data and the first survey of the copy); (**below**) the second one (original data and the second survey). Differences are in millimeters.

Based on the tests described above, we conclude that the discrepancy between verisimilitude and metric precision is evident: the colored object is much closer to reality from a qualitative and descriptive point of view, but from a metric point of view it is the one that presents the greatest deviations from the original geometry.

3.2. Dionysus with Satyr

The object surveyed is a group of sculptures preserved in the National Archaeological Museum of Venice: the "Dionysus with Satyr," a Roman copy of a group in Hellenistic style of the second half of the second century BC (Figure 9). The statue is part of the well-known legacy of 1587 by Giovanni Grimani, who with his donation contributed to the founding of the Public Statuary of the Serenissima, the precursor to the modern National Archaeological Museum.

Figure 9. (**left**) The "Dionysus with Satyr." (**right**) A detail of the marble statue.

The object in question is made of marble and is 2.17 m tall, so it is an interesting case study for research purposes, because it is considerably larger than the previous one. In fact, the different scale of the object entails logistical difficulties in the acquisition phase and usually higher tolerances in terms of precision of the data acquired.

3.2.1. Data Acquisition and Processing

The sculptural group was acquired through different techniques: triangulation-based laser scanning and digital photogrammetry. We intended to test different methods of acquisition, acting on the assumption that, being the starting point of the process that leads to the replica, the precision of methods and tools used in surveying an object must play a fundamental role in assessing the adherence of the copy to the original.

Despite the fact that the whole object has been acquired in its entirety, given its size we thought it would be appropriate to focus for this test only on a portion to avoid problems related to the processing time of such a complex model. Therefore, we decided to analyze a part of the head of the Satyr, which has fairly smooth surfaces on the face, but also has sufficiently complex on the hair.

3.2.2. Photogrammetric Survey

The photogrammetric survey was carried out using a Sony RX100 camera mounted on a telescopic carbon rod. It was remotely controlled for the acquisition of the highest parts of the statue. As the group reaches more than 2 m in height, this solution has proved to be effective for the acquisition of complete metric data.

Two hundred sixty-four images were acquired in seven strips with converging optical axes, turning around the sculptural group and setting the correct exposure from time to time (Figure 10). The statue, in fact, could not be moved, nor could a photographic set be built for the acquisition of photographs suitable for photogrammetric purposes, and the lighting of the environment in which the group is exposed, caused strong variations in light and shadow.

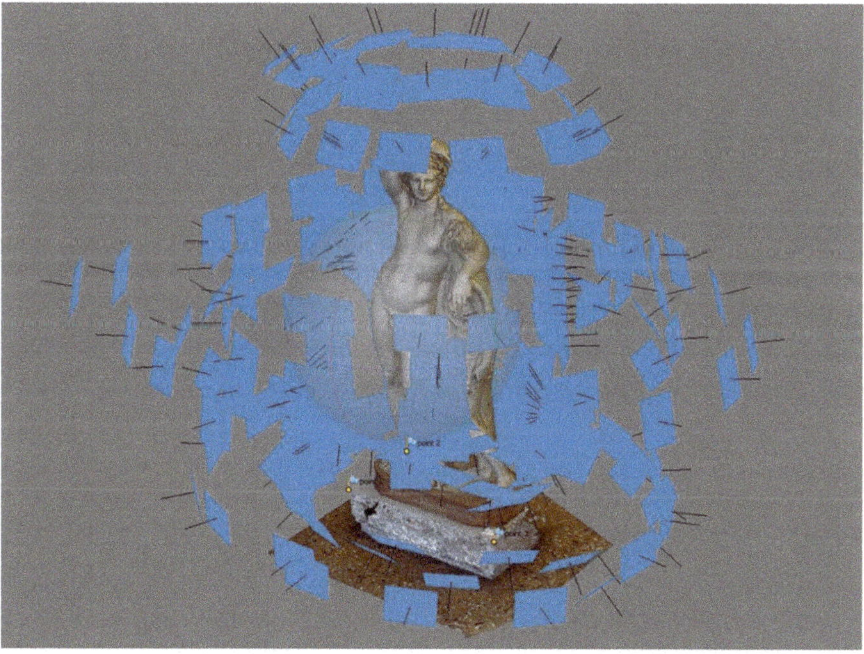

Figure 10. The photogrammetric model.

The images were processed with Structure from Motion (SfM) Agisoft Photoscan [74] software, which allowed us to directly obtain a surface textured model. The absolute orientation of the photogrammetric model was carried out using targets positioned on the basement that supported the statue.

However, the photogrammetric model could often be noisy, because the technique is less stable than laser scanning and the accuracy of the result depends on many factors, such as the geometry of the acquisition and the quality of the images captured.

In order to eliminate as much noise as possible, aggressive filters were applied during the creation of point and surface models, among those available in the software options (in fact, to sort out the outliers, Photoscan has several inbuilt filtering algorithms that respond to different needs, and the aggressive depth filtering mode sorts out most of the outliers if the area of investigation does not contain small significant details). At the end of the process, a cloud of about 26 million points and a mesh of 5 million triangles were created.

Given the size of the object, however, the number of triangles was not sufficient to describe the object with a degree of detail adequate for a representation at a 1:1 scale. As we can see from Figure 11, the details of Satyr's face and curls appear extremely smooth, such that they are sometimes almost imperceptible. To overcome this problem, we decided to separately process only the images that acquired the chosen portion. Therefore, 45 pictures were selected for the construction of a dense cloud of the head of the Satyr creating a point cloud of 1 million points, and a mesh of 240,000 triangles (in the photogrammetric model of the entire statue, the same portion was described by a dense cloud of 500,000 points). The resulting product is a much more detailed surface model than the previous one, and the effect of noise is much more visible. Noise filtering is strictly connected to details of objects, because both are high frequencies: reducing noise can cause a loss of detail, and, vice versa, increasing detail could lead to increasing noise.

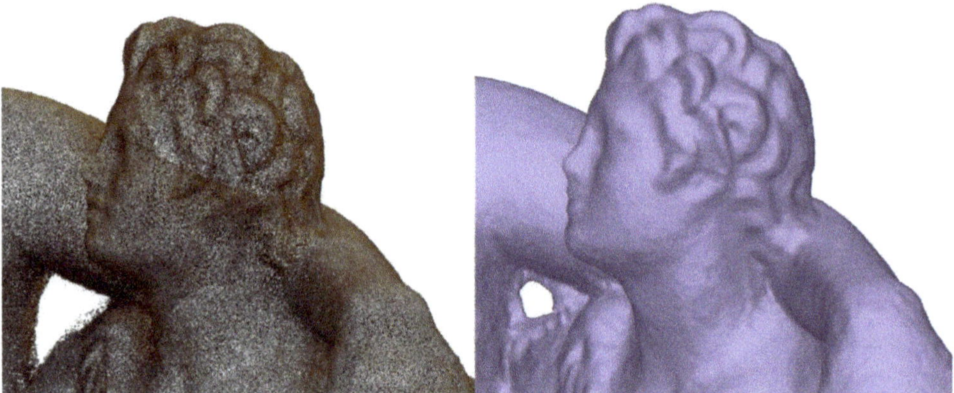

Figure 11. Detail of the point cloud (**left**) and of the mesh (**right**).

3.2.3. Laser Scanning Survey

The instrument used for the acquisition is Konica Minolta Vivid 9i [75], a triangulation-based laser scanner with projection of a laser light blade, which permits one to vary the scanning resolution using different interchangeable lenses. We chose this instrument because it allows for the acquisition of larger areas for each scan compared to Range 7 (used for the other case study) [76], at the expense of instrumental accuracy (Table 1). Since this is a large-scale sculpture, the tolerances required were greater than those required for the reproduction of the artefact presented in the previous chapter. In addition, the use of Range 7 would have meant a much longer acquisition phase.

Table 1. Technical characteristics of the two triangulation-based laser scanners.

Instrument	Scanning Range (mm)	Accuracy (mm)	Maximum Scanning Size (mm)
Vivid 9i	600–1000 500–2500 * (* extended mode)	±0.008 Tele ±0.016 Medium ±0.032 Wide	Tele 463 × 347 × 680 * Med 823 × 618 × 1100 Wide 1495 × 1121 × 1750 *
Range 7	450–800	±0.040	Tele 141 × 176 × 97 Wide 267 × 334 × 194

The instrument, with a medium lens (14 mm focal length), was calibrated before the acquisition phase; the standard process also includes a white balance, as the instrument allows one to acquire RGB data (at low quality), but some colors or materials may create noise in the acquisition phase.

The scanner directly produced a surface model (mesh). Each scan lasts on average 2.5 s. To acquire details and undercuts, many scans are often necessary, and the instrument needs to be moved on a tripod around the object, which is complex and time-consuming. For the head of the Satyr alone, nine scans were acquired.

A first rough alignment was performed during the acquisition phase, identifying three homologous points for each overlapping scan, while a more precise and metrically controlled procedure was carried out a posteriori, through Geomagic Studio, following the same process performed for the Bearded Man of Vado All'Arancio:

- a cleaning of every single scan from elements that do not belong to the object or from the portions mostly affected by noise;
- a global alignment of scans using the ICP algorithm (average distance 0.154 mm; standard deviation ± 0.168);
- a mesh union into a single surface;
- a filling of gaps with a closure of holes and correcting topology.

At the end of the process, a model consisting of about 200,000 triangles was obtained, comparable to the one obtained by photogrammetry.

3.2.4. Solid Printing

At the end of data acquisition and processing, the models were prepared for printing. Three models were physically reproduced: one created from the laser data and two created by the photogrammetric process.

Since the model to be printed was not a 3D all-round closed surface, we performed an offset of 5 mm in thickness (Figure 12). In this way, we were able to create a unique, closed model while limiting its size and avoiding a waste of material.

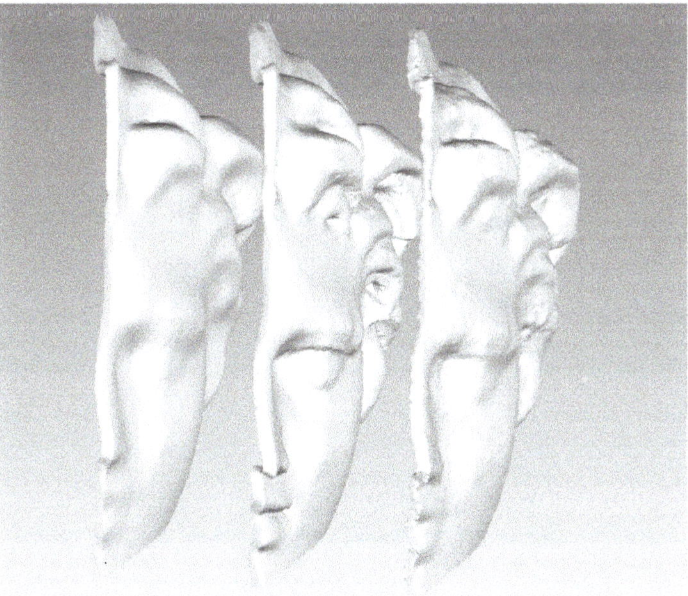

Figure 12. The three digital models: from the left: the first photogrammetric model, the laser-scanning model, and the second photogrammetric model.

The digital data were not further processed because the purpose was to highlight the quality of the final result in relation to the methodology used.

The printing was entrusted to HSL (HSL), a company founded in 1988 that deals with mechatronics and, among other things, 3D printing. The models were reproduced using a sintering technique with an EOSINT P395 printer, using a polyamide (nylon) and subjecting the external surfaces to a subsequent sandblasting process, to completely remove the excess of dust. The tolerances on sintered nylon are $\pm$ 0.3% with a minimum of $\pm$ 0.3 mm, while the tolerance on maximum deformation is $\pm$ 1%.

At first glance, without the need for further analysis, it was clear that the laser scanner produced the best physical copy, as it had the greatest level of detail and the noise of the acquired data was so low that it did not affect the quality of the result (Figure 13). Nevertheless, it was considered appropriate to also analyze the data from a metric point of view, in order to make considerations on the accuracy of the different acquisition methods.

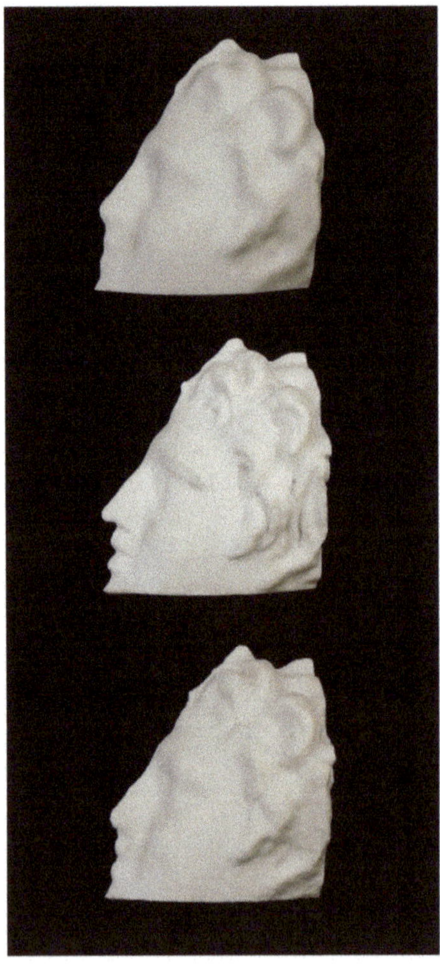

Figure 13. The three printed replicas: from above, first photogrammetry, laser-scanning, and second photogrammetry.

3.2.5. Replicas' Data Acquisition

The printed models were detected using the already mentioned Range 7 of Konica Minolta, with "tele" optics, chosen for its level of precision and for the resolution it manages to achieve: the digital data had to match the printed model as faithfully as possible. In fact, the precision that this instrument guarantees allows for the evaluation of very small discrepancies. Using different tools and methods would have led to a decay in precision and consequently would not have admitted for rigorous considerations of different acquisition methods. Moreover, using the same tool for the survey of printed objects has let us to obtain comparable data, even if the original ones were acquired by different techniques.

As in the previous case study, a rotating stage was used, and each replica was acquired in three successive steps.

1. Twelve scans placing the object in a standing position and imposing an angular step of 30° on the rotating stage.
2. Six scans with the main face upwards, setting the angular pitch to 60°.
3. Six scans with the main face downwards, setting the angular pitch to 60°.

In this way, we obtained a large amount of overlap and redundant data, allowing us to eliminate areas acquired at an inappropriate distance from the instrument and therefore slightly out of focus.

In Geomagic Studio, after the cleaning process, a global alignment process was carried out, refining the raw scan alignment already carried out by the Range Viewer software and obtaining an average distance of 0.03 mm with a standard deviation of about ± 0.07 mm for all datasets.

The three meshes obtained for each replica have been aligned and merged into a single file (see Table 2 for the accuracy obtained on the alignments).

Table 2. Statistical indices obtained as a result of global alignments of the three datasets.

Object	Average Distance (mm)	Standard Deviation (mm)
Laser	0.05	±0.06
First Photogrammetry	0.05	±0.10
Second Photogrammetry	0.05	±0.13

No further processing was carried out, except for small corrections of topological errors, such as the flipping of some normals on the edges, which were noisier.

3.2.6. Comparisons

Each digital model obtained from the survey of the printed replicas was compared with the one that generated it, i.e., the one surveyed. The analyses carried out showed very similar results between the three models: a maximum distance of about 1 mm and an average distance of about 0.1 mm, with a standard deviation of about ± 0.1 mm (Table 3).

Table 3. Statistical indices obtained from the comparisons between the printed models and the original data.

Object	Maximum Distance (mm)	Average Distance (mm)	Standard Deviation (mm)
Laser	1.428	0.130	±0.110
First Photogrammetry	0.753	0.097	±0.079
Second Photogrammetry	0.996	0.174	±0.133

Based on these first analyses, the objects, printed with the same machine, have roughly the same precision, which is in agreement with other tests carried out on other types of objects printed with different techniques. Figure 14 shows the main results of the analysis carried out: the figure compares the three replicas with the original datasets that generated them. The highest fidelity is found in the replica of the first photogrammetric model, and this is due to the fact that the mesh produced is less detailed (fewer triangles), allowing for a greater adherence between the numerical model and the printed copy. On the other hand, in the model obtained with the triangulation laser scanner, which is considered more accurate, there are greater deviations in correspondence with the more complex geometry of curls due to the difficulty in the printing process of reproducing the high detail.

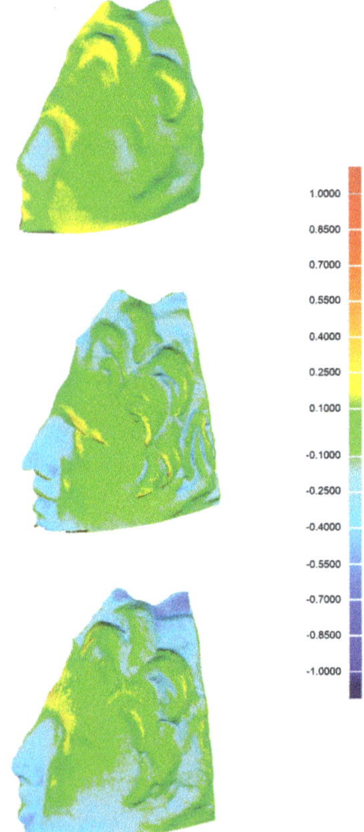

Figure 14. Results of the comparisons between printed models and the three datasets that produced them. From top to bottom, the first photogrammetric model, the laser-scanning model, and the second photogrammetric model. Distances are in millimeters.

Using the same method of analysis, we previously examined a small vase 11 cm high and 8 cm in diameter, printed at a 1:1 scale, and the bust of Francesco II Gonzaga, measuring $70 \times 56 \times 30$ cm, printed at a 1:5 scale. Both objects were reproduced with two different additive techniques: SLA (EnvisionTEC Ultra printer with photopolymer resin) and FDM (the first with a CraftBot in PLA and the second with a Makerbot Replicator 2x in ABS).

Metric analyses carried out on FDM-printed objects showed a maximum distance of about 1 mm, an average distance of about 0.1 mm, and a standard deviation of about ± 0.15 mm. In those printed with the SLA technique, on the other hand, an average distance of about 0.25 mm was found, probably caused by the post-printing process, which involves the use of an ultraviolet oven to solidify the resin, reducing the size of the replica by a few tenths of a millimeter.

Interestingly, the greatest differences are found at the edges of objects, which are most affected by noise introduced during data acquisition and processing.

Moreover, there is a clear systematism in the portion of nose and chin and in the upper part of the head. This deformation is evident in all three models and was certainly caused both by the printing process and by the thickness of the object: sintered nylon is a rather resistant material, and using thicknesses greater than 2–3 mm can generate unpredictable deformations.

Finally, as in the previous case, there is a loss of detail in the geometrically more complex part, i.e., the one relating to Satyr's curls, which are less engraved in the more inward areas and less pronounced in the more protruding areas (Figure 15).

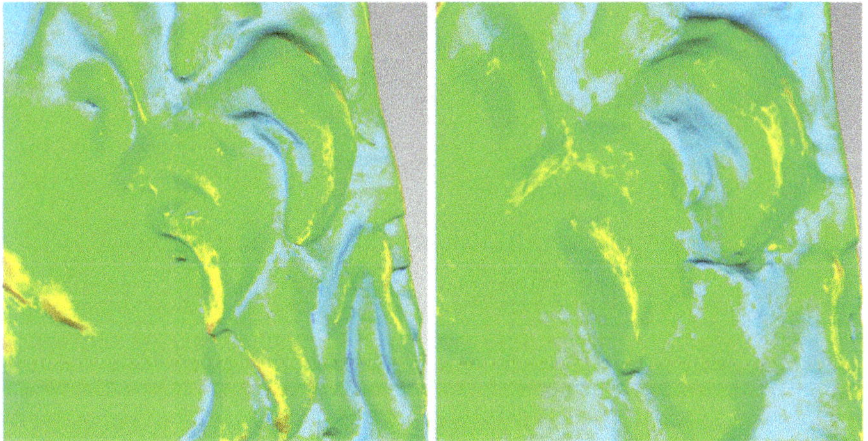

Figure 15. Details of the Satyr's curls. (**left**) laser-scanning; (**right**) second photogrammetry (scale bar is the same of Figure 14).

Nevertheless, in the light of the various tests carried out, it seems clear that the 3D printing process makes it possible to achieve sub-millimeter accuracies regardless of the technique and the printer used. What we want to confirm, however, is the different quality of results that comes from the use of different methods of acquisition. Considering the fact that it had the highest precision, the laser model was used as the basis for the comparison of those generated by the photogrammetric process. The statistical indices do not differ much from those highlighted in the previous analyses (Table 4), but it is evident from Figures 16 and 17 that the ICP algorithm used in the analyses was not able to calculate the distances between some points (in grey in the images), considered too far from each other to identify the nearest neighbor.

Table 4. Statistical indices obtained from the comparisons between the laser-scanning model and the photogrammetric ones.

Compared Objects	Maximum Distance (mm)	Average Distance (mm)	Standard Deviation (mm)
Laser—First Photogram	2.664	0.652	±0.581
Laser—Second Photogram	2.756	0.498	±0.537

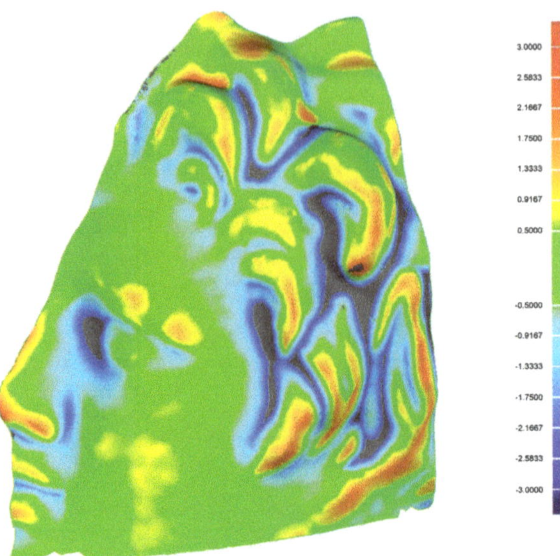

Figure 16. Comparison between the first photogrammetric model and the laser-scanning model.

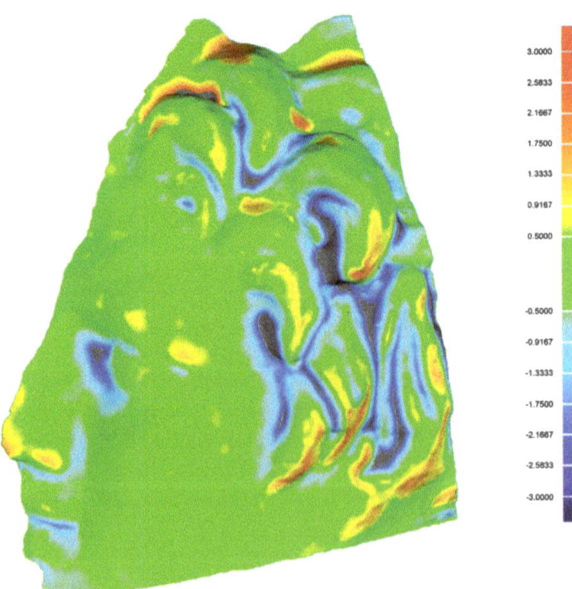

Figure 17. Comparison between the second photogrammetric model and the laser-scanning model.

Finally, it should be pointed out that the photogrammetric process can lead to much more precise results than those obtained from this test. However, our test was intended to highlight the lower stability of this technique compared to laser scanning, which is influenced by fewer variables. In fact, in photogrammetry, factors such as the image quality, the geometry of the acquisition scheme, the average frame scale, and the correct calculation of interior orientation parameters of the cameras can lead to a rapid decay in the accuracy, if they are not taken into account. The first photogrammetric model was appropriate for a representation scale far below a 1:1 scale. In fact, the group as a whole has been printed at scales of 1:10 and 1:20 (Figure 18): the noise is not evident, and the level of detail achieved is absolutely adequate to these dimensions. The second photogrammetric model, on the other hand, shows a greater level of detail, almost adequate for the 1:1 scale, but requires a long smoothing process, which can ultimately result in the elimination of the smallest details along with the noise.

Figure 18. The fused deposition modeling (FDM) replica (a scale of 1:20).

4. Conclusions

3D printing technologies have opened up a wide range of new possibilities in the CH sector, both in terms of museum usability and in terms of cataloguing and study, providing a basis for visualization on the one hand and shape analysis on the other. The tangible representation of an artefact born from digital manufacturing can be a vehicle of information, a particularly interesting characteristic, especially in CH. Replicas also have the advantage of being touched without damaging the state of preservation of the original object, thus providing a new way of interacting with works of art and therefore learning.

Precisely because of this dual nature of the replica, which is used for popular and scientific purposes, a number of factors must be taken into account: on the one hand, the realism and verisimilitude of the replica, whose texture, weight, and appearance must be consistent with the original; on the other hand, the adherence of the shape, which is expressed, in Geomatics, in terms of the accuracy and precision of the printed model. After all, 3D printing is now becoming part of the surveying sciences, positioning itself at the end of the surveying process alongside the more traditional orthogonal projection representations, and like all geomatics products, the path leading to its formation must be validated.

Within this process, the techniques and methods of data acquisition and processing may lead to a loss of adherence of the copy to the original. First, the acquisition systems have different accuracies, which, depending on the technique used, lead to the formation of a model by points that moves away from the real object. Secondly, the data processing also uses filters or procedures that, while optimizing the model for printing or viewing, lead to changes in the geometry of the copy. The main ones are decimation and smoothing, which are often inbuilt filtering algorithms responding to different needs, but no analytical formulations are declared, and their use is inevitable: the metric data acquired by photogrammetry or laser scanning is often redundant with respect to the purposes of physical reproduction and noise. From time to time, it is necessary to make choices based on the evaluation of the number of points (and consequently the size of polygons) and on the filters to be applied for noise reduction, in relation to the quality of the final model.

These considerations are linked to a further aspect related to the question of adherence of the model to reality: Which modeling process allows for the greatest mimesis of the real object? Surface models created by triangulation procedures let us obtain meshes based entirely on the points acquired by the cloud, even if they pass through the processes described above. After all, on the one hand, these operations distance the digital model from the real one; on the other hand, they are inevitable steps to obtain a qualitatively good copy.

In the case studies presented, we analyzed two different scenarios, both linked to access in a museum perspective. These two objects are very different in size, and both had to be printed at a 1:1 scale.

The first object, the archaeological find called "Uomo Barbuto di Vado all'Arancio" did not present any particular problems in the data acquisition and processing phase. Thanks to the scale of the object, we were able to use an extremely precise instrument. Therefore, we decided to focus in this case on the last step of the process: the precision of 3D printing. In fact, comparisons were made between the digital model of the printed object and the already processed one of the original object. Beyond the instrumental precision and the alignment of the individual scans, the only variable to be analyzed was the metric quality of 3D printing. Moreover, when evaluating the adherence of the copy to the original, we have to consider an inevitable uncertainty of measurement at the time of the acquisition of digital data. This is part of the acquisition process, but we tried to reduce it as much as possible by using a very precise instrument. Digital models will never be a perfectly faithful copy of reality, but statistical analysis and probability theory provides us with tools to assess uncertainty in acquisitions. One interesting element in the results of the tests is the discrepancy between verisimilitude and metric precision: the colored replica was much closer to reality from a qualitative and descriptive point of view, but from a metric point of view it was the one that showed the greatest deviations from the original.

In the second case study, the one concerning the replica of the sculptural group "Dionysus with Satyr," we decided to focus on the acquisition process, analyzing the results of the application of different techniques and instruments. The object, in fact, for size and location involved greater logistical difficulties than the previous one.

Both techniques used, triangulation-based laser scanner and photogrammetry, have their pros and cons: laser surveying is more stable than photogrammetric surveying, which is influenced by a greater number of variables (image quality, acquisition geometry, calibration methods, number of control points, etc.). When used by inexperienced users, photogrammetry can provide seemingly good products (at the descriptive level), but they are not correct from a metric point of view. On the other hand, the photogrammetric process is much faster, especially in the campaign phase. In our experience, the data processing times in the laboratory, however, are more or less similar: two to three working days, depending on the computing power of the computer. However, it is inevitable that the accuracy and/or acquisition times vary according to the scale of the objects to be acquired.

Compared to the previous case, the results of the first analyses carried out show greater deviations between the original digital model and the digital model of the printed object (an average distance in the order of tenths of a millimeter). However, we also wanted to focus on the validation of the process leading to the formation of the digital model: data acquisition and processing. For this reason, the results of the various techniques were printed, recorded, and compared with each other. First of all, it is important to underline that the photogrammetric survey carried out is not considered adequate for the scale of representation required and that it is not the best result obtainable from the application of this technique. We have here chosen to highlight some phenomena that may occur during data acquisition and processing, in particular the application of smoothing and decimation filters by the software that generates the three-dimensional data from photographs. In fact, in the photogrammetric models, the different level of detail achieved in the two cases is evident. When we processed the object in its entirety, the detail was not adequate at the representation scale chosen, while processing the single portion of the statue under examination increased the level of detail, but also the noise. As already mentioned, both the details and the noise are part of the high frequencies, and the application of filters may not be able to identify the difference between the two. The filters applied to the first photogrammetric model were suitable at a greater scale of representation (1:10).

We can therefore conclude that, if we talk about accuracy, the most important moment of the process, which from surveying leads to 3D printing, lies in the first two steps: acquisition and data processing. Despite an inevitable decay of precision at the time of printing, precisions obtained are more than adequate for any type of application within CH. The results of the tests carried out, which are comparable with others developed within the Photogrammetry Laboratory, show sub-millimeter or, at most, millimeter deviations.

The rapid evolution involving this technology, moreover, makes it possible to suppose that the limits related to the material and precision will be overcome quickly. In particular, what needs more attention in the future is the level of the likelihood of the copy, i.e., the adherence to the original in qualitative and descriptive terms (weight, texture, material, color, etc.); however, even in this aspect, the accuracy of 3D printing reaches excellent levels.

Finally, in the light of what has been analyzed, we think that it is not possible to identify a single valid workflow for every context, but it is necessary to evaluate from time to time methods and tools according to the type of object to be reproduced and the purpose of reproduction. As for the survey, and for solid printing, the great variety of shapes, sizes, and materials that characterize the world of CH (and the great variety of purposes, needs, and timing) has a significant impact on the choice of the most appropriate methods of data acquisition and management for each individual case study.

Likewise, moving from surveying to 3D printing, the protocols that need to be followed will not always be the same, since no operational protocol is suitable for every situation. Instead, we make a choice that takes into account a number of factors, such as:

- the type of object;
- the methodology used for surveying;
- the 3D printing technology used.

However, we believe that the choices made during the acquisition and processing phases can simplify the printing phase. To obtain the best products, a triangulation-based laser scanner, which allows one to obtain higher resolution and less noisy data than those acquired by other techniques, such as time-of-flight- and phase-based laser scanners and photogrammetry, can be used. It is clear, however, that this can only be used for small objects. The high redundancy in the number of scans, which guarantees the acquisition of an almost continuous model, can also be used. The presence of small holes does not compromise the general geometric description of the object.

Author Contributions: Methodology, C.B. and M.B.; validation, Christina Balletti and M.B.; formal analysis, Christina Balletti and M.B.; investigation, Christina Balletti and M.B.; resources, C.B.; data curation, M.B.; writing—original draft preparation Christina Balletti and M.B.; writing—review and editing, Christina Balletti and M.B.; supervision, Christina Balletti and M.B.; funding acquisition, C.B.

Funding: PRIN: Progetti di Ricerca di Rilevante Interesse Nazionale—Bando 2015, Prot. 2015HJLS7E.

Acknowledgments: This work has been carried out under the GAMHer project: Geomatics Data Acquisition and Management for Landscape and Built Heritage in a European Perspective. PRIN: Progetti di Ricerca di Rilevante Interesse Nazionale—Bando 2015, Prot. 2015HJLS7E.

Conflicts of Interest: The authors declare no conflict of interest.

References

1. Balletti, C.; Ballarin, M.; Guerra, F. 3D printing: State of the art, considerations and future perspectives. *J. Cult. Herit.* **2017**, *26*, 172–182. [CrossRef]
2. Gibson, I.; Rosen, D.; Stucker, B. *Additive Manufacturing Technologies: 3D Printing, Rapid Prototyping, and Direct Digital Manufacturing*, 2nd ed.; Springer Science & Business Media: New York, NY, USA, 2015.
3. Gaubatz, W.A. Rapid prototyping. In Proceedings of the IEEE Aerospace Applications Conference Proceedings, Aspen, CO, USA, 10 February 1996; Volume 3, pp. 303–311.
4. Scopigno, R.; Cignoni, P.; Pietroni, N.; Callieri, M.; Dellepiane, M. Digital fabrication technologies for cultural heritage (STAR). In Proceedings of the Eurographics Workshop on Graphics and Cultural Heritage, Darmstadt, Germany, 6–8 October 2014; pp. 75–85.
5. 3D Printing vs. CNC Machining. Available online: https://www.3dhubs.com/knowledge-base/3d-printing-vs-cnc-machining (accessed on 1 March 2019).
6. Rahman, I.A.; Adcock, K.; Garwood, R.J. Virtual fossils: A new resource for science communication in paleontology. *Evol. Educ. Outreach* **2012**, *5*, 458. [CrossRef]
7. Ioannides, M.; Quak, E. *3D Research Challenges in Cultural Heritage. A Roadmap in Digital Heritage Preservation*; Springer: Berlin/Heidelberg, Germany, 2014.
8. Ioannides, M.; Fellner, D.; Georgopoulos, A.; Hadjimitsis, D.G. Digital heritage. In Proceedings of the Third International Conference, EuroMed, Lemessos, Cyprus, 8–13 November 2010.
9. Comes, R.; Buna, Z.; Badiu, I. Creation and preservation of digital cultural heritage. *J. Anc. Hist. Archaeol.* **2014**, *1*, 50–56.
10. Choromański, K.; Łobodecki, J.; Puchała, K.; Ostrowski, W. Development of virtual reality application for cultural heritage visualization from multi-source 3D data. *Int. Arch. Photogramm. Remote Sens. Spat. Inf. Sci.* **2019**, *XLII–2/W9*, 261–267.
11. Bruno, F.; Bruno, S.; De Sensi, G.; Luchi, M.L.; Mancuso, S.; Muzzupappa, M. From 3D reconstruction to virtual reality: A complete methodology for digital archaeological exhibition. *J. Cult. Herit.* **2010**, *11*, 42–49. [CrossRef]
12. Fowles, P.S.; Larson, J.H.; Dean, C.; Solajic, M. The laser recording and virtual restoration of a wooden sculpture of Buddha. *J. Cult. Herit.* **2003**, *4*, 367–371. [CrossRef]
13. Schindler, K.; Grabner, M.; Leberl, F. Fast on-site reconstruction and visualization of archaeological finds. In Proceedings of the CIPA Symposium, Antalya, Turkey, 30 September–4 October 2003.
14. Beraldin, J.A.; Blais, F.; Cournoyer, L.; Rioux, M.; El-Hakim, S.H.; Rodella, R. Digital 3D imaging system for rapid response on remote sites. In Proceedings of the Second International Conference on 3-D Imaging and Modelling (3DIM'99), Ottawa, ON, Canada, 4–8 October 1999; p. 34.
15. Bernardini, F.; Rushmeier, H.; Martin, I.M.; Mittleman, J.; Taubin, G. Building a digital model of Michelangelo's Florentine Pieta. *IEEE Comput. Graph. Appl.* **2001**, *22*, 59–67. [CrossRef]
16. Marc, L.; Rusinkiewicz, S.; Ginzton, M.; Ginsberg, J.; Pulli, K.; Koller, D. The digital Michelangelo project: 3D scanning of large statues. In Proceedings of the ACMSIGGRAPH Conference on Computer Graphics, New Orleans, LA, USA, 23–28 July 2000; Addison Wesley: Boston, MA, USA, 2000; pp. 131–144.
17. Suveg, I.; Vosselman, G. 3D reconstruction of building models. *Int. Arch. Photogramm. Remote Sens.* **2000**, *33*, 538–545.

18. Akca, D.; Remondino, F.; Novàk, D.; Hanusch, T.; Schrotter, G.; Gruen, A. Recording and modeling of cultural heritage objects with coded structured light projection systems. In Proceedings of the 2nd International Conference on Remote Sensing in Archaeology, "From Space to Place", BAR International Series 1568, Rome, Italy, 4–7 December 2006; pp. 375–382.

19. Boulanger, P.; Rioux, M.; Taylor, J.; Livingstone, F. Automatic replication and recording of museum artifacts. In Proceedings of the 12th International Symposium on the Conservation and Restoration of Cultural Property, Tokyo, Japan, 29 September–1 October 1988; pp. 131–147.

20. Castagnetti, C.; Giannini, M.; Rivola, R. Image-based virtual tours and 3D modeling of past and current ages for the enhancement of archaeological parks: The visual Versilia 3D project. *ISPRS Arch. Photogramm. Remote Sens. Spat. Inf. Sci.* **2017**, *XLII–5/W1*, 639–645. [CrossRef]

21. Yilmaz, U.; Özün, O.; Otlu, B.; Mulayim, A.; Atalay, V. Inexpensive and robust 3D model acquisition system for three-dimensional modeling of small artifacts. In Proceedings of the CIPA Symposium, Antalya, Turkey, 30 September–4 October 2003; pp. 286–291.

22. Guidi, G.; Frischer, B.; Russo, M.; Spinetti, A.; Crosso, L.; Micoli, L.L. Three dimensional acquisition of large and detailed cultural heritage objects. *Mach. Vis. Appl.* **2006**, *17*, 349–360. [CrossRef]

23. Pavelka, K.; Dolansky, T. Using non-expensive 3D scanning instruments for cultural heritage documentation. In Proceedings of the CIPA Symposium, Antalya, Turkey, 30 September–4 October 2003; pp. 534–536.

24. Reilly, P. Data visualization in archaeology. *IBM Syst. J.* **1989**, *28*, 569–579. [CrossRef]

25. Tsioukas, V.; Patias, P.; Jacobs, P.F. A novel system for the 3D reconstruction of small archaeological objects. In Proceedings of the XXth Congress of ISPRS, Istanbul, Turkey, 12–23 July 2004; Volume XXXV. Part B5.

26. Nocerino, E.; Remondino, F.; Uccheddu, F.; Gallo, M.; Gerosa, G. 3D modelling and rapid prototyping for cardiovascular surgical planning-two case studier. *Int. Arch. Photogramm. Remote Sens. Spat. Inf. Sci.* **2004**, *XLI–B5*, 887–893.

27. Hinton, T.J.; Jallerat, Q.; Palchesko, R.N.; Park, J.H.; Grodzicki, M.S.; Shue, H.J.; Feinberg, A.W. Three-dimensional printing of complex biological structures by freeform reversible embedding of suspended hydrogels. *Sci. Adv.* **2015**, *1*, e1500758. [CrossRef] [PubMed]

28. Murphy, S.V.; Atala, A. 3D bioprinting of tissues and organs. *Nat. Biotechnol.* **2014**, *32*, 773–785. [CrossRef] [PubMed]

29. Adami, A.; Balletti, C.; Fassi, F.; Fregonese, L.; Guerra, F.; Taffurelli, L.; Vernier, P. The bust of Francesco II Gonzaga: From digital documentation to 3D printing. *ISPRS Ann. Photogramm. Remote Sens. Spat. Inf. Sci.* **2015**, *II–5/W3*, 9–15. [CrossRef]

30. Balletti, C.; Guerra, F. The survey of cultural heritage: A long story. *Rend. Lincei* **2015**, *26*, 115–125. [CrossRef]

31. Balletti, C.; D'Agnano, F.; Guerra, F.; Vernier, P. From point cloud to digital fabrication: A tangible reconstruction of Ca' Venier dei Leoni, the Guggenheim Museum in Venice. *ISPRS Ann. Photogramm. Remote Sens. Spat. Inf. Sci.* **2016**, *III–5*, 43–49. [CrossRef]

32. Brunetaud, X.; De Luca, L.; Janvier-Badosa, S.; Beck, K.; Al-Mukhtar, M. Application of digital techniques in monument preservation. *Eur. J. Environ. Civ. Eng.* **2012**, *16*, 543–556. [CrossRef]

33. Guidi, G.; Remondino, F.; Russo, M.; Menna, F.; Rizzi, A.; Ercoli, S. A multi-resolution methodology for the 3D modeling of large and complex archaeological areas. *Int. J. Archit. Comput.* **2009**, *7*, 40–55. [CrossRef]

34. Koutsoudis, A.; Vidmar, B.; Ioannakis, G.; Arnaoutoglou, F.; Pavlidis, G.; Chamzas, C. Multi-image 3D reconstruction data evaluation. *J. Cult. Herit.* **2013**, *1*, 73–79. [CrossRef]

35. Pavlidis, G.; Koutsoudis, A.; Arnaoutoglou, F.; Tsioukas, V.; Chamzas, C. Methods for 3D digitization of cultural heritage. *J. Cult. Herit.* **2007**, *8*, 93–98. [CrossRef]

36. Remondino, F. Heritage recording and 3D modeling with photogrammetry and 3D scanning. *Remote Sens.* **2011**, *3*, 1104–1138. [CrossRef]

37. Remondino, F.; Rizzi, A. Reality-based 3D documentation of natural and cultural heritage sites—Techniques, problems and examples. *Appl. Geomat.* **2010**, *2*, 85–100. [CrossRef]

38. Rivola, R.; Castagnetti, C.; Bertacchini, E.; Casagrande, F. Digitalizzazione e stampa 3D di un mosaico a tecnica bizantina a scopo documentativo e conservativo. *Archeomatica* **2016**, *7*, 1.

39. Tucci, G.; Bonora, V. Geomatics and management of at-risk cultural heritage. *Rend. Lincei* **2015**, *26*, 105–114. [CrossRef]

40. Bigliardi, G.; Dioni, P.; Panico, G.; Michiara, G.; Ravasi, L.; Romani, M.G. Restauro e innovazione al Palazzo Ducale di Mantova: La stampa 3D al servizio dei Gonzaga. *Archeomatica* **2015**, *3*, 40–44.

41. Arbace, L.; Sonnino, E.; Callieri, M.; Dellepiane, M.; Fabbri, M.; Idelson, A.I.; Scopigno, R. Innovative uses of 3D digital technologies to assist the restoration of a fragmented terracotta statue. *J. Cult. Herit.* **2013**, *14*, 332–345. [CrossRef]
42. Louvre Abu Dhabi. Available online: https://www.louvreabudhabi.ae/ (accessed on 19 June 2018).
43. Tate Modern. Available online: http://www.tate.org.uk/whats-on/tate-modern/exhibition/modigliani (accessed on 19 June 2018).
44. Petrelli, D.; Ciolfi, L.; Van Dick, D.; Horneker, E.; Not, E.; Schmidt, A. Integrating material and digital: A new way for cultural heritage. *Interactions* **2013**, *20*, 58–63. [CrossRef]
45. Wilson, P.F.; Stott, J.; Warnett, J.M.; Attridge, A.; Smith, M.P.; Williams, M.A. Evaluation of touchable 3d-printed replicas in museums. *Curator Mus. J.* **2017**, *60*, 445–465. [CrossRef]
46. Brown, S.J.; Collins, A.; Duguid, P. Situated cognition and the culture of learning. *Educ. Res.* **1989**, *18*, 32–42. [CrossRef]
47. Hölzel, B.K.; Lazar, S.W.; Gard, T.; Schuman-Olivier, Z.; Vago, D.R.; Ott, U. How does mindfulness meditation work? Proposing mechanisms of action from a conceptual and neural perspective. *Perspect. Psychol. Sci.* **2011**, *6*, 537–559. [CrossRef] [PubMed]
48. Di Franco, P.D.G.; Camporesi, C.; Galeazzi, F.; Kallmann, M. 3D prnting and immersive visualization for improved perception of ancient artifacts. *Presence Teleoperatoris Virtual Environ.* **2015**, *24*, 243–264. [CrossRef]
49. Mc Ginnis, R. Islands of stimulation: Perspectives on the museum experience, present and future. In *The Multisensory Museum: Cross-Disciplinary Perspectives on Touch, Sound, Smell, Memory, and Space*; Levent, N., Pascual-Leone, A., Eds.; Roman & Littlefield: Plymouth, UK, 2014; pp. 319–329.
50. Shwandt, H.; Weinhold, J. 3D technologies for museums in Berlin. In Proceedings of the EVA London 2014 on Electronic Visualisation and the Arts, London, UK, 8–10 July 2014; pp. 255–261.
51. Sportun, S. The future landscape of 3D in museums. In *The Multisensory Museum: Cross-Disciplinary Perspectives on Touch, Sound, Smell, Memory, and Space*; Levent, N., Pascual-Leone, A., Eds.; Roman & Littlefield: Plymouth, UK, 2014; pp. 319–329.
52. Candling, F. *Art, Museums and Touch*; Manchester University Press: Manchester, UK, 2010.
53. Chatterjee, H.; Vreeland, S.; Noble, G. Museopathy: Exploring the healing potential of handling museum object. *Mus. Soc.* **2009**, *7*, 164–177.
54. Kuo, C.-W.; Lin, C.-T.; Wang, M.C. New media display tecnologiy and exhibition experience. *ACSIJ Adv. Comput. Sci. Int. J.* **2016**, *5*, 103–109.
55. Jansson, G.; Bergamasco, M.; Frisoli, A. A new option for the visually impaired to experience 3D art at museums: Manual exploration of virtual copies. *Vis. Impair. Res.* **2003**, *5*, 1–12. [CrossRef]
56. Brewster, S. The impact of haptic 'touching' technology on cultural applications. In *Digital Applications for Cultural and Heritage Institutions*; Hemsley, J., Cappellini, V., Stanke, G., Eds.; Routledge: London, UK, 2001; pp. 1–12.
57. Dima, M.; Hurcombe, L.; Wright, M. Touching the past: Haptic augmented reality for museum artefacts. In *Lecture Notes in Computer Science, Proceedings of the Virtual, Augmented and Mixed Reality: Applications of Virtual and Augmented Reality, VAMR 2014, Crete, Greece, 22–27 June 2014*; Shumaker, R., Lackey, S., Eds.; Springer: Cham, Switzerland, 2014; Volume 8526, pp. 3–14.
58. D'Agnano, F.; Balletti, C.; Guerra, F.; Vernier, P. TOOTEKO: A case study of augmented reality for an accessibile cultural heritage. Digitization, 3D printing and sensors for audio-tactile experience. *Int. Arch. Photogramm. Remote Sens. Spat. Inf. Sci.* **2015**, *XL–5/W4*, 207–213.
59. Neumüller, M.; Reichinger, A.; Rist, F.; Kern, C. 3D printing for cultural heritage: Preservation, accessibility, research and education. In *Lecture Notes in Computer Science*; Springer: Cham, Switzerland, 2014; Volume 8355, pp. 119–134.
60. Neumüller, M.; Reichinger, A. From stereoscopy to tactile photography. *Photo Res.* **2014**, *19*, 59–63.
61. Rossetti, V.; Furfari, F.; Leporini, B.; Pelegatti, S.; Quarta, A. Enabling access to cultural heritage for the visually impaired: An interactive 3D model of a cultural site. *Procedia Comput. Sci.* **2018**, *130*, 383–391. [CrossRef]
62. Callieri, M.; Pingi, P.; Potenziani, M.; Dellepiane, M. Alchemy in 3D: A digitization for a journey through matter. *Proc. Digit. Herit. Int. Congr.* **2015**, *1*, 223–231.

63. Balletti, C.; Galeazzo, L.; Gottardi, C.; Guerra, F.; Vernier, P. New technologies applied to the history of the Venice Lagoon. In Proceedings of the 11th ICA Conference on Digital Approaches to Cartographic Heritage, Thessaloniki, CartoGeoLab-Laboratory of Cartography & Geographic Analysis, Xi'an, China, 18–27 November 2016; Livieratos, E., Ed.; AUTH CartoGeoLab: Thessaloniki, Greece, 2016; pp. 182–3893.

64. Calabi, D.; Galeazzo, L. *Acqua e Cibo a Venezia. Storie Della Laguna e Della Città*; Marsilio: Venezia, Italy, 2015.

65. Galeazzo, L. Mapping change and motion in the lagoon: The island of San Secondo. In *Visualizing Venice: Mapping and Modeling Time and Change in a City*; Huffman, K.L., Giordano, A., Bruzelius, C., Eds.; Routledge: London, UK, 2017; pp. 43–50. ISBN 978-11-3828-599-6.

66. Allard, T.T.; Sitchon, M.L.; Sawatzky, R.; Hoppa, R.D. Use of hand-held laser scanning and 3D printing for creation of a museum exhibit. In Proceedings of the 6th International Symposium on Virtual Reality, Archaeology and Cultural Heritage, Pisa, Italy, 8–11 November 2005; Mudge, M., Ryan, N., Scopigno, R., Eds.; Eurographic Association: Goslar, Germany, 2005; pp. 182–196.

67. Scopigno, R.; Cignoni, P.; Pietroni, N.; Callieri, M.; Dellepiane, M. Digital fabrication techniques for cultural geritage: A survey. *Comput. Graph. Forum* **2017**, *36*, 6–21. [CrossRef]

68. Capurro, C.; Nollet, D.; Pletinckx, D. Tangible interfaces for digital museum applications: The virtex and virtex light systems in the keys to Rome exhibition. In Proceedings of the 2015 Digital Heritage International Congress, Granada, Spain, 28 September–2 October 2015; Guidi, G., Scopigno, R., Torres, J.C., Grafs, H., Eds.; Springer: New York, NY, USA, 2015.

69. Van Gogh. Available online: https://www.theguardian.com/artanddesign/2013/aug/24/3d-replicas-van-gogh (accessed on 15 May 2018).

70. Ballarin, M.; Balletti, C.; Vernier, P. Replicas in cultural heritage: 3D printing and the museum experience. *Int. Arch. Photogramm. Remote Sens. Spat. Inf. Sci.* **2018**, *42*, 55–62. [CrossRef]

71. Martini, F. L'arte paleolitica e mesolitica in Italia, Millenni. *Studi Archeol. Preist.* **2016**, 12.

72. Geomagic. Available online: https://it.3dsystems.com/software (accessed on 3 April 2018).

73. Sculpteo. Available online: https://www.sculpteo.com/en/ (accessed on 3 April 2018).

74. PhotoScan. Available online: http://www.agisoft.com/ (accessed on 6 July 2018).

75. Vivid. Available online: https://www.konicaminolta.com/instruments/download/instruction_manual/3d/pdf/vivid-9i_vi-9i_instruction_eng.pdf (accessed on 6 July 2018).

76. Range7. Available online: https://www.konicaminolta.eu/it/strumenti-di-misura/prodotti/misurazioni-3d/range-7/introduzione.html (accessed on 5 July 2018).

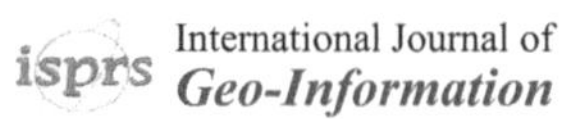 International Journal of *Geo-Information*

Article

Novel Method for Virtual Restoration of Cultural Relics with Complex Geometric Structure Based on Multiscale Spatial Geometry

Miaole Hou [1,2], Su Yang [3,*], Yungang Hu [1,2,*], Yuhua Wu [4], Lili Jiang [5], Sizhong Zhao [1,2] and Putong Wei [1,2]

[1] School of Geomatics and Urban Information, Beijing University of Civil Engineering and Architecture, Beijing 100044, China; houmiaole@bucea.edu.cn (M.H.); 210813j116013@stu.bucea.edu.cn (S.Z.); 210813j117014@stu.bucea.edu.cn (P.W.)
[2] Beijing Key Laboratory for Architectural Heritage Fine Reconstruction & Health Monitoring, Beijing University of Civil Engineering and Architecture, Beijing 102616, China
[3] North China Institute of Computing Technology, Beijing 100083, China
[4] Historic Cultural Research Institute, Beijing 100029, China; wuyuhua@cahch.org.cn
[5] Beijing Digsur Science and Technology Co. Ltd., Beijing 100012, China; jiangll18810970196@outlook.com
* Correspondence: cinder0711@hotmail.com (S.Y.); huyungang@bucea.edu.cn (Y.H.);
  Tel.: +86-138-1100-7050 (S.Y.); +86-136-4134-0555 (Y.H.)

Received: 18 July 2018; Accepted: 9 August 2018; Published: 27 August 2018

Abstract: Because of the age of relics and the lack of historical data, the geometric forms of missing parts can only be judged by the subjective experience of repair personnel, which leads to varying restoration effects when the geometric structure of the complex relic is reconstructed. Therefore, virtual repair effects cannot fully reflect the historical appearance of cultural relics. In order to solve this problem, this paper presents a virtual restoration method based on the multiscale spatial geometric features of cultural relics in the case of complex construction where the geometric shape of the damaged area is unknown, using the Dazu Thousand-Hand Bodhisattva statue in China as an example. In this study, the global geometric features of the three-dimensional (3D) model are analyzed in space to determine the geometric shape of the damaged parts of cultural relics. The local geometric features are represented by skeleton lines based on regression analysis, and a geometric size prediction model of the defective parts is established, which is used to calculate the geometric dimensions of the missing parts. Finally, 3D surface reconstruction technology is used to quantitate virtual restoration of the defective parts. This method not only provides a new idea for the virtual restoration of artifacts with complex geometric structure, but also may play a vital role in the protection of cultural relics.

Keywords: cultural heritage; virtual restoration; spatial geometric features; skeleton line; regression analysis; Dazu Thousand-Hand Bodhisattva statue

---

1. Introduction

Computer-aided restoration of cultural heritage relics has made substantial progress, yet it still faces major challenges when the geometry of the damaged parts of artifacts is unknown. Due to the lack of reliable evidence for restoration, the geometry of the missing parts of many artifacts cannot be specified. As a result, for the same piece of an artifact, different persons may repair its incomplete parts differently, and its true historical appearance cannot be determined. As shown in Figure 1, in a part of the Dazu Thousand-Hand Bodhisattva statue, due to its long history and erosion, the index finger is missing. In order to provide a reference for the actual repair of this finger, a computer may be used to perform virtual restoration. However, due to a lack of reliable evidence for the virtual restoration, different restoration methods can lead to different repair results (see Figure 1b–d).

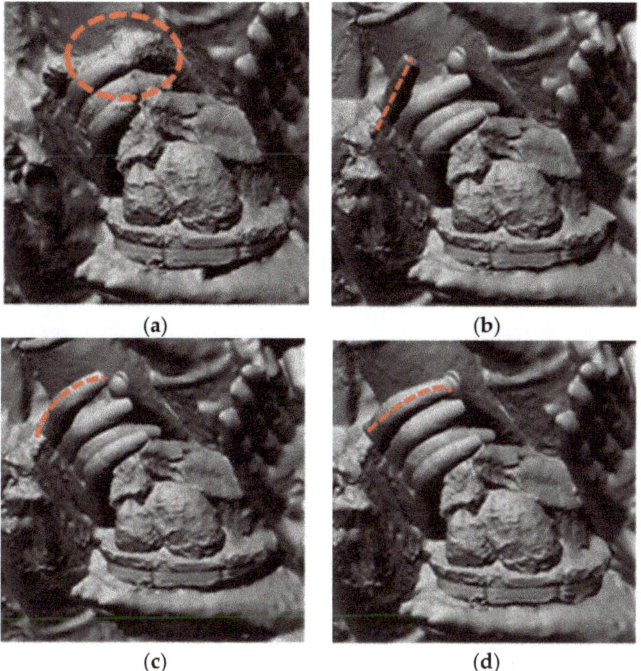

Figure 1. Different restoration results due to lack of reliable restoration evidence: (**a**) damaged finger; (**b**) possible result 1; (**c**) possible result 2; and (**d**) possible result 3.

Therefore, for damaged and missing parts of an artifact, it is essential to find evidence for virtual restoration by analyzing its basic geometry in the same period or of the same type, so that the effect of virtual restoration can be close to its true historical appearance. This not only helps with the three-dimensional display of artifacts, but also provides a more accurate repair plan for artifact protection, so that valuable historical artifacts can be continuously preserved. Hence, this problem has become an important and timely issue that needs to be solved in the field of virtual artifact restoration.

In response to these problems, this paper presents a virtual restoration method for artifacts with complex geometric structure based on multiscale spatial geometric features, by using virtual restoration of defective fingers in the Dazu Thousand-Hand Bodhisattva statue as an example. In this study, the basic geometric features of the missing parts are determined by analyzing the global geometric features, such as the symmetry and similarity of the three-dimensional models. High-precision three-dimensional models of cultural relics are extracted from the skeleton lines as the local geometric features of the relics, and the geometric dimensions of the missing parts are determined by mathematical analysis.

The paper is organized as follows. After this introduction, some related works are reviewed in Section 2, and Section 3 introduces the overall process of the proposed method and the key methods in the process. Section 4 describes the experimental application of the proposed method using the Dazu Thousand-Hand Bodhisattva statue. Section 5 provides some discussion about the method. Section 6 concludes the paper with observations obtained from this study.

2. Related Works

In the field of virtual artifact restoration, some progress has already been made with regard to the technologies used for virtual restoration of broken artifacts. For example, Su [1] used historical records, pictures, and an analysis of the style of the Qingzhou Buddha statues to restore one of the statues by computer. However, due to their long history, no reliable information is available for many artifacts. Many researchers [2–4] have investigated techniques for extraction and matching of surface boundaries of artifact fragments to provide more reliable repair information. For example, Fan [5] applied shape matching of complementary three-dimensional (3D) polygonal arcs under the constraint of tangent vector across boundaries and achieved automated restoration of fragmented artifacts. Huang et al. [6] used spatial surface-matching techniques in a study of automated assembly of artifact fragments. Wang [7] developed an improved Hausdorff distance-matching algorithm for assembling fragmented artifacts, and applied this algorithm in virtual artifact restoration. Li [8] proposed an algorithm for the restoration of broken rigid bodies based on the matching of fracture surfaces, and showed that this method could be used for accurate assembly of relatively complex artifact fragments for restoration. Geoffrey et al. [9] achieved automated assembly of thin artifact fragments under the condition that the geometry of the artifact was unknown. This method is of important reference value to the virtual restoration of hollow artifacts, such as ceramic containers, bowls, and vases. However, all the above methods only apply to situations where all fragments of the broken artifacts are available and substantially preserved, and it is still difficult to perform virtual restoration of artifacts with missing parts.

In order to recover artifacts with missing parts in damaged areas, some scholars have proposed methods to estimate the geometry of the missing parts based on the symmetry of the artifact, in order to ensure the reliability of virtual restoration results. Li et al. [10] proposed a geometric method that estimates the symmetry axis of broken artifacts to obtain information on the generatrix of the artifact fragments, and applied this method to successfully restore an artifact model with inner and outer surfaces. Yang et al. [11] discussed the framework of a computer-aided artifact restoration system. They used quadratic surface fitting to calculate the normal vectors, and applied straight-line geometric methods to calculate the rotary axis of the entire object, in order to improve the precision of the calculation on the rotary axis. With this method, they successfully restored artifact fragments in the shape of common rotating bodies. Willis et al. [12] performed recombination on ceramic fragments with axial symmetry, and applied this method in the restoration of artifacts that show the property of axial symmetry. However, the application of the above methods is limited to broken artifacts with axial symmetry, such as bowls and pots. For artifacts in which the missing parts do not have a regular geometry, it is still challenging to obtain evidence for restoration.

Among the studies of virtual restoration of damaged artifacts with complex geometric configurations, researchers have used the constraints on local geometric features of the surface of the artifact model to create a smooth repair of missing parts [13–15]. In order to conduct virtual restoration of artifacts with relatively abundant geometric texture on the surface, Henning et al. [16] proposed an interactive cut-and-paste method based on multiresolution subdivision surfaces for texture restoration of three-dimensional artifact models. Yu et al. [17] used Poisson's equation as the theoretical basis to provide mesh editing, deformation, consolidation, and smoothing operations of artifact models. Wei [18] applied carving on the surface of the artifact model, as well as cutting and pasting the geometric texture and other operations to repair the damaged area of the artifact model. However, although these methods achieve free editing and operations on the surface of the artifact's model so that the artifact could be restored to a shape that met the restorer's expectations, it could not be reliably repaired by this method. In response to this, one research institute recombined the fragments of a terracotta statue damaged in a 2009 earthquake and used the geometric shape of the right hand of the statue as a reference for virtual restoration of the missing left hand, followed by exploring the restoration of its polychrome decoration (2013). Lu et al. [19] and Lu et al. [20] applied a method based on the matching of rigid surfaces, and performed a virtual restoration of damaged Buddha faces in artifact restoration and protection of the Bayon Temple in Cambodia, utilizing the characteristic that the four faces in the same spire bear similarities. Wu et al. [21] conducted virtual restoration on damaged hands of the rock carving of the Dazu Thousand-Hand statue, but their study mainly focused on damage to the surface area of the three-dimensional artifact modeler for incomplete statue hands that have missing parts, and these studies found no reliable evidence for the restoration.

3. Methodology

3.1. Overall Process of the Proposed Method

This research aims to achieve accurate and objective virtual restoration of damaged cultural relics with complex geometry. To solve the problem of virtual restoration to diversify the results, the proposed method was used to analyze the spatial geometric features of 3D cultural relic models under multiscales. That analysis can obtain historical information on the geometric shapes and sizes of the missing parts. On this basis, the result can show the historical origin of the relics. The whole process of virtual repair is shown in Figure 2.

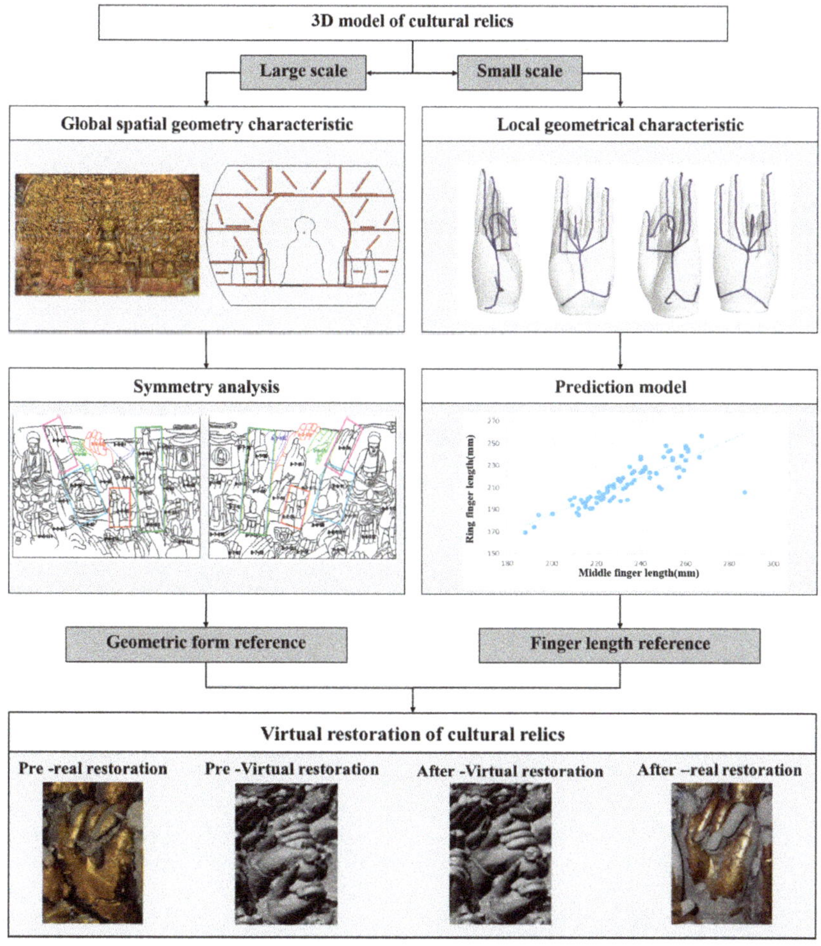

Figure 2. Overall technical process of the proposed method.

The overall process is divided into two aspects: the first is to analyze the global geometric characteristics of the cultural relics on a large-scale model, and the second is to analyze the local geometric features on fine models with small scale. In the global space geometry analysis, the purpose is to find the geometric information of the defective part in the relic, then analyze the symmetry of the whole object. Using this method, the geometric reference samples of the defect object are usually obtained.

Further, in order to solve the condition of similar geometric shape but different proportion, the local geometric features of the cultural relic needs to be analyzed. This study uses the skeleton line of fine model to express the local geometric characteristics, because the skeleton line can express the geometric shape of each finger and its topologic relation while ignoring unwanted model surface noise. Through the statistics and analysis of artifacts in geometric characteristics, especially the skeleton line, the goal of local space geometry analysis is to establish a prediction model of the size of missing parts in relics. It is important to note that the limitations of the predictive model lie in the fact that artifacts are required to be grouped together or that subregions are grouped in a whole artifact.

Through these steps, the geometric shape and actual size of the missing part can be found. These are the constraints of virtual restoration of cultural relics. It is important that the constraint conditions are analyzed by the geometric characteristics of the relics so that the constraint conditions are in accordance with the history. After that, 3D reconstruction technology is used for virtual repair.

3.2. Global Spatial Geometry Characteristics Analysis

The analysis of global spatial geometry characteristics is mainly divided into two steps, spatial geometry, and similarity analysis. The geometric construction of cultural relics divides the relics into several subregions through a certain kind of spacing mechanism. The purpose of the geometric feature analysis of the entire cultural relic space is to examine the premise of whether there is any connection between geometries shown in relics. If this assumption holds, the inherent relationships between geometric constructs can be used to support the virtual restoration of artifacts. For example, after the cultural relics are divided into regions, they may exhibit spatial symmetry. In this way, the geometric similarity between structures can be utilized to obtain geometric references of artifacts. Geometric analysis of the space can vary according to the geometry of different artifacts, but the main considerations include the overall size of the artifacts, the overall axis of symmetry, and the center of the whole artifact.

3.3. Local Geometric Characteristics Analysis

3.3.1. Extraction of Skeleton Line

The local geometric features of cultural relics can usually be calculated by geometric calculation on the 3D model, but simple calculation cannot get accurate results on the skeleton line. Furthermore, the skeleton line of a 3D model of a cultural relic refers to the one-dimensional representation of the three-dimensional model geometry and its topologic structure. The three-dimensional model of complex geometric structure is expressed by the skeleton line, which not only reflects its local geometric features, but also preserves its topologic relation. The 3D model of the relic is expressed in this form, which makes it easier to analyze the local geometric features. Therefore, the skeleton line is of great significance to the virtual restoration of cultural relics.

For the extraction of skeleton lines of 3D models, the method based on the Laplace operator, proposed by Au [22], is adopted and improved by taking into account the topology and centrality of the skeleton lines.

The algorithm is divided into three steps: geometric contraction, skeleton simplification, and node fine-tuning.

1. The geometric contraction process can be expressed as follows:

Step 1: Solve $\begin{bmatrix} W_L^t & L^t \\ W_H^t & 0 \end{bmatrix} V^{t+1} = \begin{bmatrix} 0 & 0 \\ W_H^t & V^t \end{bmatrix}$ and obtain the solution V^{t+1}, where L is an $n \times n$ weighted Laplace operator of curvature flow, V is a vertex of the manifold, and W_L^t and W_L^t are both $n \times n$ weighted diagonal matrices. These two matrices are used to balance the contraction strength and constraint of shape preservation during the iterative contraction process. The ith element in the diagonal matrix $W_L^t(W_H^t)$ is defined as $W_{L,i}^t(W_{H,i}^t)$.

Step 2: Use the newly obtained solution V^{t+1} to update L^{t+1}.

Step 3: Update the contraction weight and the constraint weight. A is the area of the first-ring triangle mesh with i as the vertex in the triangulation model, and S_L is the contraction step from the origin to the L vertex, where

$$W_L^{t+1} = S_L W_L^t \tag{1}$$

$$W_{H,i}^{t+1} = W_{H,i}^0 \sqrt{A_i^0 / A_i^t} \tag{2}$$

Step 4: The condition for terminating the iteration is reached when the ratio of the model volume after contraction and the original model volume is less than 1×10^{-5}. If the termination condition is not reached, go back to Step 1; otherwise, the iteration is terminated.

2. Skeleton simplification is a method for further converting this mesh model of approximately zero volume to one-dimensional skeleton lines.

In this process, we define a quadratic error matrix K_{ij} for each edge (i, j) in the model (see Equation (3)). A matrix method based on the distance between the midpoint and the line of space can be obtained. The quadratic error measure of point p to its opposite side is $P^T\left(K_{ij}^T K_{ij}\right)P$.

$$K_{ij} = \begin{bmatrix} 0 & -a_z & -a_y & -b_x \\ a_z & 0 & a_x & b_y \\ -a_y & a_x & 0 & -b_z \end{bmatrix} \tag{3}$$

where a is the normalized edge vector of edge (i, j), and $b = a \times \hat{V}_i$.

As shown in Equation (4), the initial quadratic error of vertex i is the sum of squares of the distance from the point to its associated edge

$$F_i p = P^T \sum_{(i,j) \in E} \left(K_{ij}^T K_{ij}\right) P \tag{4}$$

In the process of simplification, in order to guarantee the shape of the skeleton line as much as possible, the quadratic error measure produced by folding an edge (i, j) is

$$\varepsilon_a(i,j) = F_i\left(\widetilde{V_j}\right) + F_j\left(\widetilde{V_j}\right) \tag{5}$$

In this simplification algorithm, the method based on the quadratic error may easily cause oversimplification of the skeleton nodes when the geometry of the finger is close to a straight line. Hence, to ensure that the skeleton nodes with relatively even density on fingers with different curvatures in the same statue hand are obtained, the weights to appropriately amplify the quadratic errors in the model where the finger is approximately a straight line are added.

$$\varepsilon_b(i,j) = \sum_{(i,k) \in E}^{n} \|\widetilde{V_i} - \widetilde{V_j}\| \tag{6}$$

Thus, the overall quadratic error is the weighted sum of the above two parts of quadratic errors

$$\varepsilon(i,j) = a\varepsilon_a(i.j) + b\varepsilon_b(i,j) \tag{7}$$

3. In the node fine-tuning process, the generated skeleton nodes are divided into three categories: regular nodes, branch nodes, and terminal nodes. Assuming the upper and lower boundaries of the region is set and each boundary contains a vertex set S_j that includes vertices in the original model, the weighted average displacement of the vertex d_j in the boundary can be expressed by Equation (8)

$$d_{j=} \frac{\sum_{i \in set(j)} l_{i,j}(\widetilde{v}_i - v_j)}{\sum_{i \in set(j)} l_{j,i}} \tag{8}$$

where $l_{j,i}$ is the sum of the two adjacent sides converging to vertex i in the first ring boundary j.

For ordinary nodes, there are only two adjacent edges in the skeleton point. Since the skeleton nodes correspond to the two ring boundaries d_1 and d_2 in the original model, the spatial coordinates of the skeleton points are adjusted to

$$V = V - \frac{d_1 + d_2}{2} \; V = V - \frac{d_2 + d_2}{2} \tag{9}$$

where V is the spatial coordinates of the skeleton point.

For branch nodes that have three or more adjacent edges, in order to make the skeleton line of the model more in line with the topological relations among the geometric structures of the model ontology, it is necessary to merge the branch nodes that are too close. It should be noted that for models with different geometric constructions, the thresholds for branch node merging need to be adjusted according to the actual situation. Moreover, the centrality of the skeleton point of the branch node is adjusted to

$$V = \frac{d_1 d_2 + \dots d_n}{N} \tag{10}$$

The last kind of skeleton nodes are called end nodes, which are characterized by one point with only one adjacent edge. The fine-tuning method for this class of vertices is

$$V = V - d \; V = V - d \tag{11}$$

3.3.2. Prediction Model

In this study, regression analysis is used to fit the prediction model, which is used to calculate the length of broken relics (in this study, the finger length), to obtain reliable references for the restoration. The equation of the model is

$$Y = \hat{a} + \hat{b}x \tag{12}$$

where $\hat{b} = \frac{n\sum_{i=1}^{n} x_i y_i - \left(\sum_{i=1}^{n} x_i\right)\left(\sum_{i=1}^{n} y_i\right)}{n\sum_{i=1}^{n} x_i^2 - \left(\sum_{i=1}^{n} x_i\right)^2}$, $\hat{a} = \frac{1}{n}\sum_{i=1}^{n} y_i - \frac{\hat{b}}{n}\sum_{i=1}^{n} x_i$, x and y represent the lengths of two fingers on the same hand, i is the ith sample, and n is the number of samples.

4. Results and Analysis

4.1. Experimental Artifact

To test the proposed method, the Dazu Thousand-Hand Bodhisattva statue in China was chosen as the study artifact, as shown in Figure 3a. This artifact is located in section no. 8 of the South Cliff of the Giant Buddha Bay in Baodingshan, Dazu District, Chongqing City. The Thousand-Hand statue was carved in the Southern Song Dynasty, dating back over 800 years. It is one of the most exquisite parts of the Dazu rock carvings. The Dazu Thousand-Hand Bodhisattva statue is famous for having nearly 1000 hands in different forms and holding different ritual instruments, positioned in a radial arrangement. The entire statue is covered with gold foil, showing a magnificent appearance, vivid forms, and resplendent gold color. However, due to its long history and environmental factors, damage such as rock weathering and peeling of the gold foil have occurred (Figure 3b). Because of this damage, many fingers in the hands are completely or partially missing, so that the rock statue can no longer faithfully represent its historical appearance, seriously affecting archaeological research. Hence, using three-dimensional laser scanning to digitize the geometry of the statue and perform virtual restoration of the incomplete fingers, aiming to precisely restore the statue's historical appearance, is important to both the protection of this artifact and archaeological research.

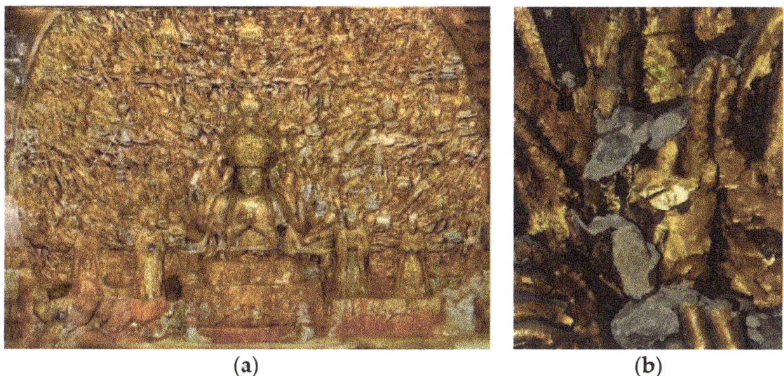

Figure 3. Dazu Thousand-Hand Bodhisattva statue: (**a**) orthophoto map; (**b**) one damaged hand.

4.2. Experimental Data

The data used in our research included different scales of three-dimensional point clouds. The three dimensional point clouds of the whole shrine of the Dazu Thousand Hand Bodhisattva statue were acquired using a FARO LS420 scanner (Figure 4a) with high measurement accuracy (position 3–6 mm, sampling density 1 mm). These data were mainly used to construct a triangular mesh model of the whole niche. During the construction process, the original point cloud data needed to be preprocessed, including point cloud denoising and station. Because of the large size of the Guanyin statue, multistation data were required to be able to scan it intact. The complete model was formed by splicing the repeated parts between the stations. Finally, after repairing some holes caused by disease or occlusion of cultural relics, the three-dimensional model was built by the point cloud model encapsulation as shown in Figure 5a.

A CimCore Infinite 2.0 articulating arm (Figure 4b) was used to obtain finer scanning data (position 0.045 mm, sampling density 0.1 mm) of the hands and Dharma instruments by doing meticulous scanning one by one. These data were used to construct a high-precision triangular mesh model of the statue hands. These data show the detailed geometry of the Guanyin hands and the geometric texture of the surface. The difference with the above process is that after preprocessing the point cloud data, some redundant data caused by disease on the model surface, as the gold foil is missing from the surface, needed to be deleted.

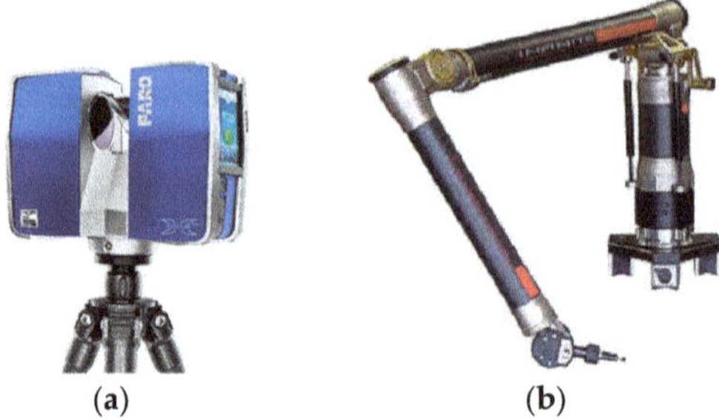

Figure 4. Equipment used in data acquisition: (a) FARO LS420 scanner; (b) CimCore Infinite 2.0 articulating arm.

4.3. Global Spatial Geometry Characteristics Analysis

4.3.1. Spatial Partition

The purpose of partitioning the Thousand-Hand Bodhisattva statue was to facilitate spatial analysis. According to the partitioning, each hand was given a unique identification. In the experiment, the whole statue was divided into a nine-by-nine grid, ending up with 81 areas covering the whole area. The lower left corner was the origin. The serial number of each hand was a combination of line number, column number, and number of hands in the area (Figure 6a). For example, the ID 9-7-s3 represents the third hand from the spectator at row 9, column 7 (Figure 6b).

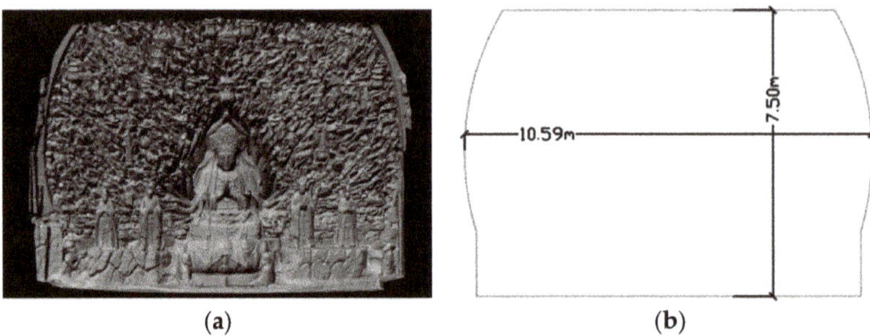

Figure 5. Geometric information of the Dazu Thousand-Hand Bodhisattva statue: (a) orthophoto quad; and (b) dimensions of statue model.

4.3.2. Spatial Geometry Analysis

It was important to determine the center point of the Dazu Thousand-Hand Bodhisattva statue to analyze the spatial distribution of the hands. Through three-dimensional measurement and statistics on the triangular mesh models that were reconstructed by point clouds, the results showed that the statue is about 7.5 m high and 10.59 m wide (Figure 5). Furthermore, there are 830 hands, of which 599 are relatively well maintained.

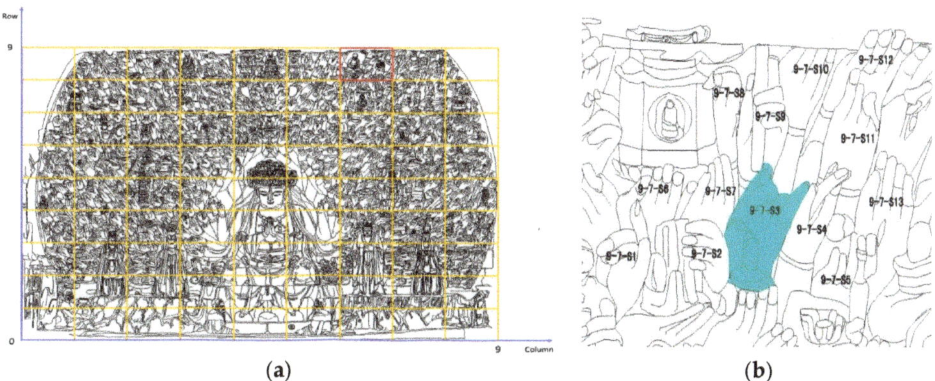

Figure 6. Spatial partitioning of the Dazu Thousand-Hand Bodhisattva statue: (**a**) partitioning; (**b**) line drawing of hand number 9-7-s3.

In the three-dimensional model shown in Figure 5a, the curved boundary on both sides of the statue can fit into an approximate ellipse. The basic parameters of the ellipse are as follows: long axis (vertical direction) 12.5 m, short axis (horizontal direction) 10.59 m, and flat curvature 0.85. The center of the ellipse is measured by the origin. The objects are measured at the center of the ellipse. The result is that the upper boundary of the center is 3.70 m, the lower boundary of the statue is 3.80 m, and the boundary between the left and right sides is 5.30 m (see Figure 7a). In addition, the center of the ellipse is basically coincident with an extremely important and very special place called the "sky eye" on the Dazu Thousand-Hand Bodhisattva statue. Figure 7b shows the positions of the center and "sky eye". In this study, the "sky eye" is considered to be the center of the whole statue. It is very important to analyze the symmetry of cultural relics. Through this position, the cultural relic can be divided vertically into left and right halves. On the two sides of the vertical bisector, the spectator can appreciate the characteristics of spatial symmetry and similar geometric morphology.

(a)

Figure 7. *Cont.*

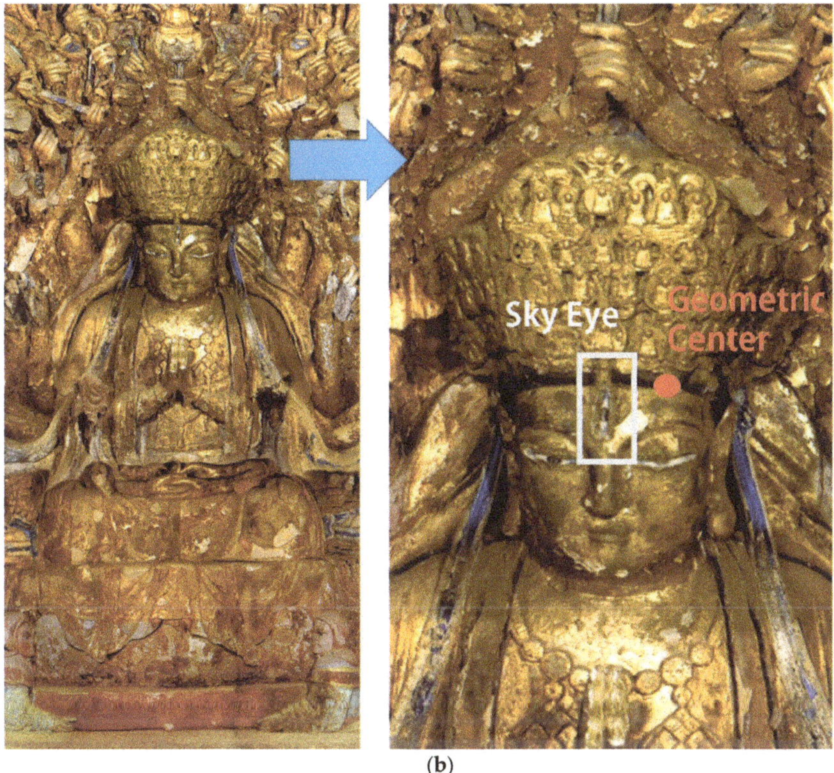

Figure 7. Global spatial geometric characteristics analysis of the Dazu Thousand-Hand Bodhisattva statue: (**a**) example of spatial geometric feature fitting on the object; (**b**) positions of geometric center point and "sky eye".

4.3.3. Analysis of Global Geometric Similarity

The purpose of analyzing the symmetries of the Dazu Thousand-Hand Bodhisattva statue is to find similar hands in the statue. The approximate geometric shape of a damaged hand can be obtained through similarities detected by the spectator. This study identifies similar hands through global geometric features and related religious knowledge, mainly including the following:

1. Symmetry of hand positions, whether the positions of both hands are based on the centerline symmetry of the statue.
2. Geometrical morphology, the main factor that determines whether the similarity of the geometric form of a symmetrical hand is the geometric form of each finger. Usually, each hand represents a gesture, which has a special religious meaning. Therefore, it is possible to check whether symmetrical hands express the same gesture. Finally, it can be judged by whether the hands hold the same instrument.
3. Neighboring relations, whether the hand adjacent to the target hand is symmetrical and similar in shape.

Based on the above principles, two corresponding hands can be found as the left hand and the right hand. If one of them is well preserved, it can provide a geometric reference to the corresponding broken one.

4.4. Analysis of Local Geometric Characteristics

4.4.1. Extraction of Skeleton Line

The method for extracting skeleton lines was applied to the statue. After several experimental rounds, the iteration parameters for the three-dimensional model of the statue were obtained as follows: contraction strength $W_L^0 = 77.56$, contraction step $S_0 = 2.0$, shape preservation constraint $W_H^0 = 1.0$.

Figure 8 shows the process of the iterative contraction of the triangulation model of this statue hand. Each iteration is, in essence, one mesh contraction of the model. The shape preservation constraint of each vertex was then updated to ensure that the model contraction continues while preserving the key geometric details on the surface.

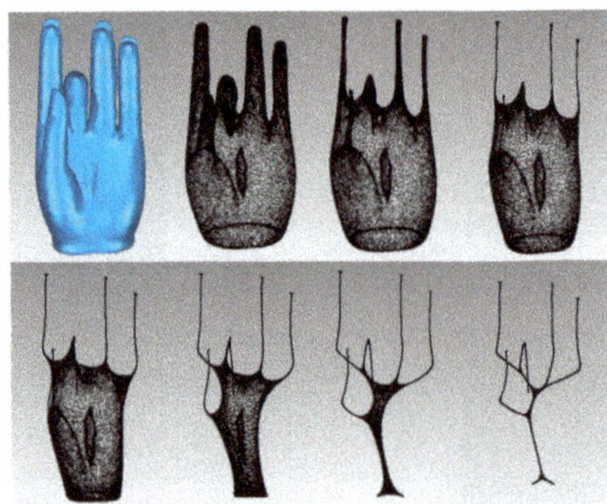

Figure 8. Iteration contraction process of cultural relic model.

According to Equations (3)–(7), when $a = 1$ and $b = 0.1$, 130 skeleton nodes can be obtained, as shown in Figure 9a; when $a = 1$ and $b = 1$, 149 skeleton nodes can be obtained, as shown in Figure 9b. Under the condition that the fingers are curved in the original hand model, adding weights does not have a large impact on the number of skeleton nodes. However, in areas where the geometric shape of the finger is relatively straight, adding weights increases the density of the skeleton nodes.

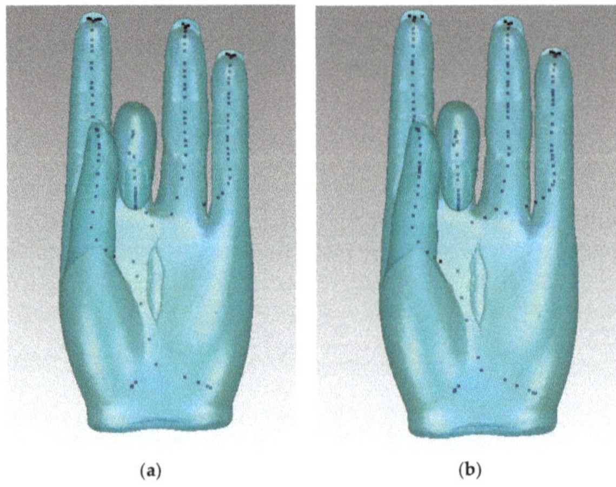

(a)   (b)

Figure 9. Simplification result of QEM (Quadric error metrics) skeleton nodes under different weights: (a) $a = 1$, and $b = 0.1$, 130 skeleton nodes; (b) $a = 1$ and $b = 1$, 149 skeleton nodes.

Figure 10 shows the results of node fine-tuning. In view of the characteristics of the three-dimensional model of the hand in the statue, the topologic relation and the separation of some skeletons from the surface of the model are improved. The red dots and yellow lines are key skeleton nodes and lines, respectively, before taking into consideration tuning with centrality. The white dots and blue lines are the key skeleton nodes and lines, respectively, after the fine-tuning. As can be seen from the figure, after applying the fine-tuning algorithm that takes centrality into account, the key skeleton nodes at the thumb junction and between the little finger and the palm are moved to the correct positions. In addition, after the fine-tuning, the topologic relationships between the skeleton lines are more in line with the biological structure of human hands. Figure 11 illustrates the multiangle display of the skeleton lines extracted from the three-dimensional model using the skeleton line extraction algorithm that takes centrality into account.

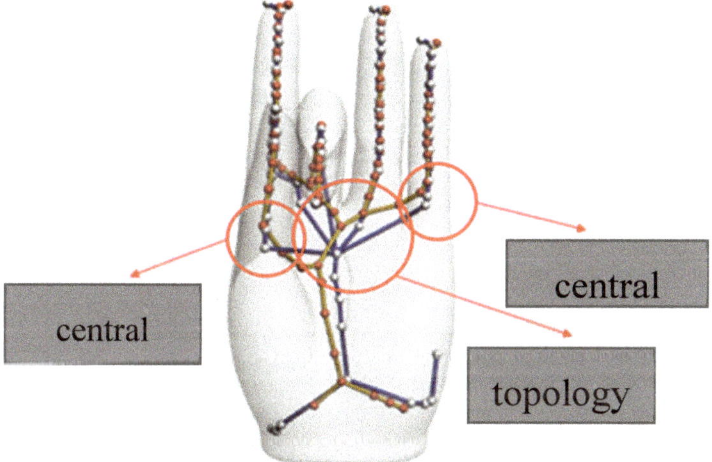

Figure 10. Comparison of skeleton nodes before and after fine-tuning.

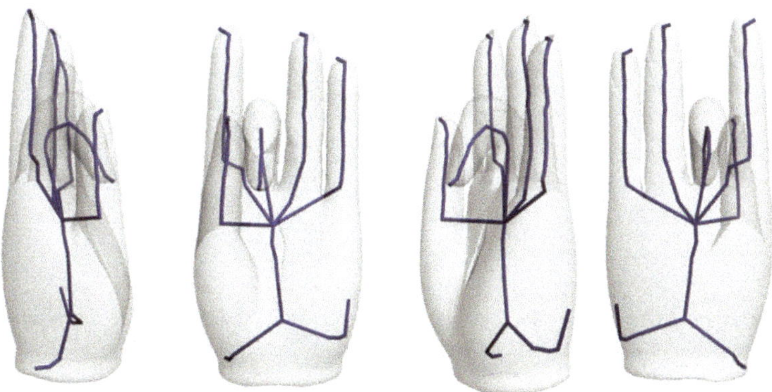

Figure 11. Multiangle displays of skeleton lines.

4.4.2. Prediction of Finger Length

Through the skeleton line extraction algorithm, the lengths of the skeletons of fingers indicate the real lengths of the fingers. Then, calculated finger lengths of the well-preserved hands are classified by thumb, forefinger, middle finger, ring finger, and little finger. Finally, the regression analysis method can be used to construct a mathematical model, which is a computational model of mathematical relations between various fingers. For the same hand, using the prediction model, the length of the broken finger can be estimated.

In the experiment, 66 well-preserved hands were identified on the Dazu Thousand-Hand Bodhisattva statue. Taking the index and middle fingers as an example to extract the length of the skeleton line, regression analysis was performed to check whether there was a linear relationship between different fingers.

4.4.3. Hypothesis Testing and Anomaly Detection

As shown in Figure 12a, a linear relationship can be observed. However, it is also obvious that there is a small number of outlier points (red dots). These unusual points may be caused by the hand grip, or by the natural form of the stone wall. Therefore, in order to ensure the accuracy of the proposed model, the anomaly points need to be excluded. In this paper, three statistical indices are used to detect abnormal points: centered leverage value, studentized deleted residual, and Cook's distance. Among them, the first two indices mainly detect a single finger length anomaly, the experimental selection centered leverage value is less than 0.03, and the studentized deleted residual is between $(-1, 1)$. The final index is used to detect the imbalance of the finger ratio, which is less than 0.01. After the anomaly detection, six sets of data were excluded, shown by red dots in Figure 12b.

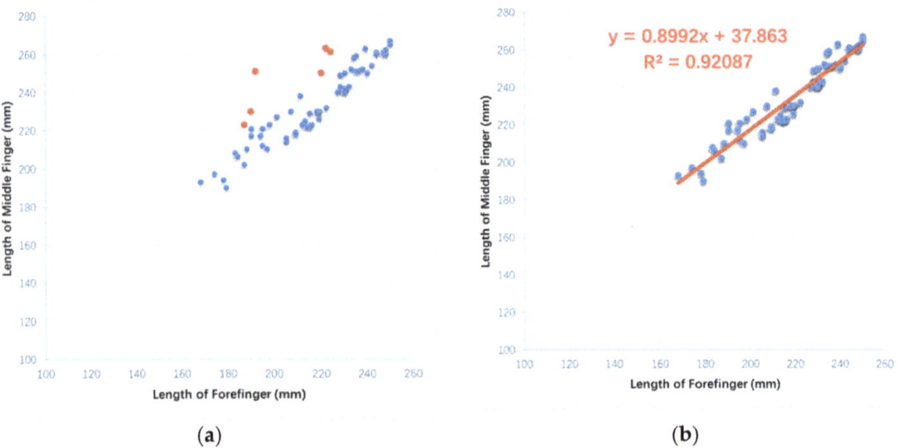

Figure 12. Prediction model: (**a**) hypothesis testing and outlier analysis of the data; (**b**) prediction model of incomplete finger length.

4.4.4. Model Fitting

After the hypothesis testing and outlier point exclusion, the lengths of the index fingers and middle fingers of 60 hands were fitted to the prediction model (Figure 12). The prediction model is

$$Y = 0.8992x + 37.863 \tag{13}$$

where x is the length of the index finger and Y is the length of the middle finger.

In this model, $R^2 = 0.92087$, and *t* test *p* value is 0.02, indicating that the prediction model has significant statistical significance. Two standard deviations were the length of the prediction interval, about 6.23 mm. This length can be fine-tuned as an aesthetic.

4.5. Example of Virtual Restoration of Incomplete Hands

This section describes virtual restoration of the hand ID 9-7-s4 (Figure 13a) on the Bodhisattva statue as an example. In this hand, the index finger is broken, and the middle finger, ring finger, and small finger have undergone weathering.

First, a similar hand, ID 9-5-s6 (Figure 13b), was found by analyzing the spatial geometry. Hands ID 9-7-s4 (broken finger) and ID 9-5-s6 (basically intact fingers) are geometrically similar, and their spatial location is symmetric based on the statue's center (Figure 14a). Furthermore, the hands adjacent to those two hands are also similar (Figure 14b). Therefore, the geometric shape of the missing parts of hand 9-7-s4 can refer to hand 9-5-s6.

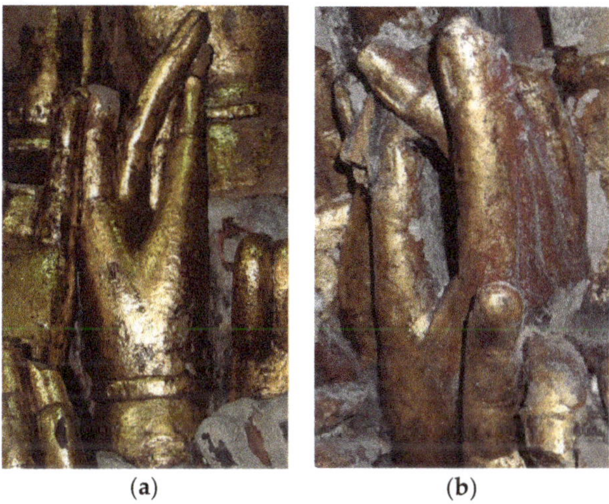

(a) **(b)**

Figure 13. Photographs of missing fingers and their reference hands: (**a**) hand 9-7-s4; (**b**) hand 9-5-s6; which is the hand symmetric to 9-7-s4.

Second, the skeleton lines of the fingers in hand ID 9-7-s4 were extracted. Among them, the length of the middle finger is 241 mm, and the length of the current index finger (broken) is 110 mm. Through Equation (13), it can be concluded that the length of the broken index finger should be 226 mm. Therefore, the finger length needs to be increased by 116 mm based on the geometric form of the index finger of 9-5-s6.

Finally, based on the geometric shape of the reference hand and the length of the finger calculated by the prediction model, the broken hand is repaired in a computer simulation. Geomagic 2017 commercial software was used to virtually recover the model by human–computer interaction. Figure 15 shows the results of this virtual restoration.

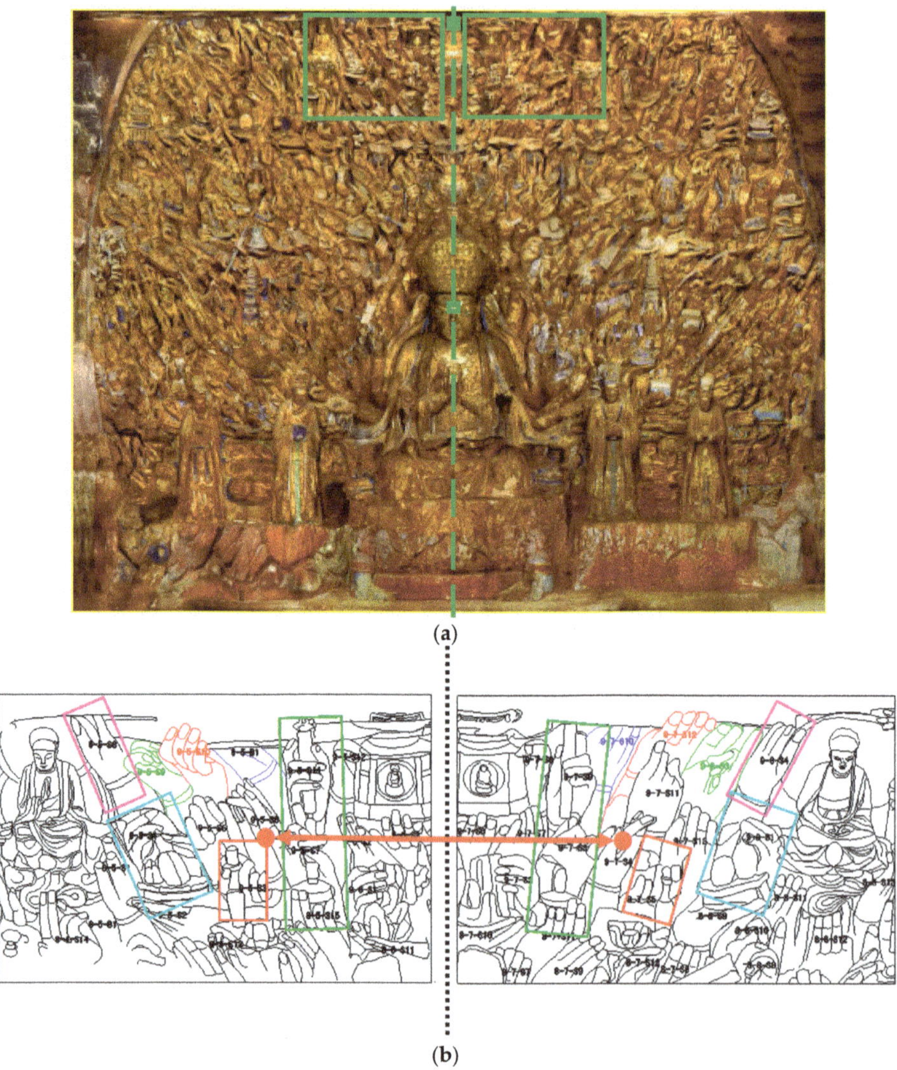

Figure 14. Characteristics of symmetry in the Dazu Thousand-Hand Bodhisattva statue: (**a**) space areas of hands 9-7-s4 (right side) and 9-5-s6 (left side); (**b**) local enlargement of line drawing with the geometrically similar hand.

There are altogether 231 incomplete Guanyin hands. Some of them can be repaired directly because of their few incomplete parts and obvious geometric shape. According to the above virtual repair process, a total of 58 broken fingers were repaired.

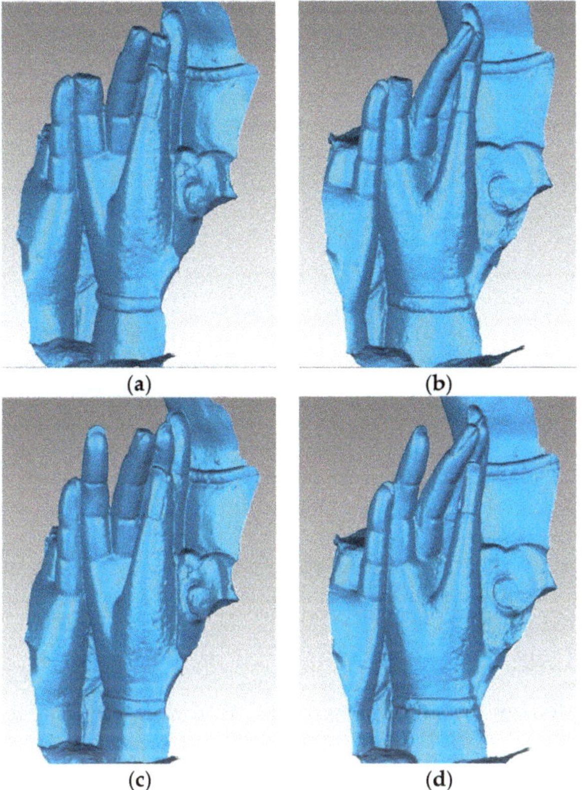

Figure 15. Virtual restoration effect of ID 9-7-s4 hand: (a,b) original model; (c,d) virtual restoration effect.

5. Discussion

In the virtual restoration of cultural heritage relics, the unknown geometries of damaged parts of artifacts present a huge challenge. How to make the virtual repair effect close to the historical appearance is a difficult problem in the absence of relevant historical records. This research used the Dazu Thousand-Hand Bodhisattva statue, part of China's world heritage, for the study. The virtual restoration effects were successfully applied to the conservation project of this cultural relic. The restoration workers refer to the virtual restoration effect provided by this method.

The method proposed in this paper not only can be used for virtual restoration of the Dazu Thousand-Hand Bodhisattva statue, but also can be applied to many other cultural relics. This is because many artifacts were designed and measured, not randomly constructed. Therefore, in most cases, cultural relics are regulated. If the tacit knowledge can be extracted by the geometric features of the relics and analyzed, this information can be used to recreate their historical shape and also provide a reference for the actual repair work. Therefore, if a cultural relic has a certain spatial distribution, or if there is a regression relationship between the different geometric parts, it can be repaired with the methods and ideas proposed in this paper. It is important to remember that the global geometric features are not limited to the cultural relic space symmetry, and the skeleton line is only one of the typical features of local geometry. In practical applications, according to the characteristics of cultural relics, different spatial geometry characteristics can be selected at different scales. Through the process of spatial and mathematical analysis, the implicit regulations of the spatial geometry of cultural relics can be discovered. At the same time, virtual restoration based on the geometric features of different scales can complement and verify each other. This also has great significance for archaeology. In summary, this method has strong reference and repeatability potential in the field of cultural relic restoration.

In future work, researchers could consider creating a 3D semantic database of the hands' 3D models and richly describe each one—such as geometric position, shape of gesture, etc.—in order to achieve semantics-based spatial analysis and retrieval. The advantage of this method is that on the basis of a geometric morphologic analysis of cultural relics, it can also use semantics to retrieve the symmetry and similarity of cultural relics, and it is easier to analyze the degree of matching.

6. Conclusions

This paper presents a practical method for virtual restoration of artifacts with complex geometric structures based on multiscale spatial geometry, especially those artifacts that have missing and broken parts and no proper documented information. In this paper, for the first time, the geometric shape and size of cultural relics are determined by analyzing the global and local geometric features. Based on this, objective virtual restoration of cultural relics is carried out. In the repair process, the spatial symmetry of relics is utilized. The skeleton line was first proposed to express the three-dimensional models. The adaptive adjustment of the skeleton line extraction algorithm improves the central and topological relationship of the skeleton line. In the end, the length prediction model of the relic is established by regression analysis.

This method mainly solves the problem that the complex geometry of relics is difficult to find in the evidence of virtual restoration and is more scientific and objective than subjective judgment. The effect of virtual restoration can be explained by the results of multiscale spatial geometric analysis of cultural relics. The results of this experiment have been successfully applied to the virtual restoration of the Dazu Thousand-Hand Bodhisattva stone carvings. In addition, the methods and ideas provided in this paper also apply to the virtual restoration of other cultural relics, which has great reference value. Therefore, the method is of great significance to the virtual restoration of complex geometric structures.

Author Contributions: Conceptualization, M.H. and S.Y.; Methodology, Y.H. and Y.W.; Software, L.J., S.Z., and P.W.; Validation, S.Y.; Formal analysis, M.H. and S.Y.; Investigation, S.Z. and P.W.; Resources, M.H. and Y.H.; Data curation, L.J.; Writing—Original draft preparation and review, all authors.

Funding: The research work conducted in this paper was supported by the research fund of the National Key Research and Development Program (no. 2016YFB0501404 and no. 2017YFB1402525) and the Beijing Advanced Innovation Center for Future Urban Design (no. UDC2016030200), and by Supporting Plan for Cultivating High Level Teachers in Colleges and Universities in Beijing (no. IDHT20170508).

Acknowledgments: The authors would like to acknowledge and thank the Dazu Museum for its support of this work. The authors also would like to thank Ian Dowman from University College London, UK, and Songnian Li from Ryerson University, Canada, for their valuable comments on improving the manuscript, Changfa Zhan from Dazu Rock Carvings Museum in Chongqing for his novel suggestions, and Fangyin Li and Huili Chen from Dazu Rock Carvings Museum in Chongqing for their work on the capturing and processing of the data.

Conflicts of Interest: The authors declare no conflict of interest.

References

1. Su, Y.Q. A Comprehensive Study on Three-Dimensional Digital Restoration of Qingzhou Buddha Statues. Master's Thesis, Yantai University, Yantai, China, 2008.
2. Yang, L.B. The Application of Shape Matching Technology in Artifact Restoration. Master's Thesis, Northwest University, Xi'an, China, 2002.
3. Ru, S.F. A Study on the Restoration of Fragmented Objects and Computer-Aided Artifact Restoration. Ph.D. Thesis, Northwest University, Xi'an, China, 2004.
4. Pan, R.J. Several Problems in Computer-Aided Artifact Restoration. Ph.D. Thesis, Henan University of Science and Technology, Luoyang, China, 2005.
5. Fan, S.R. Methods for Shaping Matching and Assembly of Complementary Parts of Rigid Bodies. Master's Thesis, Northwest University, Xi'an, China, 2005.

6. Huang, Q.X.; Flöry, S.; Gelfand, N.; Hofer, M.; Pottmann, H. Reassembling fractured objects by geometric matching. *ACM Trans. Graph.* **2006**, *25*, 569–578. [CrossRef]

7. Wang, Y.T. The Application of Contour Matching Technology in Artifact Restoration. Master's Thesis, Henan University of Science and Technology, Luoyang, China, 2011.

8. Li, Q.H.; Zhou, M.Q.; Geng, G.H. Reassembly of broken 3D solids based on fractured surfaces matching. *J. Image Graph.* **2012**, *17*, 1298–1304.

9. Oxholm, G.; Nishino, K. A flexible approach to reassembling thin artifacts of unknown geometry. *J. Cult. Herit.* **2013**, *14*, 51–61. [CrossRef]

10. Li, C.L.; Zhou, M.Q.; Cheng, X.; Cheng, R.B. Virtual restoration of axisymmetric relic fragments. *J. Comput.-Aided Des. Comput. Graph.* **2006**, *18*, 620–624.

11. Yang, C.L.; Zhang, Z.X.; Pan, R.J. A study on the framework of computer-aided artifact repair system and key technologies. *J. Syst. Simul.* **2006**, *18*, 2003–2006.

12. Willis, A.R.; Cooper, D.B. Bayesian assembly of 3D axially symmetric shapes from fragments. In Proceedings of the 2004 IEEE Computer Society Conference on Computer Vision and Pattern Recognition (CVPR 2004), Washington, DC, USA, 27 June–2 July 2004; Volume 81, pp. I-82–I-89.

13. Chalmoviansky, P.; Juettler, B. Filling holes in point clouds. *Lect. Notes Comput. Sci.* **2003**, *2768*, 196–212.

14. Nooruddin, F.S.; Turk, G. Simplification and repair of polygonal models using volumetric techniques. *IEEE Trans. Vis. Comput. Graph.* **2003**, *9*, 191–205. [CrossRef]

15. Ju, T. Robust repair of polygonal models. *ACM Trans. Graph.* **2004**, *23*, 888–895. [CrossRef]

16. Biermann, H.; Martin, I.; Bernardini, F.; Zorin, D. Cut-and-paste editing of multiresolution surfaces. *ACM Trans. Graph.* **2002**, *21*, 312–321. [CrossRef]

17. Yu, Y.Z.; Zhou, K.; Xu, D.; Shi, X.H.; Bao, H.J.; Guo, B.N.; Shum, H.Y. Mesh editing with poisson-based gradient field manipulation. *ACM Trans. Graph.* **2004**, *23*, 644–651. [CrossRef]

18. Wei, T. Data Integration and Interactive Repair Techniques for 3D Geometric Modeling. Master's Thesis, Peking University, Beijing, China, 2007.

19. Lu, M.; Kamakura, M.; Zheng, B.; Takamatsu, J.; Nishino, K.; Ikeuchi, K. Clustering bayon face towers using restored 3D shape models. In Proceedings of the Second International Conference on Culture and Computing, Kyoto, Japan, 20–22 October 2011; pp. 39–44.

20. Lu, M.; Zheng, B.; Takamatsu, J.; Nishino, K.; Ikeuchi, K. 3D shape restoration via matrix recovery. *Lect. Notes Comput. Sci.* **2010**, *6469*, 306–315.

21. Wu, Y.H.; Hou, M.L.; Zhang, Y.M. Progress and future direction of the application of 3D laser scanning technology in rock artifact protection. *Geomat. World* **2011**, *4*, 53–57.

22. Au, K.C.; Tai, C.L.; Chu, H.K.; Cohen-Or, D.; Lee, T.Y. Skeleton extraction by mesh contraction. *ACM Trans. Graph.* **2008**, *27*, 1–10. [CrossRef]

 International Journal of
Geo-Information

Article

A Methodology for Planar Representation of Frescoed Oval Domes: Formulation and Testing on Pisa Cathedral

Andrea Piemonte [1,*,†], Gabriella Caroti [1,†], Isabel Martínez-Espejo Zaragoza [1,†], Filippo Fantini [2,†] and Luca Cipriani [2,†]

1   Civil and Industrial Engineering Department (DICI), University of Pisa, Largo Lucio Lazzarino 1, 56122 Pisa, Italy; gabriella.caroti@unipi.it (G.C.); isaroju@gmail.com (I.M.-E.Z.)
2   DA–Dipartimento di Architettura, Alma Mater Studiorum Università di Bologna–Bologna, Viale del Risorgimento, 2, 40136 Bologna, Italy; filippo.fantini2@unibo.it (F.F.); luca.cipriani@unibo.it (L.C.)
*   Correspondence: andrea.piemonte@unipi.it; Tel.: +39-050-221-7773
†   These authors contributed equally to this work.

Received: 8 June 2018; Accepted: 3 August 2018; Published: 7 August 2018

Abstract: This paper presents an original methodology for planar development of a frescoed dome with an oval plan. Input data include a rigorous geometric survey, performed with a laser scanner, and a photogrammetry campaign, which associates a high-quality photographic texture to the 3D model. Therefore, the main topics include the development of geometry and, contextually, of the associated textures. In order to overcome the inability to directly develop the surface, an orthographic azimuthal projection is used. Starting from a prerequisite study of building methodology, the dome is divided into sectors and bands, each linked with the maximum acceptable deformations and the actual geometric discontinuities detectable by the analysis of Gaussian curvature. Upon definition of the development model, a custom automation script has been devised for geometry projection. This effectively generates a (u,v) map, associated to the model, which is used for model texturing and provides the planar development of the fresco.

Keywords: frescoed vault; planar representation of vault; cultural heritage documentation

---

1. Introduction

The availability of large-scale planar representations of frescoes painted on curved surfaces provides useful support for restoration and maintenance operators. During operations related to diagnostics and restoration and/or maintenance of frescoes, specialized operators document the status quo and each intervention in technical forms, providing rationale for the decisions and stating the steps of the intervention that leads to the final result. Such documentation includes an explanatory report and 2D documents, both graphical and photographical, for each step at an adequate scale. In addition, these documents form a relevant archival knowledge base which, when used in conjunction with the parent 3D model, can provide the foundation for reconstruction of disrupted or destroyed heritage. In order to generate 3D models incorporating metrically exact geometries and high-quality photo-realistic textures, it is possible to integrate two well-established surveying technologies—Terrestrial Laser Scanning (TLS) and photogrammetry [1,2]. TLS enables fairly consistent geometric precision for any given scanner surface range, irrespective of the surface texture of the object. On the other hand, photogrammetry exploits cameras and image collection techniques so as to achieve high-resolution, high chromatic quality textures.

Advances in digital photogrammetry, in particular Structure from Motion (SfM) and Multi-View Stereo (MVS), ensure good results, even with extreme image taking conditions that are otherwise unusable in analogue or analytical photogrammetry [3].

The potential offered by current photogrammetric methodologies allows the exploitation of historic photographic archives in order to register archive images on up-to-date models and reconstruct destroyed or severely damaged frescoes [4] as well as supplying visitors of cultural heritage sites with augmented reality solutions capable of overlaying contents on a digital representation of a work of art [5].

Upon the generation of textured models of curved surfaces, a further problem involves their best planar representation, reducing or at least controlling the ensuing deformations.

This problem involves two distinct issues:

- performing the development (more appropriately termed as projection and pseudo-development for non-directly developable surfaces) of geometric shapes;
- applying a reference system (u,v) to geometric shapes in order to retain their association with textures.

While the literature provides several suggestions to solve the first issue, some of which have already been implemented in CAID (Computer Aided Industrial Design) software, there are only a few implementations that partially solve the second problem with applications such as Technodigit3DReshaper and the open-source Cloud Compare [6]. Planar representation of geometric shapes must therefore differentiate between developable and non-developable curved surfaces. Developable surfaces can be "unfolded" on a plane without any deformation. This operation is bijective and can be reversed with no loss of accuracy [7].

Planar representation of non-directly developable surfaces requires their projection on a plane or other developable auxiliary surfaces. This operation is typical of cartographic representations, where the Earth's surface is approximated to an ellipsoid and projected on a plane. Such projections, however, entail some linear, angular, or area deformations. While no projection can completely remove these deformations, it is possible to check their dimension and keep it within given tolerance ranges by means of adequate precautions, which often provide reduction of the projected surface area and repetition of the projection with different auxiliary surfaces.

Unlike previous contributions, the present work aims to define a methodological approach for planar pseudo-development of a frescoed dome with oval-plan design and polycentric sections.

The object of study is the dome of Pisa Cathedral (Figure 1), figuring a fresco depicting the Assumption of the Virgin Mary with several saints, painted between 1627 and 1630 by Orazio Riminaldi [8].

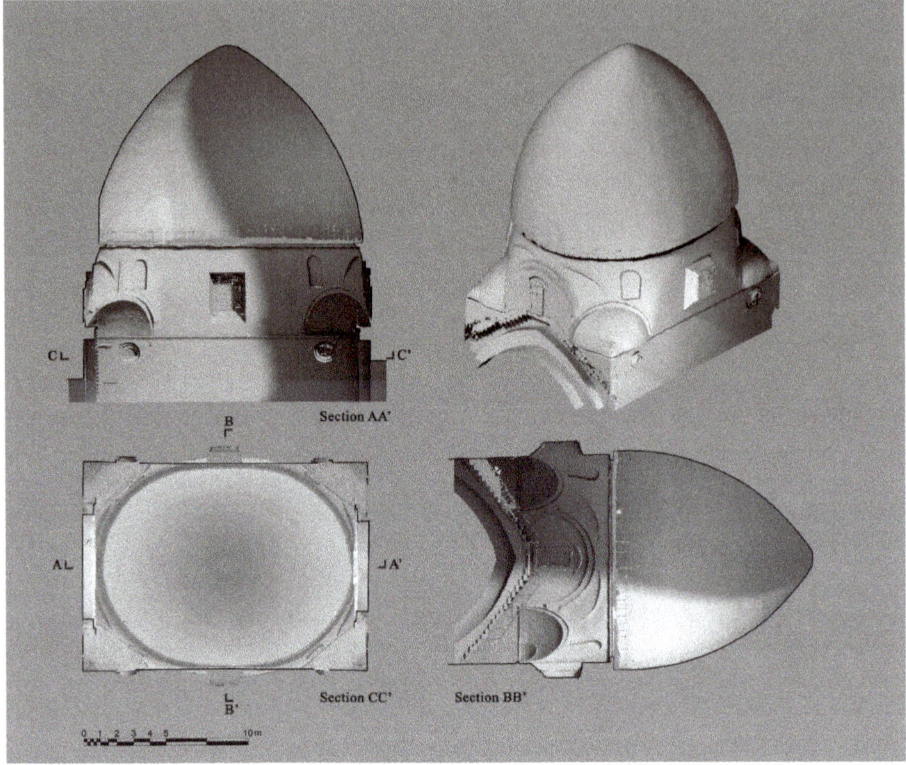

Figure 1. Frescoed dome, Pisa Cathedral.

2. Materials

The dome of Pisa Cathedral dates back to the early twelfth century and presents an original shape compared to the European panorama of the time. As pointed out by Piero Sanpaolesi [9], who worked as superintendent in Pisa between 1943 and 1960, a clear typological interpretation is a prerequisite when carrying out studies on the domes of historic buildings. There are two types of domes: extrados ones (*cupole estradossate*) and the pseudo-cupulas (or "corbelled", or "horizontal rings"). Pisa Cathedral dome belongs to the first category, which has a curvilinear profile of the extrados similar to that of the intrados (the volume of the cupula can be considered as the result of a "shift" of the intrados surface outside).

This is a departure from the technical solutions adopted in other buildings such as the Pantheon in Rome which, seen from the outside, hides for the most part the hollow spherical volume of the intrados inside a heavy cylinder made of bricks. In other terms, in the first case, the extrados is dimensioned by means of a simple offsetting operation of the intrados surface, i.e., the internal and external shapes coincide, with the respective positions differing by a value homogeneously defined in the design and execution phase.

In the second case, inside and outside shapes present different, independent shapes; an inside semi-sphere externally corresponds to a "lowered", flatter profile, such as in the Pantheon in Rome. According to Sanpaolesi, the example supplied by the Pisan dome is one of the first examples in Europe and in the Mediterranean Basin in which the form perceived internally is also visible externally, forgoing the heavy and massive stereometry that characterizes the drums of many previous buildings, in particular of the classical era. The domes belonging to the latter category, in fact, base their strength on the progressive decrease in thickness as well as the weight of the construction materials, ring after ring. In addition to the already mentioned and iconic Pantheon in Rome, numerous mausoleums built in the Imperial era and late Antiquity [10], such as the Mausoleum of Sant'Elena, are characterized by the same feature, i.e., the internal and external shapes do not match.

In the specific case of the Cathedral of Pisa, the shape of the plan at the impost height is oval; this feature makes this constitutive element incompatible with a revolved surface, as occurs in numerous previous examples, also belonging to the classical age, e.g., the Temple of Diana in Baia [11–13].

The original impetus that characterized the original shape of the Pisan dome was reduced by ancient restorations, in particular by the addition of a small loggia built starting from 1389 for the next seven years by Puccio di Landuccio and Lapo di Gante based on a model [9]. The small loggia appears to have had the function of a reinforcing hoop aimed at containing horizontal thrusts through an elegant colonnade that increased the vertical component of the forces. Similar strategies were also adopted by architects for the restoration of ancient buildings as in the case of the previously mentioned Mausoleum of Sant'Elena with the raising of the upper drum, which incorporated the original tiered cornice [10]. The loggia follows the irregular octagonal profile of a drum but leans on a series of coeval arches, whose purpose was the unloading of forces along the eight angular pillars.

Surveying

Currently, 3D architectural surveys mainly rely on two modeling methodologies—range-based and image-based [14]. The former exploits TLS to provide 3D point cloud models made of millions of points framed in a common reference system, which can represent the survey object geometry with subcentimeter accuracy [15]. The second technique—through the integration of efficient computer vision algorithms such as SfM and MVS with classic photogrammetry procedures—can provide both a geometric model and the radiometric information needed for model texturing [16].

Given the need to obtain a model of the dome ensuring both the strict geometric definition of its shapes and the analysis of the fresco, it was necessary to proceed with both surveying methods and to integrate their output [1,2].

The possibility of placing surveying instrumentation in close proximity of the survey object, together with the visual interference of the internal scaffolding during its staging, led to planning distinct steps for the geometry survey:

1. Topographical survey of a support network, at ground level.
2. Laser scanning and photogrammetry of the intrados, at ground level.
3. Laser scanning and photogrammetry of the intrados, at matroneum level.
4. Topographical survey of an internal support network, at springer level.
5. Laser scanning and photogrammetry of the intrados, at springer level.

Completion of steps 1 to 3 took place prior to the staging of scaffolding. Steps 4 and 5 were completed during a temporary work stoppage, with the internal scaffolding at springer level.

As previously stated, while the scaffolding does provide a privileged access point to upper levels, it intrinsically lacks the stability required for precision surveys. For this reason, the support survey carried out by total station is of utmost importance. Throughout its execution, measuring stations were consistently placed on steady spots, such as ground, matroneum floor, and masonry landings along stair flights linking service doors at drum and springer level.

Points surveyed via total station acted as either Ground Control Points (GCPs) or Check Points (CPs) in photogrammetric processing and point cloud registration; GCPs were used for scaling and georeferencing the model, and CPs were used to control its accuracy.

As for TLS surveys—performed via Leica's C10 ScanStation with a point cloud density of about 70 pts/cm^2 on average—point clouds from lower level scans included both the area of interest, i.e., the dome, and portions closer to the ground. These clouds were orientated using ground-level benchmarks, which provided higher reliability than those at dome level. Registration of clouds collected in close proximity to the dome relied on both dome-level benchmarks and cloud-matching algorithms with previously registered clouds.

Photogrammetric surveying—carried out by means of Nikon's D700 DSLR fitted with a 50 mm lens at dome level and an 85 mm lens at matroneum and ground levels—yielded a mean Ground Sampling Distance (GSD) of 3 mm. Image processing via SfM algorithms require an overlap of more than 70%. Acquisition of images taken from different distances (12 m at dome drum level, 45 m at ground level) allowed for model checking and increased rigidity. Bundle adjustment was performed with 10 GCPs and yielded a standard deviation of 2 mm based on 15 CPs.

Based on these values, images were correctly orientated in the relevant reference system and therefore could be used for texturing TLS models registered in a common reference system.

3. Methods

3.1. Geometry Pseudo-Development

Rieck et al. [7] provided an in-depth study of unfolding methodologies for high-resolution 3D meshes of objects with rotation symmetry. Previous photogrammetry works used cylinders for the development of 3D surfaces having a precise analytical representation [17]. The use of map-like projections has already proved effective for the achievement of raster images of frescoes painted on arches or spherical surfaces [18,19]. A thorough discussion of the development methods for spherical objects can be found in the cartographic literature [20]. In later works, arbitrary surfaces were interpolated with triangle strips prior to their development [21]. Bevilacqua et al. [22] provided an example of planar representation of a fresco painted on an *"a schifo"* vault for which the error for planar projection was checked.

The methodology adopted for the planar representation of the dome can be described as an orthographic azimuthal projection (Figure 2). In this kind of projection, linear deformation Δ_L for a given curvature radius is directly linked to the distance of the projected point relative to the tangency point of the plane and therefore to the angle α, according to Equation (3):

$$|\hat{C}P| = R \cdot \sin \alpha \tag{1}$$

$$\left|\overline{CP'}\right| = R \cdot \sin \alpha \cdot \cos \alpha \tag{2}$$

$$\Delta_L = R \cdot \sin \alpha \cdot (1 - \cos \alpha) \tag{3}$$

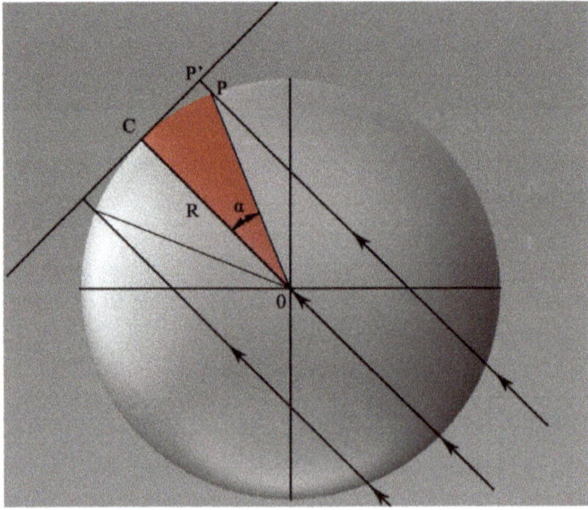

Figure 2. Geometric diagram of orthographic azimuthal projection.

Due to its double curvature, together with variable curvature radius for both meridian and horizontal sections, the surface was split into zones and bands in order to reduce linear deformations for the points farthest away from the tangency point on the projection plane, resulting, in fact, as a whole in a polycentric orthogonal azimuthal projection. As regards the fuses, the choice of their number directly affects the fragmentation of the final representation; for this reason, it is advisable to keep the greatest amplitude compatible with the final scale and the related allowable deformation. As regards the bands, reducing their extent does not affect representation quality or processing time, thanks to a custom script, which will be discussed later.

Processing included use of algorithms belonging to distinct application fields (photogrammetry, reverse modeling, polygonal modeling) and the development of a script aimed at the automation of iterated operations required for the pseudo-development of the dome intrados.

A rapid, intuitive way to assess the developable nature of surfaces requires the use of Gaussian maps, which evaluates the curvature at any point for the main directions of the surface. Given any planar curve, it is possible to define the tangent, the normal, and the osculating circle for any of its points, with the inverse of the radius of the osculating circle being the curvature k [23]:

$$k = \frac{1}{R} \tag{4}$$

For 3D surfaces, these concepts apply as follows: For any point, it is possible to define the normal, the tangent, and infinite curvatures, i.e., one for each section curve obtained by the intersection of a plane containing the normal. Notable among these are the main curvatures $k1$ and $k2$—the maximum and the minimum—which can be positive or negative, based on the accordance with the normal. Gaussian curvature is defined by:

$$k = k1 \cdot k2 \tag{5}$$

It is therefore possible to obtain the following cases:

1. simple curvature surfaces: $k1 = 0, k2 \neq 0$; or $k1 \neq 0, k2 = 0$.
2. double curvature surfaces: $k1 \neq 0, k2 \neq 0$.

Developable surfaces can only have one curvature, i.e., when $k = k1 \cdot k2 = 0$.
The sign of Gaussian curvature is also useful for shape definition:

1. $k > 0$ define ellipsoids, which can be concave ($k1 > 0$ and $k2 > 0$) or convex ($k1 < 0$ and $k2 < 0$);
2. $k = 0$ further distinguishes between $k1 = k2 = 0$ (plane) and either $k1$ or $k2 \neq 0$ (cylinder);
3. with $k < 0$, the two main curvatures have opposite signs, identifying a saddle surface.

In 3D modeling applications based on Non Uniform Rational Basis-Splines (NURBS), it is possible to visualize the Gaussian curvature of a shape in order to immediately perceive if, and in which areas, this is developable.

In the case of a morphologically complex digital model—obtained by laser scanning—it is logical to expect that the numerous irregularities due to superficial alterations, structural settlements, and restorations lead to an unclear reading of the Gaussian curvature. For these reasons, the Pisa dome is a case in point, as it requires particular arrangements in order to undergo Gaussian curvature analysis and understand which parts of its shape is developable and which are not, to provide a possible solution for the development.

The essential first phase is the achievement of a NURBS representation from the TLS mesh in order to carry out the Gaussian curvature analysis expressed through a visual output in false colors directly on the model. This step of the process involves reverse modeling techniques to convert the intrados—made of triangles—into a boundary representation (B-Rep) formed by parametric patches. False colors on a connected set of NURBS patches supply the user with precise information on the location of developable areas—displayed in green by the application used in this work (approximately equivalent to cones or cylinders, $k = 0$). The aim is to obtain "heuristic" clues to be used in the following steps.

3.2. Geometry Pseudo-Development

There are numerous ways for obtaining a mathematical model from a high-detail mesh. Each system has its pros and cons regarding lead times and overall metric reliability of the model obtained through reverse modeling techniques [24] compared to the original mesh and, last but not least, an easy portability of the model towards a NURBS application.

(1) Direct patching: Direct patching—present in all the reverse modeling applications, such as 3D Systems Geomagic Design X or InnovmetricPolyWorks—was deemed not suitable, although they are highly automated and faster than the other strategies examined (Figure 3).

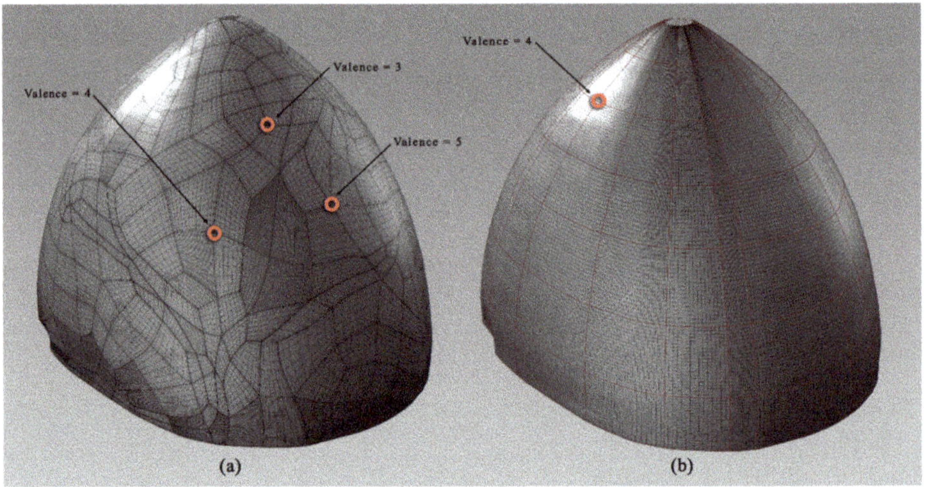

Figure 3. (a) Automated direct patching; (b) supervised 3D patching.

The main reason is that the set of NURBS patches "covering" the high-density mesh is obtained through a series of customizable parameters that are unreliable for both constructive and design features of the dome. In Figure 3a, the automatic procedure results in discontinuous and interrupted sequences of loops of edges; in addition, patch organization is neither horizontal, nor meridian. In fact, boundary vertices belonging to the patches do not have constant valence equal to four, which is the optimum for the next steps of the procedure. Regular horizontal and vertical rows and columns of patch would have matched the natural arrangement of bricks and ribs of the centring adopted for the dome, as shown in Figure 3b. The cupola was apparently built by arranging rows of horizontally laid bricks (consistent with the parallels) or normal to the surface of the intrados. Nevertheless, direct patching leads to an inconsistent arrangement with respect to the natural division into parallels and meridians of the dome. In fact, the automatic and seemingly random arrangement of the patches and the relative isoparametric curves (isoparms) makes it extremely difficult to obtain a 2D version of the whole B-Rep that is able to visually and intuitively respond to the needs of this study, i.e., interdisciplinary approach, collaborative work with restorers, etc.

In other words, the isoparametric curves, namely the graphic lines running along the surface in the U and V directions and giving shape to NURBS patches, do not fit with a constructive reading of the artifact.

(2) 3D Patch Network: This is an alternative strategy to obtain a NURBS model inside the already mentioned applications, and it consists of the combined use of reverse modeling techniques (section extraction and features) and is a procedure which, depending on the situation, can be highly automated or manual. Firstly, the technique involves the operation called "de-featuring", which leads to removal from the model the areas that hinder the continuity of the shape by integrating them with mesh in continuity with neighboring areas, in particular, holes and tiny dangling elements (like windows, niches, or overhanging parts). Once the surface has been normalized, the automatic, manual, or assisted identification of feature curves (generatrix and directrix) leads to the determination of a wireframe representation of the mesh model. In general, contiguous patches are constructed in such a way as to intersect in points of valence equal to 4, defined by the intersection of parallels and meridians.

(3) Direct modeling: This starts with a series of sections/feature curves. It presents some problems, among which the main one is the reliability of the final NURBS model in relation to the original, since the deviation between mesh and NURBS cannot be evaluated step-by-step as in reverse modeling techniques. Once a limited number of sections—characterized by a large number of poles—are imported inside applications such as McNeel Rhinoceros and Autodesk Alias, they have to undergo a simplification process since fewer poles are always a better solution for the achievement of a nice curvature of patches. Nevertheless, CAID applications implement less specific controls with respect to those present in reverse modeling software, when simplifying a spline. On the contrary, if the number of poles on splines is kept high, without any particular intervention, the generated model can be locally very reliable (around the curves extracted from the mesh) but overall less reliable in the areas of connection between the sections (patch quality can be poor and conflicting with development tools).

(4) Retopology: This solution is implemented inside several entertainment applications (Foundry Modo, PixologicZBrush, etc.), and it can be a manual or an automatic process [25,26] aimed at remeshing highly detailed triangular meshes into "quads", namely four-sided polygons. Certain advances that have recently been implemented in modeling tools, both NURBS and polygonal, allow the conversion of such quad-polygonal meshes into mathematical models [27].

This set of techniques and opportunities—all evaluated by the research group—led to the use of a personalized approach. The first step involved the use of the 3D Patch Network to obtain information on the Gaussian curvature; the second phase is based on the use of the spline curves extracted from the model in the previous step and used as generatrices and directrices in order to obtain a second simplified geometric model based on NURBS technology through commands able to keep low and uniform the number of isoparametric curves. Reverse modeling techniques enable a full control on the deviation introduced during the simplification process while maintaining the intersection of the curves in points of valence equal to four. This model is constructed in such a way as to allow a particular type of linear development that is functional to verifying the validity of the approach used with the subsequent steps that will be based on discrete models in quadrangular polygons.

3.3. Design Analysis and Constructive Issues

The reverse modeling steps (3D Patch Network) were based on a series of studies and verifications concerning design analysis and ancient construction techniques. Hypotheses on the oval-plan outline of the dome (ichnographia) benefit from the new instrumental survey outcomes and verified hypotheses by Sanpaolesi who had already identified the centers and the curvature radii at the dome's impost plane—a polycentric curve identified by a series of modules (Figure 4).

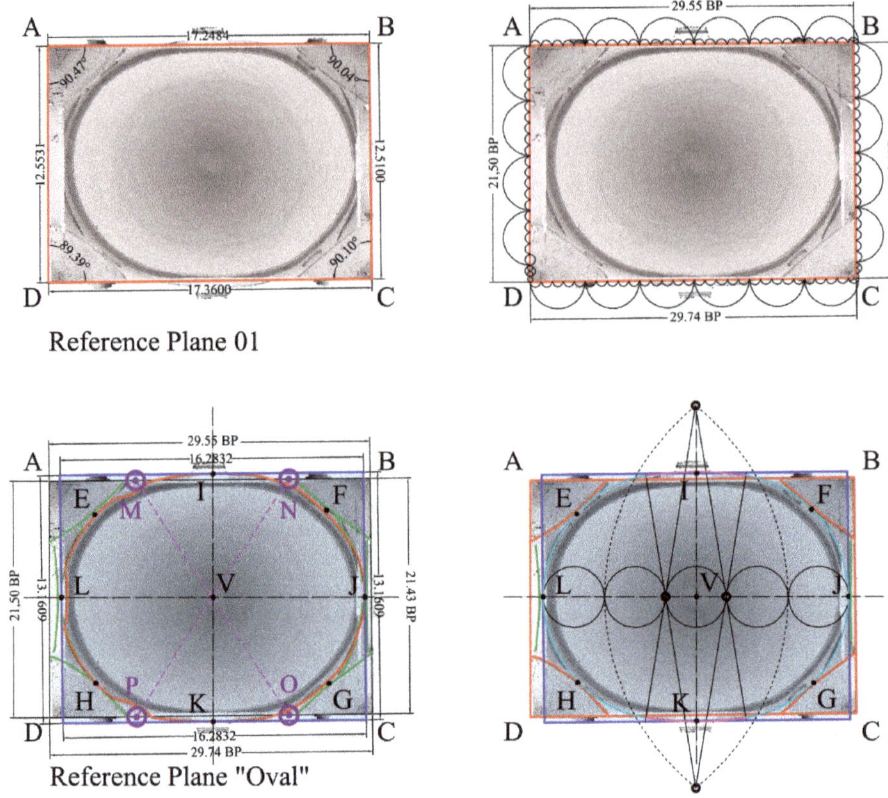

Figure 4. Design analysis and hypotheses on oval-plan outline. The main lengths are in both meters and ancient measuring system Braccia Pisane (BP).

The use of similar techniques for obtaining oval curves had been known since antiquity to define the plan design of amphitheatres [28]. This "geometric algorithm" led to the achievement of a closed curve with continuity of position (G0), of tangency (G1), but not of curvature (G2). The problem of reverse modeling therefore needs to be carried out bearing in mind these assumptions of ancient mathematics and geometry applied to design and construction.

Designing and construction issues can be considered in these terms (limiting for the moment the analysis to the intrados); the irregular quadrilateral (ABCD in Figure 4) constitutes the starting shape from which the irregular octagon was found out. This form will contain the oval of the intrados, but the octagon does not inscribe its curvilinear shape.

The individual sides of the quadrilateral (ABCD) measure in meters: AB = 17.248, BC = 12.510, CD = 17.360, DA = 12.553. In Pisan arms (Braccia Pisane, BP) measuring 0.5836 m [29], the following measurements are obtained: AB = 29.55; BC = 21.43; CD = 29.74; DA = 21.50.

Observing the major axis and the minor axis of the oval, it is possible to notice how the first (LJ) is shorter than the long side of the underlying rectangle (ABCD), while the second (IK) comes out of the same "boundary" drawn by the quadrangular perimeter.

Therefore, the oval is not exactly inscribed; instead, its layout is obtained through a tangency relationship solely affecting points E, F, G, and H (belonging to the oblique sides of the irregular octagon). The tangency in E, F, G, and H is accurately achieved, taking into account the constructive difficulties and irregularities due to deformations induced by phenomena of deterioration and subsidence of the foundation and elevated structures occurring over nine centuries of life of the building [14].

These considerations have a possible construction-based explanation; when looking at the longitudinal section of the cathedral, overlapped with the plan, it is possible to understand the reason of this "shift" between the octagon and the bounding box of the oval (in blue in Figure 4).

When locating the main support points for temporary structures (centring), the architects of the cathedral should have opted for the minor axis of the oval. At the same time, they had to exploit the structural continuity of the piers from the first order to the compact connection masonry of the drum. Possibly, they placed the main ribs of the wooden centring in the areas between the oval and the ABCD rectangle, in particular near the points M, N, O and P (Figure 4).

It is presumable that additional ribs, probably of secondary importance, were laid on these two main arches following the same operating logic—the exploitation of intersection areas between the irregular octagon and the blue square in Figure 4.

For the ichnografia, it is still valid to assume what was claimed by Sanpaolesi [9], who proposed a polycentric curve based on the division of the major axis of the oval into five modules (Figure 4).

3.4. Segmentation of the Cupola

By executing a series of horizontal contour lines of the intrados surface, it is possible to obtain further information that is useful for carrying out a geometric and structural reading functional to the development of the intrados.

The field of geometric modeling provides two types of surfaces—primary and secondary—or blends [30]. In this study, this criterion was applied to perform a segmentation of the model useful for the purposes of this investigation, i.e., to subdivide a complex shape into several subsets coherent with the design and construction of the dome.

The primary surfaces are attributable to geometric primitives (plane, box, sphere, cone, etc.), while the secondary ones are used to connect two primary surfaces with zones of "passage", also referred to as blends [24]. By restricting the analysis to contour lines and not to the entire morphology, major discontinuities visible to the naked eye are evident, and they lead to the identification of two sets of curves: "primary" (type 1 and type 2 in Figure 5a) in continuity with the Sanpaolesi hypothesis and "secondary" (joints 1, 2, 3, 4). This is a first segmentation of the model to detect the most suitable sectors for development:

1. Two sectors (type 1 in Figure 5a): shapes formed by contour lines with minor curvature.
2. Two sectors (type 2 in Figure 5a): shapes formed by contour lines with greater curvature.
3. Four connecting sectors ("Joint" in Figure 5a): shapes formed by contour lines approximated to linear segments.

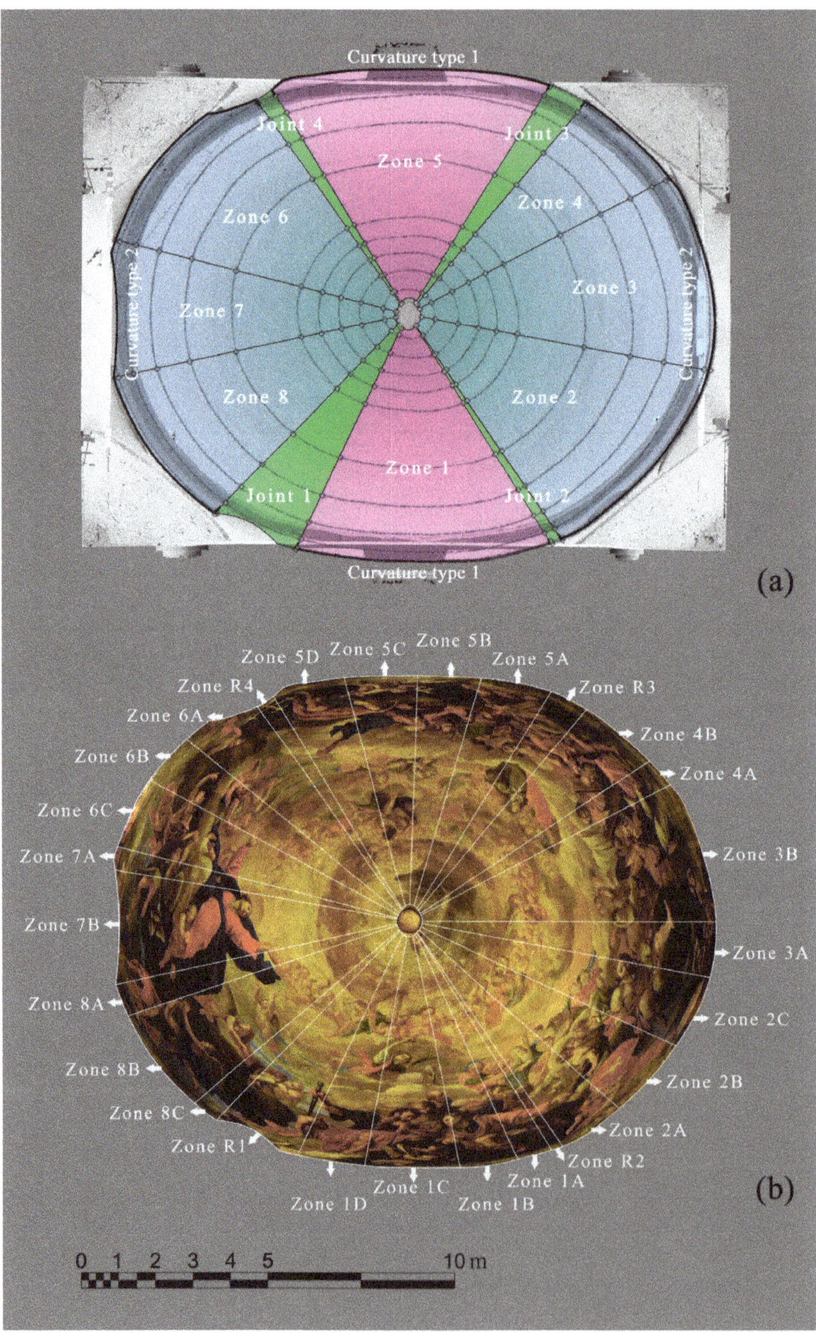

Figure 5. (**a**) Dome main zones subdivision; (**b**) Dome secondary zones subdivision.

In turn, these sectors can be further divided (second partition) on the basis of the variation of the Gaussian curvature (Figure 6) that identifies the overall trend of the surface and provides information about the areas to be subjected to specific shape analysis by means of the curvature graph.

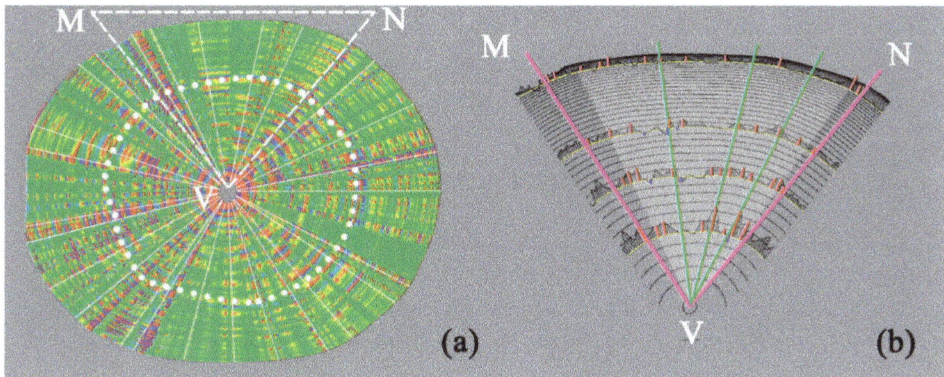

Figure 6. (a) Gaussian curvature analysis; (b) Zoom on MNV sector.

The subdivision of the main sectors into further parts of smaller extension responds to the need to reduce the deformations which, in an orthographic projection, increase as the distance of the projected element increases from the point of tangency of the plane on which the projection takes place.

Instead of making a subdivision based solely on the acceptable size of the area to be developed, the method chosen was aimed at defining additional discontinuity lines by comparing the trends deduced by curvature flow.

In general terms, the dome is composed of "locally developable" surfaces (green colored areas in Figure 6). However, at the height of 8.70 m—with respect to the impost line—a white-dotted contour line underlines a sudden change of Gaussian curvature sign. Above this curve, developable areas decrease significantly.

To illustrate the segmentation method used and applied in all sectors, the portion of the dome included in the MNV triangle is examined (Figure 6b).

The method is based on the analysis of 4 sample curves. One of them is the white-dotted contour line, which represents the "watershed" between areas that are more suitable for development and those that are not; two curves are above this threshold line; and just one curve is below since the area closer to the drum is for the most part colored in green (developable). The method to identify the "fine" partition is based on the following criteria shown in Figure 6:

1. identification in the macro-sector of positive "peaks" of the curvature (red color);
2. identification of areas with low and constant curvature (blue);
3. identification of inflection point areas (change from concave to convex, or vice versa): blue color abruptly alternated with red.

From these considerations, there derives a segmentation proposal for the oval dome into 8 main parts, in turn subdivided into 26 parts (Figure 5b). These 26 secondary zones have extension characteristics able to guarantee linear deformations compatible with a representation scale of 1:100, once having undergone an azimuthal-orthographic-polycentric projection (according to Equation (3) for a radius of curvature, equal to the maximum one found equal to 17 m).

Moreover, since these areas can be approximated in accordance with Gaussian curvature values to cylindrical or conical surfaces, some of the tools have already been implemented in geometric modeling software for the achievement of a pseudo-development of these geometric shapes.

For the modeling aimed at development, the widely used McNeel's Rhinoceros application (https://www.rhino3d.com) was used, since it provides the user with tools useful for the tests carried out.

The first is the possibility of developing continuous NURBS "patches" with Gaussian curvature equal to zero.

On the other hand, the development implemented in the application leads to the creation of a new 3D model belonging to the horizontal reference plane (X, Y), presenting a different parameterization with respect to its "original" 3D version (different number of isoparametric in the u and v direction). The reparametrization of the developed model represents a significant variation of the initial model structure—a variation in the number of nodes that govern the flow of the NURBS surface. In fact, it is a different model, with some properties in common with the original one.

However, the development tool has some tolerances and therefore manages to develop—within certain, not clearly definable, limits—surfaces with curvature $k \neq 0$ unless the structure of the isoparametric curves is regular (low number and homogeneous distribution of poles).

An additional tool gives the possibility of measuring models to evaluate the degree of reliability of the development operation. This essentially means measurement of lengths, areas, and angles in 2D and 3D.

Another aspect that must be pointed out is that it does not make NURBS modeling compatible with rigorous development intended for texturing operations aimed at supplying the user with some bitmap-type 2D output. In fact, the native-born parameterization of NURBS models does not correspond to a geometric development of the model into the (u,v) parameter space; the parametrized version of the NURBS model into the u and v dominion cannot be modified or altered since it is strictly functional to modeling. On the contrary, in polygonal models (born without parametrization), the creation of a one-to-one function that correspond a 3D shape makes to a 2D representation in (u,v) parameter space is facilitated by a range of tools [23].

Therefore, the NURBS application will be used only and exclusively for obtaining a boundary representation of the intrados of the dome; this will be realized starting from 26 curves lying on vertical planes obtained within the application of reverse modeling and mutually connected through the loft tool, which keeps the number of piece parameters low (Figure 7).

As already mentioned, the texturing systems implemented in the Rhinoceros software allow a very simplified parametrization aimed at texturing with respect to the familiar tools implemented in other applications for entertainment. Therefore, it was necessary to continue the work by introducing a program belonging to this field.

The 26 NURBS patches were converted into the same number of polygonal surfaces in such a way that the sampling step is constant across the whole region (along u and v) and compatible with the reliability necessary for the present study. This was measured on the basis of deviation tools inside the reverse modeling application used in this paper.

The method developed for the pseudo-development by projection of patches is as follows:

1. Use of Rhinoceros to obtain the measurable surface for validation of the process and to determine whether it is developable or not. If it is not, a further partition of the piece is carried out according to the stated method.
2. Export of the single NURBS patches into a modeler equipped with a robust polygonal modeling kernel. The model is tessellated so as to maintain quadrangular topology, namely with all vertex valences equal to 4, with the only exception of the vertices forming part of the outer edge of the patch (valence equal to 2 or 3).
3. Projection of each piece in the parameter space (u,v) through a script.
4. Conversion of the (u,v) parameterized version into a mesh belonging to the reference system (x, y, z) and export towards NURBS modeling application.

5. 2D polygon mesh conversion—obtained from parameterization—into NURBS model. This process takes place through the transformation of each cell (polygonal) of an equivalent patch (NURBS). This way, it will be possible to check the metric reliability of the projection operation.
6. Model comparison and error evaluation— areas and lengths of the edges of the patch.

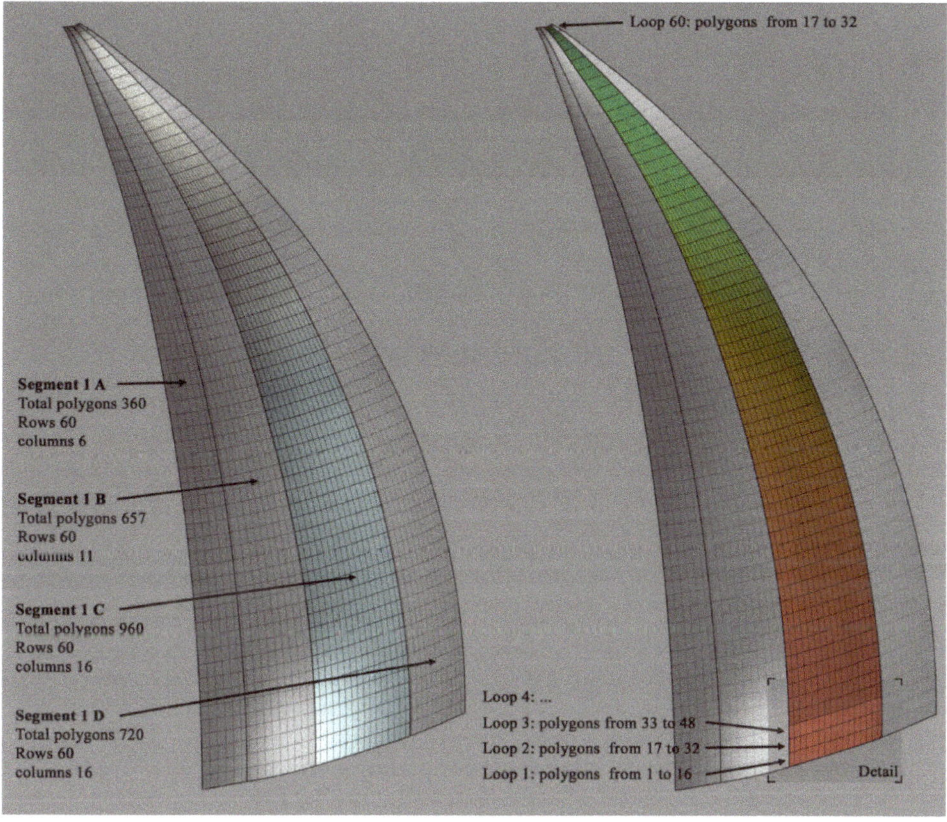

Figure 7. Portion of the dome included in MNV triangle.

3.5. Automatic Multicenter Orthographic Projection (AMOP) Script

The series of instructions aimed at developing has been carried out within the scripting system of The Foundry Modo and is based on the reiteration of a series of commands:

```
#LXMacro#
select.loop
workPlane.fitSelect
viewport.fit
tool.setuv.viewProj on
tool.apply
select.loopprev
workPlane.fitSelect
viewport.fitAlignSelected
tool.setuv.viewProj on
```

tool.apply
select.loopprev
workPlane.fitSelect
viewport.fitAlignSelected
tool.setuv.viewProj on
tool.apply

The purpose of the script is to automate the series of operations listed below and schematized graphically in Figure 8.

0. The necessary starting condition is that the model of each segment is displayed autonomously and in a single viewport, which must be strictly the orthographic view indicated with the name TOP (Top View). The first polygon in the lower left corner of the structured mesh of quadrangular polygons must be selected.

1. The script is launched, and it automatically selects the entire horizontal band of polygons (Belt 1): select.loop.

2. A drawing reference plane (UCS) is automatically set up; its normal vector is determined by the program through the average of the set of normals belonging to Belt 1, which we will call UCSBA1, i.e., UCS Belt Average 1: workPlane.fitSelect.

3. The initial TOP view (which allowed visualization of the segment from above) is now oriented in space in a congruent way with the normal outgoing from the plane defined by UCSBA1, which is centered on the selection together with the zoom value: viewport.fit.

4. A command is launched that makes a projection (central or parallel; in this case parallel) of the set of polygons selected in (x, y, z) onto the parameter space (u,v) using the view determined in the previous step: tool.setuv.viewProj on.

5. A one-to-one correspondence is established between the set of polygons of Belt 1 in (x, y, z) and an island in space (u,v): tool.apply.

6. The next band of polygons is automatically selected (Belt2): select.loop prev. From this point on, the sequence is repeated (Belt3, ... , Beltn) until all the horizontal bands of polygons forming the segment are projected. These will appear superimposed in the parameter space (u,v) and therefore do not result as one-to-one correspondence with the whole set of points of the model in R3: a necessary and sufficient condition for correct texturing.

7. The biunique relationship is re-established by assigning $\forall$ Belt1, ... , n a specific space in (u,v): uv.pack true false true auto 0.2 false false nearest 1001 0.0 −1.0 2.0 2.0.

Once this first phase is over, it is necessary to use a second script that automates the "recomposition" of islands corresponding to Belt1, ... , n.

8. The first polygon on the bottom left is selected by the user that launches the second script that selects the entire horizontal band of polygons (Belt 1): select.loop.

9. The selection of Belt 1 is converted from polygons into edges: select.convert edge.

10. The one-to-one correspondence between the bands in R3 (Belt1, ... , n) and their counterparts in (u,v) is not verified along the edges of the single islands of the Belts. The outlines of the islands are partly duplicated (see image) since they will belong to the previous and the next band. Since the program highlights this duplication, it is possible to exploit it to automate the "stitching" of the adjacent belts through the command: uv.sewMove select true.

11. From here on, the script repeats the operation of the previous point, changing first the selection filter: select.NextMode.

12. The next band is selected: select.loop next.

13. Upon merging all the belts into one island, a fitting is necessary in order not to extend the patch beyond the boundaries of the (u,v) space: uv.fit false.

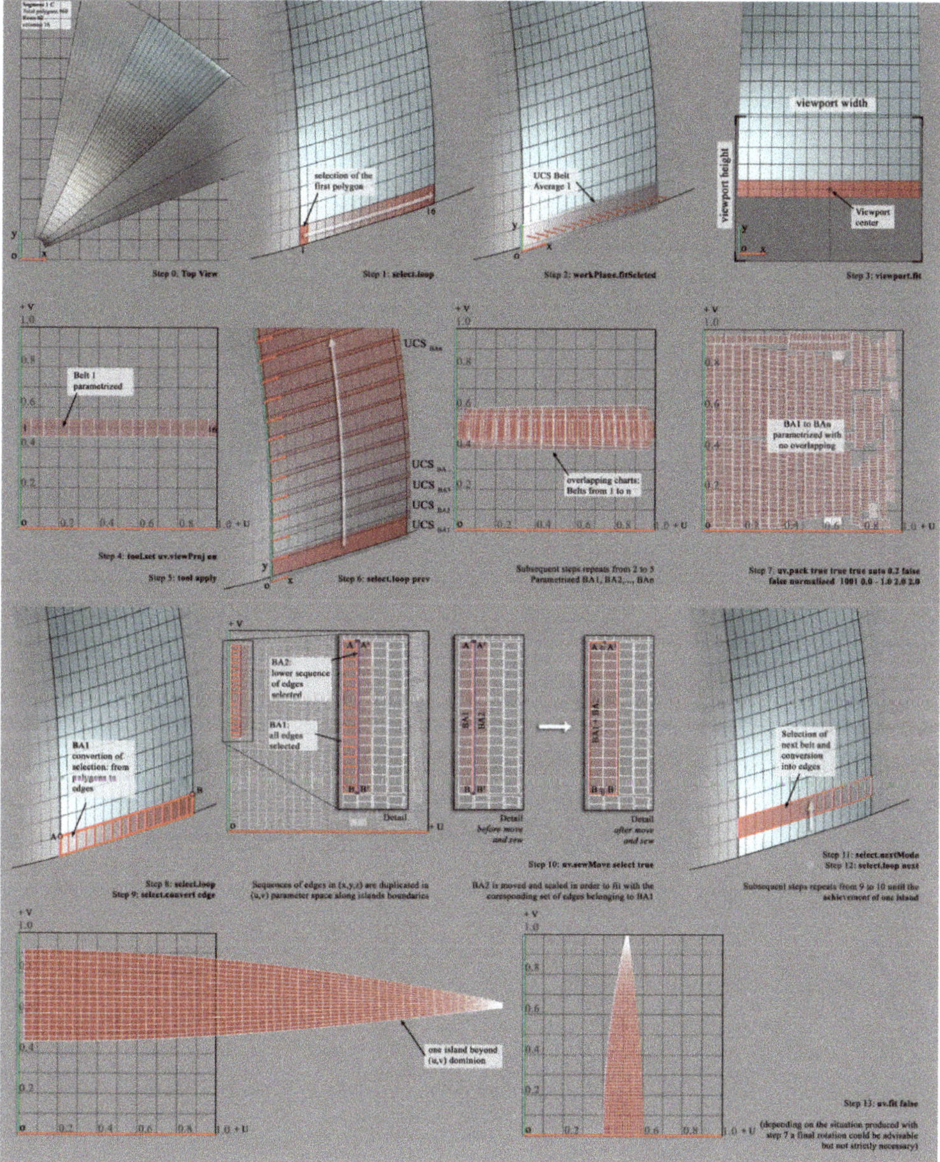

Figure 8. Automatic Multicenter Orthographic Projection (AMOP) script graphical description.

3.6. Associating Textures with Geometry Pseudo-Development

Execution of the above script allows parametrization of each section resulting by dome subdivision. Upon association of the UV map with each section, it is possible to import them into the photogrammetry software used for image modeling, subsequently allowing definition of image orientation parameters. In this environment, it is possible to import each section with the associated UV map and recalculate model textures based on the new (u,v) map, which effectively acts as the pseudo-development by projection of geometry and texture [31].

The final step allows scaling of the (u,v) map, defined in the image space by coordinates ranging from 0 to 1, against the actual dimensions of boundary representation. For this purpose, boundary lines of the sections—derived by geometry development—are defined as reference for image scaling.

4. Results and Discussion

The accuracy of geometry development was checked by comparing, in each zone, the boundary length on the original 3D model and on projections obtained either with the Rhinoceros software or by scaling the (u,v) map at real dimensions. Results, shown in Table 1, are compatible with the 1:100 project scale and in some cases, with definitely greater scales. In Table 1, the following zones are checked:

Table 1. Accuracy assessment of boundary representation planar development. La: length of left zone's boundary along meridian direction measured on the original 3D model. Ld: length of the bottom zone's boundary along horizontal direction measured on the original 3D model. Las: length of left zone's boundary along meridian direction measured on planar projection. Lds: length of the bottom zone's boundary along horizontal direction measured on planar projection.

Zone / Boundary Segment	La [m]	Las [m]	La-Las [m]	Ld [m]	Lds [m]	Ld-Lds [m]
R1	14.014	14.013	0.002	2.356	2.351	0.005
R2	13.954	13.954	0.000	0.304	0.304	0.000
R3	14.053	14.053	0.000	1.059	1.058	0.001
R4	13.932	13.932	0.000	0.529	0.529	0.000
S2A	13.987	14.002	−0.015	1.825	1.806	0.019
S2B	14.109	14.104	0.005	2.263	2.257	0.007
S2C	14.281	14.280	0.001	2.044	2.033	0.011
S3A	14.456	14.465	−0.010	1.505	1.501	0.004
S3B	14.440	14.443	−0.002	3.905	3.882	0.023
S4A	14.261	14.269	−0.008	0.834	0.834	0.000
S4B	14.212	14.210	0.001	2.174	2.170	0.004
S5A	13.930	13.942	−0.012	1.893	1.890	0.003
S5B	13.676	13.688	−0.012	1.744	1.746	−0.001
S5C	13.559	13.563	−0.004	1.975	1.975	0.000
S5D	13.631	13.629	0.002	2.010	2.000	0.010
S6A	14.007	14.006	0.001	2.086	2.070	0.016
S6B	14.225	14.217	0.008	1.474	1.473	0.001
S6C	14.318	14.310	0.008	1.975	1.968	0.007
S7A	14.412	14.421	−0.009	0.622	0.618	0.004
S7B	14.423	14.438	−0.015	3.104	3.103	0.000
S8A	14.396	14.404	−0.008	0.896	0.895	0.000
S8B	14.333	14.346	−0.012	2.959	2.941	0.018
S8C	14.084	14.086	−0.002	0.750	0.748	0.002

Figure 9 shows a collage of the pseudo-developments by projection of the regions in which the dome of Pisa Cathedral was divided. Considering the accuracy of the 3D survey, image resolution, the extension of the single regions, and the checks performed on geometry, the collage is suitable for 1:100 printing.

Achieving large scales depends solely on reducing region size and is therefore attained by dividing the model into a greater number of regions. At greater scales, this kind of document can provide restorers with on-site support for the reconstruction of frescoed domes. The availability of such archive documentation allows reconstruction of damaged frescoes while ensuring full respect of original geometries.

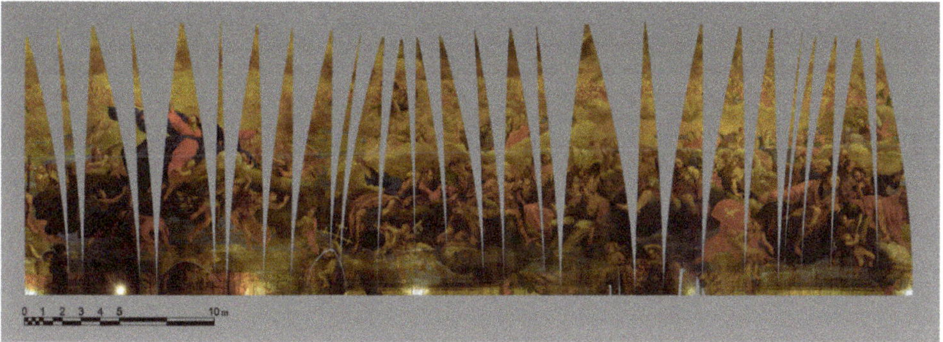

Figure 9. Pseudo-development by projection of the Pisa Cathedral dome.

By analyzing the pseudo-development, representation of the portion closer to the dome impost is relatively continuous due to the comparatively great extensions of the regions. On the other hand, planar representation of the fresco portion closer to the dome summit is quite fragmented due to the limited extension of the regions.

A similar problem is also found in cartographic projections where, in order to prevent excessive fragmentation, latitudes ranging up to 70°–80° N use the Transverse Mercator projection, while polar caps use stereographic polar projection.

This could also be effectively used for the representation of the summit portion of domes, since it allows an easier, more continuous view of frescoed areas. The enhancement of the methodology described incorporating these suggestions will be the object of future investigations.

5. Conclusions

The methodology described (shown using a flow chart in Figure 10) allowed pseudo-development by projection of the fresco painted on an oval dome, which provides the more inclusive example of vaulted or domed surfaces due to the complexity and curvature variability of its sections.

As regards surveying methodology, integrating TLS and photogrammetry enabled the best exploitation of their respective strengths, i.e., homogeneous geometric accuracy and color quality.

The pseudo-development required solving the problem of processing surfaces not approximated by a developable primitive using a polycentric orthographic azimuthal projection, dividing the dome into fuses and bands. The sections delimiting the fuses were defined considering the required representation scale, which limited the width of the regions, along with the defining sections related to the building methodology and the analysis of the Gaussian curvature. Bands were defined by identifying the normal of the different projection planes via a custom AMOP script.

In fact, a good portion of the originality of the present methodology lies in this script, which allows, via the projection of geometry in the (u,v) space, the achievement of the planar representation not only of geometric shapes but also of the texture associated with the 3D model.

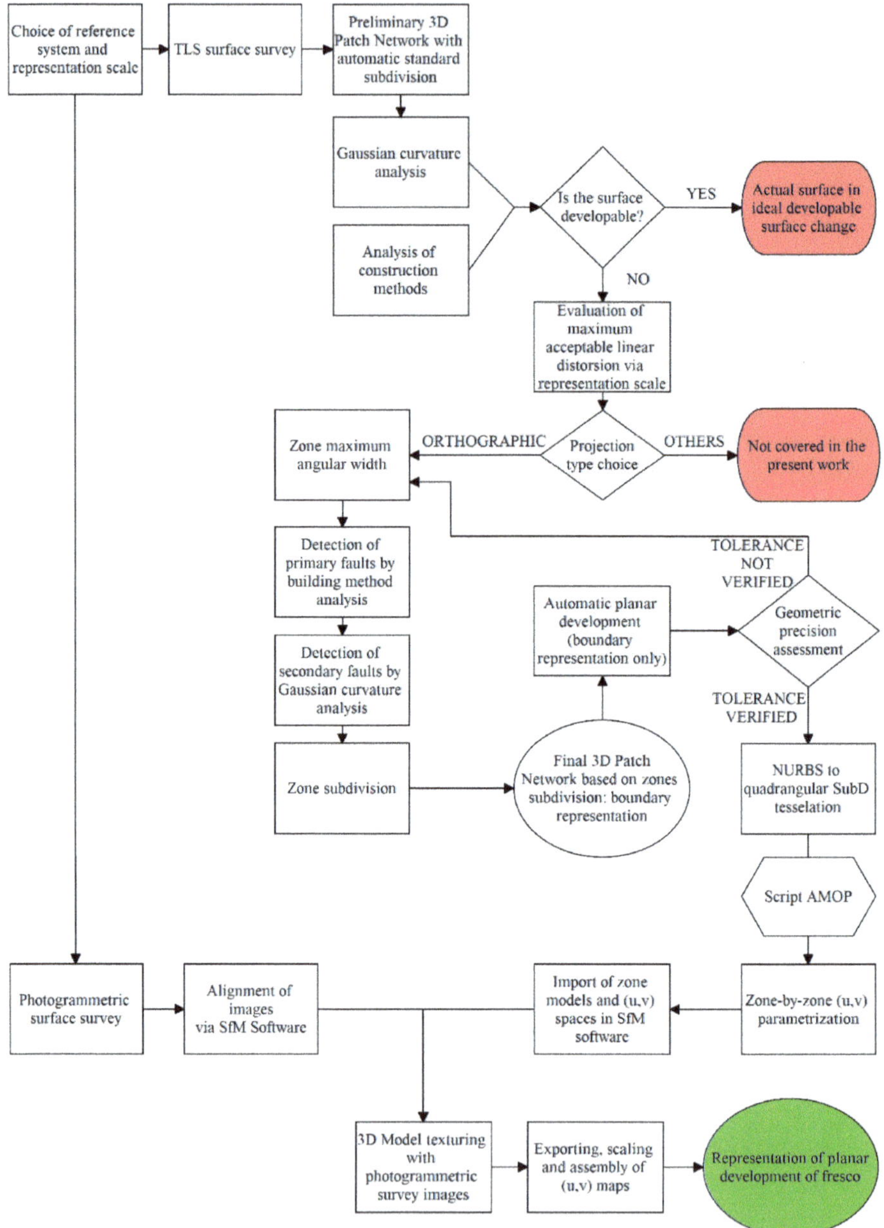

Figure 10. Flow chart of the methodology suggested.

Author Contributions: Conceptualization, A.P. and F.F.; Data curation, A.P., G.C., I.M.-E.Z., F.F., and L.C.; Funding acquisition, A.P. and G.C.; Investigation, A.P., G.C., and I.M.-E.Z.; Methodology, A.P., G.C., F.F., and L.C.; Software, F.F.; Supervision, A.P.; Validation, A.P., I.M.-E.Z., and F.F.; Visualization, I.M.-E.Z., and F.F.; Writing—Original Draft, A.P., G.C., I.M.-E.Z., F.F., and L.C.; Writing—Review & Editing, A.P., G.C., I.M.-E.Z., F.F., and L.C.

Funding: This research was partially funded by Università di Pisa grant number PRA_2017_12.

Acknowledgments: Thanks are due to Opera Primaziale Pisana for granting the required authorizations for the survey of the Pisa Cathedral dome; Eurotec Pisa srl for collaboration throughout the survey; technical staff members Andrea Bedini and Jessica Micheloni for survey execution.

Conflicts of Interest: The authors declare no conflict of interest. The funders had no role in the design of the study; in the collection, analyses, or interpretation of data; in the writing of the manuscript; and in the decision to publish the results.

References

1. Robleda, P.G.; Caroti, G.; Martínez-Espejo Zaragoza, I.; Piemonte, A. Computational vision in UV-mapping of textured meshes coming from photogrammetric recovery: Unwrapping frescoed vaults. *ISPRS Int. Arch. Photogramm. Remote Sens. Spat. Inf. Sci.* **2016**, *XLI-B5*, 391–398. [CrossRef]
2. Caroti, G.; Martínez-Espejo Zaragoza, I.; Piemonte, A. Range and image based modelling: A way for frescoed vault texturing optimization. *ISPRS Int. Arch. Photogramm. Remote Sens. Spat. Inf. Sci.* **2015**, *XL-5/W4*, 285–290. [CrossRef]
3. Chiabrando, F.; Rinaudo, F. Recovering a collapsed medieval fresco by using 3D modeling techniques. *ISPRS Ann. Photogramm. Remote Sens. Spat. Inf. Sci.* **2014**, *II-5*, 105–112. [CrossRef]
4. Bevilacqua, M.G.; Caroti, G.; Piemonte, A.; Ruschi, P.; Tenchini, L. 3D survey techniques for the architectutal restoration: The case of St. Agata in Pisa. *ISPRS Int. Arch. Photogramm. Remote Sens. Spat. Inf. Sci.* **2017**, *XLII-5/W1*, 441–447. [CrossRef]
5. Clini, P.; Frontoni, E.; Quattrini, R.; Pierdicca, R. Augmented Reality Experience: From High-Resolution Acquisition to Real Time Augmented Contents. *Adv. Multimed.* **2014**, *2014*, 597476. [CrossRef]
6. Sammartano, G.; Spanò, A. High scale 3D modelling and orthophoto of curved masonriesfor a multipurpose representation, analysis and assessment. *ISPRS Int. Arch. Photogramm. Remote Sens. Spat. Inf. Sci.* **2017**, *XLII-5/W1*, 245–252. [CrossRef]
7. Rieck, B.; Mara, H.; Krömker, S. Unwrapping highly-detailed 3d meshes of rotationally symmetric man-made objects. *ISPRS Ann. Photogramm. Remote Sens. Spat. Inf. Sci.* **2013**, *II-5/W1*, 259–264. [CrossRef]
8. Carli, E. *Il Duomo Di Pisa. Collana Chiese Monumentali d'Italia*; Nardini Collana: Firenze, Italy, 1989.
9. Sanpaolesi, P. Il restauro delle strutture della cupola della Cattedrale di Pisa. *Bolletino d'Arte* **1959**, *44*, 199–230.
10. Venditelli, L. *Il Mausoleo Di Sant'Elena. Gli Scavi*; Mondadori Electa: Firenze, Italy, 2011.
11. Rakob, F. Römische Kuppelbauten in Baiae. In *Mitteilungen Des DeutschenArchäologischenInstituts*; Römische Abteilung: Mainz, Germany, 1988.
12. Cipriani, L.; Fantini, F.; Bertacchi, S. The Geometric Enigma of Small Baths at Hadrian's Villa: Mixtilinear Plan Design and Complex Roofing Conception. *Nexus Netw. J.* **2017**, *19*, 427–453. [CrossRef]
13. WardPerkins, J.B. *Architettura Romana*; Mondadori Electa: Firenze, Italy, 1974.
14. Aita, D.; Barsotti, R.; Bennati, S.; Caroti, G.; Piemonte, A. 3-dimensional geometric survey and structural modelling of the dome of Pisa cathedral. *ISPRS Int. Arch. Photogramm. Remote Sens. Spat. Inf. Sci.* **2017**, *XLII-2/W3*, 39–46. [CrossRef]
15. Guidi, G.; Remondino, F. 3D Modelling from Real Data. In *Modeling and Simulation in Engineering*; InTech: Munich/Garching, Germany, 2012. [CrossRef]
16. Remondino, F.; Spera, M.G.; Nocerino, E.; Menna, F.; Nex, F. State of the art in high density image matching. *Photogramm. Rec.* **2014**, *29*, 144–166. [CrossRef]
17. Karras, G.E.; Patias, P.; Petsa, E. Digital monoplotting and photo-unwrapping of developable surfaces in architectural photogrammetry. *Int. Arch. Photogramm. Remote Sens.* **1996**, *31*, 290–294.
18. Karras, G.E.; Patias, P.; Petsa, E.; Ketipis, K. Raster projection and development of curved surfaces. *Int. Arch. Photogramm. Remote Sens.* **1997**, *XXXII*, 179–185.
19. Verhoeven, G.J.; Missinne, S.J. Unfolding leonardo da Vinci's globe (AD 1504) to reveal its historical world map. *ISPRS Ann. Photogramm. Remote Sens. Spat. Inf. Sci.* **2017**, *IV-2/W2*, 303–310. [CrossRef]
20. Grafarend, E.W.; Krumm, F.W. *Map Projections*; Springer: Berlin/Heidelberg, Germany, 2006.
21. Massarwi, F.; Gotsman, C.; Elber, G. Papercraft Models using Generalized Cylinders. In Proceedings of the 15th Pacific Conference on Computer Graphics and Applications (PG'07), Maui, HI, USA, 29 October–2 November 2007; pp. 148–157.

22. Bevilacqua, M.; Caroti, G.; Martínez-Espejo Zaragoza, I.; Piemonte, A. Frescoed Vaults: Accuracy Controlled Simplified Methodology for Planar Development of Three-Dimensional Textured Models. *Remote Sens.* **2016**, *8*, 239. [CrossRef]

23. Ciarloni, R. La logica delle forme. In *La Ricerca Nell'ambito Della Geometria Descrittiva*; Gangemi: Roma, Italy, 2016.

24. Gaiani, M. *La Rappresentazione Riconfigurata. Un Viaggio Lungo Il Processo Di Produzione Del Progetto Di Disegno Industriale*; POLI. Design: Milan, Italy, 2006.

25. Lai, Y.-K.; Kobbelt, L.; Hu, S.-M. An incremental approach to feature aligned quad dominant remeshing. In Proceedings of the 2008 ACM Symposium on Solid and Physical Modeling (SPM'08), Stony Brook, NY, USA, 2–4 June 2008; p. 137.

26. Jakob, W.; Tarini, M.; Panozzo, D.; Sorkine-Hornung, O. Instant field-aligned meshes. *ACM. Trans. Graph.* **2015**, *34*, 1–15. [CrossRef]

27. Shen, J.; Kosinka, J.; Sabin, M.; Dodgson, N. Converting a CAD model into a non-uniform subdivision surface. *Comput. Aided Geom. Des.* **2016**, *48*, 17–35. [CrossRef]

28. Valerio, V. La forma dell'ellisse. In *Arte e Matematica: Un Sorprendente Binomio*; Arte Tipografica: Napoli, Italy, 2006; pp. 241–262.

29. Scalzo, M.S. Agata a Pisa. In *Rotonde d'Italia. Analisi Tipologica Della Pianta Centrale*; Volta Valentino: Milano, Italy, 2008.

30. Filkins, P.C.; Tuohy, S.T.; Patrikalakis, N.M. Computational methods for blending surface approximation. *Eng. Comput.* **1993**, *9*, 49–62. [CrossRef]

31. Verhoeven, G.J. Computer graphics meets image fusion: The power of texture baking to simultaneously visualise 3D surface features and colour. *ISPRS Ann. Photogramm. Remote Sens. Spat. Inf. Sci.* **2017**, *IV-2/W2*, 295–302. [CrossRef]

 International Journal of
Geo-Information

Article

Hybrid 3D Models: When Geomatics Innovations Meet Extensive Built Heritage Complexes

Filiberto Chiabrando, Giulia Sammartano *, Antonia Spanò and Alessandra Spreafico

Department of Architecture and Design (DAD), Politecnico di Torino, Viale Mattioli, 39, 10125 Torino, Italy; filiberto.chiabrando@polito.it (F.C.); antonia.spano@polito.it (A.S.); alessandra.spreafico@polito.it (A.S.)
* Correspondence: giulia.sammartano@polito.it; Tel.: +39-0110904380

Received: 22 January 2019; Accepted: 23 February 2019; Published: 1 March 2019

Abstract: This article proposes the use of a multiscale and multisensor approach to collect and model three-dimensional (3D) data concerning wide and complex areas to obtain a variety of metric information in the same 3D archive, which is based on a single coordinate system. The employment of these 3D georeferenced products is multifaceted and the fusion or integration among different sensors' data, scales, and resolutions is promising, and it could be useful in the generation of a model that could be defined as a hybrid. The correct geometry, accuracy, radiometry, and weight of the data models are hereby evaluated when comparing integrated processes and results from Terrestrial Laser Scanner (TLS), Mobile Mapping System (MMS), Unmanned Aerial Vehicle (UAV), and terrestrial photogrammetry, while using Total Station (TS) and Global Navigation Satellite System (GNSS) for topographic surveys. The entire analysis underlines the potentiality of the integration and fusion of different solutions and it is a crucial part of the 'Torino 1911' project whose main purpose is mapping and virtually reconstructing the 1911 Great Exhibition settled in the Valentino Park in Turin (Italy).

Keywords: 3D models; multisensor; multiscale; SLAM; MMS; LiDAR; UAV; data integration; data fusion; cultural heritage

---

1. Introduction

The creation of a three-dimensional (3D) model, with the aim of documenting extensive complexes, is a great challenge where the geomatics techniques and methodology could play a key role. A 3D model, which relies on an accurate metric survey, could provide a useful base on which other analyses can be carried out, especially in the cultural heritage domain where highly detailed information that is connected to models and descriptive model data that provide a more general view of the studied object are necessary. An important aspect, which is related to the wealth of the available data after multisensor acquisition, is connected to the optimization phase, where it is important to follow a predefined pipeline to optimize the model, not only in terms of file dimension, but also according to the detail of the recorded object. High resolution of digital models must be harmonized to high accuracy and density of information, but, in contrast, this compromises the capability of visualization, requiring more efficient hardware or a simplification of the model is necessary [1]. The creation of an accurate (level of detail depends on its final use) 3D model is the first step. The management of a useful 3D model requires the creation of an ad hoc digital surface; this implies a series of reflections and choices referring to time and requirements both in acquisition and processing phases, geometric, and radiometric aspects of final products, according to resolution and quality needs, but also the size and interoperability of files.

Nowadays, digital 3D documentation of the built heritage presents a useful tool in the analysis and interpretation of the historical site. The employed sensors and methodologies for capturing reality could be different and they are mainly divided into image-based or range-based techniques. Because each of the aforementioned techniques is connected with pro and cons, it is well affirmed that the right way to carry out a correct survey, which could be considered to be an important research issue, is the 'hybridization' of the approach, in the sense of a combination or fusion of data to obtain a complete, usable, and users-driven 3D digital model.

Starting from the work that is reported in [2], where the main issues connected to the use of a Simultaneous Localization and Mapping (SLAM)-based system for surveying complex built heritage were discussed, the objective of the research that is presented in the next sections is to extend and evaluate the contribution of the multisensor approach; mostly referring to the connected techniques for carrying out suitable 3D models for the documentation of extensive built heritage complexes. The different techniques that are employed for data acquisition and processing will be discussed and the adopted strategy, with the accompanying advantages and disadvantages, will be addressed to define a good strategy for obtaining a complete 3D digital model. Furthermore, in the second part of the paper, the usability of the achieved models and the descriptive characteristics will be reported. The idea, following the actual development for 3D real-time visualization, is connected to the improvement of the use of this new digital support for enhancing the knowledge that is connected to the documented built heritage. These instruments can recreate an interactive, virtual-reality experience that is usable by students, researchers, and other users to explore the site's buildings and the related artefacts.

State of the Art and Related Works

Architectural heritage, or more generally the complex built heritage, is a very interesting test field for 3D documentation purpose. The characteristics of those objects allow for experimenting with new sensors, methodologies, and techniques, which are parallel to more consolidated ones, to evaluate the pros and cons of each followed approach. Moreover, nowadays it is generally agreed that the combination of the different techniques is the proper way to obtain a metric product that respects all of the requirements of a correct survey in terms of accuracy, completeness, and reliability.

Nowadays, the well-consolidated techniques that are able to produce a 3D recording of the built heritage can be divided into two main categories, namely image-based and range-based techniques [3,4]. The image-based techniques are related to the use of photogrammetric acquisitions, primarily close-range or Unmanned Aerial Vehicle (UAV). Nowadays, this methodology is probably one of the most used in the architectural survey field, according to the significant developments of the last decade in terms of sensors and algorithms, even more, based on a computer vision approach—the well-known Structure from Motion (SfM). Because it could be considered to be low-cost, easy to use, flexible (multiresolution and multiscale), and, thanks to the massive use of UAV, photogrammetry allows for moving the point of view from the terrain to the air. As a consequence, in every work related to the survey of a complex site, starting from the archaeological areas [5,6], up to the built heritage complex site [7–10], the use of photogrammetric techniques are used with the aim of extracting 3D point clouds, 3D models, or simple orthoprojections. Further developments in this field are connected to the use of different sensors, such as multispectral, hyperspectral, thermal or mobile devices [11,12], and in the employment in terrestrial or UAV applications of the new generation of panoramic /spherical cameras [13,14]. The range-based techniques (i.e., LiDAR, Light Detection and Ranging), dating back to the early '80s, have recently developed different strategies for data management and processing [15,16], along with several workflows and a lot of device improvements, to produce high-performance point clouds in terms of accuracy (short-range vs long-range scanners), speed of data acquisition (Time of Phase vs Phase Shift), and radiometric quality (embedded or external camera). Starting from the reliability of the LiDAR in connection with the evolution of the computer vision algorithms and the improvement of the inertial systems in mobile platforms, an emerging methodology that is even more frequently used for survey purpose is the SLAM approach. These techniques could be considered to be a Mobile Mapping System (MMS), where a moving head is equipped with a ranging measurement laser capturing two-dimensional (2D) point profiles, an inertial measurement unit with triaxial gyros, accelerometers, and three-axis magnetometers. Over the last year, different tests and applications have been carried out through the use of this system [17–19] and, according to the achieved results, it could be considered to be more than suitable for architectural scale purposes, also in articulated and compound environments [20], even indoor areas. The problems of RGB data and (geo)positioning issues are currently not completely solved.

Nowadays, it is clear that the main issues driving the research test and validation are addressed to the data processing and management in terms of data and procedures optimization. The different techniques allow for extracting or collecting 3D point clouds and the fusion of the different achieved results is an accepted methodology [21–23]; these products constitute the starting point for different following steps that could drive the surveyors to carry out a traditional drawing, an orthophoto, a 3D textured model, or to define the geometry behind an Heritage Building Information Modelling (HBIM). The next sections are dedicated to reporting the experiences of our research group, which aimed at obtaining a final hybrid model, i.e., derived from a combination of fusion and the integration of multisensor data.

2. The 'Torino 1911' Project: Focuses and Aims

1911 is a symbolic year in Italian history; in that year, Italy celebrated the 50th anniversary of national unity with the so-called 'Fabulous Exposition' that settled in the Valentino Park of Turin. The International Exhibition displayed the major works of technology and science from the entire civilized world. The spectacular architectures that were realized for this purpose were the mirror of society at that time: ephemeral elements characterized an overabundance of eclectic styles to overwhelm seven-million visitors [2,24]. The Torino 1911 exposition celebrated the industrial fervour in the western part of Italy with its robust and explosive production of 'goods, science and technology, national boundary and Capital' [25]. The architects created a fantastic city on both banks of the Po River, a city of dazzling whiteness, with staircases, pediments, steeples, and colonnades overburdened with friezes. The decoration was done with superfluous capitals like no one had ever seen before, with fountains, waterfalls, and *tapis roulant, fastigiums, porticos,* statues of Victories with rustling veils, eagles with outstretched wings, and angels playing trumpets.

The overall project *Torino 1911* aspires to virtually recreate the world that was revealed in 1911 and to facilitate constructive encounters between the old world's fair technologies and new digital technologies. Today, the monumental Valentino Park preserves its historic structure with avenues and ancient trees [26], while most of the temporary buildings are no longer present. In contrast, there are some monumental complexes, which, although not built for the occasion in 1911, constituted an integral and important part.

It is from this point that we started and that constitutes the main objective of this work. How should models of complex buildings be generated, organized, and characterized that must be subjected to many different objectives at the same time? Being not only navigated, but also investigated to know the intrinsic characteristics and the insertion in a scenario that has to be rebuilt?

Case Study

The selected case studies are those that are most directly connected with the themes of survey and model generation from real-based methods, that is, the existing building complexes. Subsequently, the analysis focuses on two castles settled in the Valentino Park of Turin: the *Rocca* (Figure 1) and the Valentino Castle (Figure 2), this last mainly referring to the noble floor.

(a)   (b)

Figure 1. Test area: (**a**) The Rocca in the '*Borgo Medievale*' and (**b**) The Valentino Castle.

The first one is part of a suggested reconstruction of a medieval hamlet (*Borgo Medievale*), which was conceived for the 1884 Italian Exhibition. The latter is one of the Savoy Royal Residences (all inscribed on the World Heritage List for their outstanding value from 1997), built by Cristina of France in the XVIIth century as the actual configuration, and now hosting the Faculty of Architecture of the Polytechnic of Turin. The *Borgo Medievale* and the related *Rocca* were designed by Alfredo d'Andrade to symbolize the medieval Italian styles and be materialized in architecture, furnishing, flourishes, utensils, and works [27]. Enclosed by an enceinte and protected by a tower with a drawbridge, the *Borgo* is composed of a church, some houses, workshops, a garden, and the Rocca, the innermost keep; all of these structures are a revisit of some real medieval architectures that are spread in two Italian regions, Piedmont and Valle d'Aosta. The *Rocca* is articulated in four levels and two towers, one circular and one square. All of the rooms revolve around the inner courtyard, a double height open space with two levels of wooden balconies, and partially covered by a roof, which very similar to the Fenis one also for the mural paintings figuring Saint George over the stone staircase and philosophers, saints, and heroes all around the balconies. Each local has an irregular plan, depicted walls and brick vault, or wooden beam ceilings [28]; moreover, furnishing and usual accoutrements are designed to represent medieval life.

(a)   (b)   (c)

Figure 2. The main staircase in the courtyard of (**a**) the Fenis Castle (AO) in 2004 and (**b**) back to 1884 [29]. (**c**) The same scene in the courtyard of the Medieval Rocca in Turin.

The second case study refers to another type of castle; developed from a XVth century suburban palace and settled alongside the river Po, the Valentino Castle [30] was transformed into a royal residence during the XVIIth century for the Savoy family. During the XVIIIth century, the Castle was involved in urban planning, in which it was the central focus for the design of the Valentino Park [31], an urban public space of 42 hectares, and a place of a series of national and international exhibitions, which dated between the end of the XIXth century and the beginning of the XXth century. The Castle began the headquarters of the Royal School of Engineers in the mid-XIXth century and subsequently of the *Politecnico di Torino*. Nowadays, the noble floor is the most representative part of the building, which is also open to the public as a museum on predefined days. The present research investigates some chambers of the Castle, in detail, the Great Salon, the Fleur-de-lis, and the Roses Chambers in the south apartment of the main floor. With its double height, the favoured view of the river Po, mural paintings with never-ending scene spreads in the walls, and sculpture masked in the corners, the Salon is the Honour chamber represents the Savoy family. The Savoy's military feats are celebrated in the Salon, whose decoration was carried out by various artists during the XVIIth century and restored in the XIXth and XXIst centuries. The Honour Salon is the core of the whole castle, facing the ancient access from the river, and the newest from the courtyard towards the city. Carlo and his son Amedeo of Castellamonte, followed the architectural works of the Valentino Castle were followed by, who transformed the Palace into a 'Pavillon' Castle with French style. Isidoro Bianchi supervised the Cristina's apartment decorations between 1633 and 1642. The restoration of the entire noble floor of the castle, in addition to the ones affecting the artistic heritage, has concerned the structures, especially the vaulted systems, which have suffered damage and water infiltration. The vaults are so-called 'fake vault' and they are realized with plaster that is applied on reed mats, which are hanging on a rib wood frame. In addition to the historical restoration that was carried out by D'Andrade in the late nineteenth century, some consolidations of the vaults—made after the second world war to recover the damage of the bombings and another in 1978—have already been studied to determine the safe conditions [32].

3. The 3D Documentation Project: Multisensor Data Acquisition

In these types of scenarios, many operative solutions are commonly deployed by endorsed workflows, as introduced, to overcome multiscale and multiresolution issues in outdoor and indoor spaces and some innovative mapping solutions have already been tested [2]. Surely, the multisensors approaches should be validated from a partial to a global viewpoint, to optimize the effectiveness of each sensor data contribution to their integration, finalized to overcome their singular applicability limits and that are aimed at making an experimental workflow effective. The test areas present a variety of characteristics and restrictions that influenced the choice of which sensor was the best solution to perform a complete and continuous 3D metric survey. Therefore, both image- and range-based sensors were used during the acquisition phase and a general overview of the raw data is presented in Table 1.

Table 1. Sensors employed in the acquisition phase and general overview of the amount of collected data for the entire project.

Type of Survey	Systems	Sensors	Medieval Rocca and Hamlet	Valentino Castle Main Floor
Range-based	TLS	FARO® Focus³ᴰ X120 & X330	93 scans	112 scans
	MMS	GeoSLAM ZEB Revo RT+ZEBCam	29 scans	8 scans
Image-based	UAV	DJI Phantom 4 Pro Obsidian DJI Mavic Pro Platinum DJI Spark	1919 images	264 images
	DSLR	Sony ILCE 7RM2 Canon EOS 5DS R	2455 images	1942 images
	Low-cost sensors	DJI Osmo+ GoPro Fusion 360	15 videos	10 videos
Topography	GNSS	Geomax Zenith 35	18 vertices 145 targets	17 vertices 162 targets
	TS	Geomax Manual TS Zoom30 Pro		

Hereby, only the main results from some of them are later referred to in detail. Moreover, another crucial aspect to be considered is that the acquisition campaigns, due to operative needs of the project, were organized at different moments, while employing various groups of surveyors to cover the entire area and to achieve different levels of detail.

3.1. Multisensors Approach for Multiscale Data

In detail, the investigated approaches are a Terrestrial Laser Scanner (TLS), an MMS, Digital Single Lens Reflex (DSLR) cameras, various UAVs, and some image-based low-cost and portable sensors, as steady cameras and action cams. Moreover, a geo-reference system was established, thanks to the integration of a principal network that was acquired with a Global Navigation Satellite System (GNSS) system and a huge amount of materialized codified contrast markers and natural points from the scene, well distributed in the test area, acquired by a Total Station (TS), in order to obtain a reliable network of control and check-points.

According to the complexity and the extension of the test areas, the phase shift laser FARO® Focus (Figure 3) was chosen as the TLS system, S120, and the superior range X330; the characteristics of which are tested in these cases of wide distance in outdoor scenes.

(a)   (b)

Figure 3. Terrestrial Laser Scanner (TLS) by FARO®: (**a**) The FARO® Focus 120 during the acquisition and (**b**) planimetric view of an outdoor coverage range of an X330 scanner.

Another type of laser-based sensor is the ZEB Revo Real Time (RT) by GeoSLAM (Figure 4); contrary to the motionless TLS system, it is a portable MMS working with a Hokuyo UTM-30LX-F scanner technology and using the SLAM algorithm and an Inertial Measurement Unit (IMU) [33]. This type of sensor is promising because of its 'go-anywhere' definition and that it is currently already quite investigated in the literature [2,17,34–42].

In fact, lightness, handheld solution, and manoeuvrability characteristics allow for rapid acquisition of a vast number of points, especially in an articulated, multilevel, narrow, and extended area—such as the interiors of furnished museums, offices, and underground spaces—where other sensors cannot be employed or would be very time-consuming. Furthermore, the tested sensor mounted a ZEB Cam, which is a commercial off-the-shelf (COTS) GoPro HERO4 Session.

The scanner works with a tablet through a specific Wi-Fi connection, allowing for the real-time view during the acquisition; these two implementations increase the potentiality of the whole system when compared with the previous versions, the first Zebedee [43], and the following ZEB1 [37] and ZEB Revo [44]. The scanner works in pair with a tablet through a specific Wi-Fi connection, allowing for the real-time view during the acquisition; these two implementations increase the potentiality of the whole system as compared with the previous versions, the first Zebedee [43] and the following ZEB1 [37] and ZEB Revo [44].

According to the aim of the work that is connected to multisensor and multiresolution data acquisition and processing, in the analysed areas, different UAVs have been employed (Figure 5) accurately to document the built heritage from different points of view.

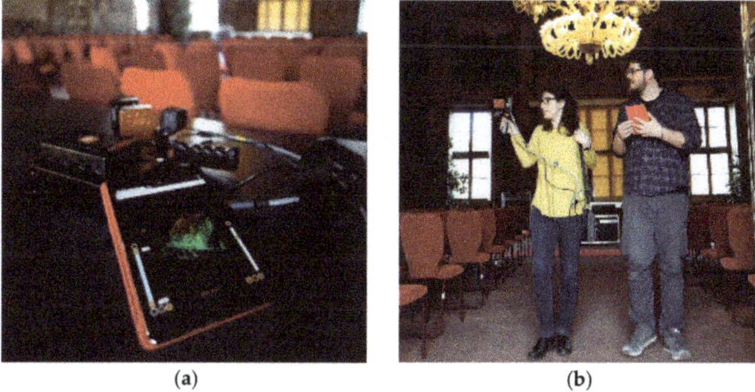

Figure 4. The ZEB Revo Real Time Mobile Mapping System (RT MMS) by Geo Simultaneous Localization and Mapping (GeoSLAM): (**a**) The data logger, the rotating head and the Go Pro mounted above and the IPad Pro employed for the real-time view during the acquisition phase; and; (**b**) Operators scanning the Great Salon with the instrument.

The flights were performed using a standard approach according to the different characteristics of the used platforms (automatically for the Phantom and Mavic and manually for the Spark). In the next sections, only the data that was acquired by the Phantom will be described because the highest resolution of the images acquired by that platform allows for better integration and/or fusion of the achieved results with the ones that were acquired by the other sensors. Furthermore, to complete the multisensor survey, a photogrammetric close-range acquisition was carried out to test different low-cost sensors and consolidated DSLR blocks and also acquire different qualities of radiometric data.

PHANTOM 4 PRO OBSIDIAN	MAVIC PRO	SPARK
Sensor 20 MP, 4096 × 2160 px	Sensor 12 MP, 4000 × 3000 px	Sensor 12 MP, 3968 × 2976 px

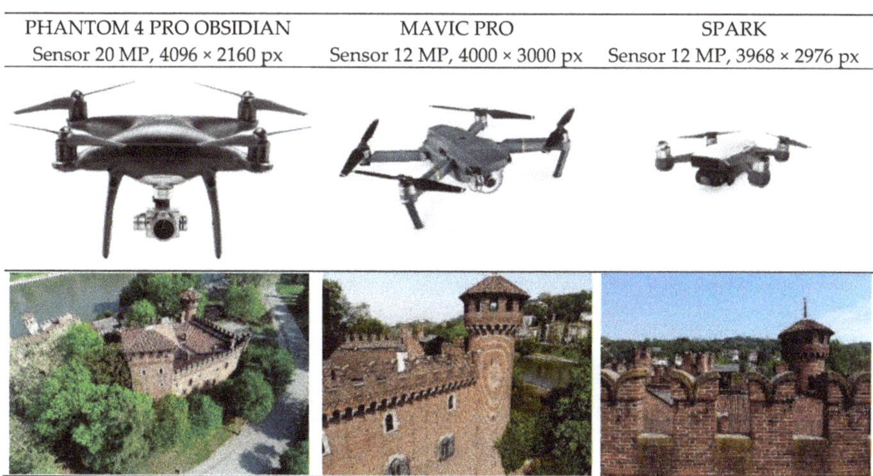

Figure 5. Different Unmanned Aerial Vehicle (UAV) platforms by DJI company and related sensors with captured images.

3.2. Data Acquisition Strategy and Data Preprocessing

The strategies that have been put in action are addressed, as introduced, to the problem of data acquisition protocols' optimization, especially if coming from different technologies use and the issues that are related to multiscale and multisensor models' creation and validation. Certainly, in such compound environments featured to a large extent in interior and exterior spaces, the variables of space-to-be-covered and time-to-be-spent in composite operations workflows require deep analysis. Therefore, different sensors and integrated strategies were tested for indoor and outdoor acquisitions. Certainly, the castle royal rooms and medieval *Rocca* apartments have been challenging environments, because furnishings indoor and trees outdoors forced the consideration of some aspects:

- Flight plan, overlap, resolution, and UAV camera axes in oblique acquisitions: the moment of the year in which the survey is performed will be influenced by the presence of the trees' foliage.
- Number, position, and resolution of TLS scans required obtaining a continuous surface with homogeneity of recovering without horizontal and vertical shadings: if they need to be increased, consequently, the acquisition and processing time are also time-consuming.
- SLAM-based trajectories in the case of MMS, deployment should be adapted to these types of locations: certainly, they will be successful in terms of time-saving, but they would bring with it an increase in the already known problems of noise errors affecting the point clouds and the drift errors that affect the trajectory.

3.2.1. The UAV Flight on the Medieval Rocca

Over the analysed area, a complete UAV survey was planned and performed using the aforementioned platforms to accurately describe the external part of the castle. Despite the use of different platforms, hereafter only the Phantom data will be considered as a general example for using the UAV for data acquisition and extracting 3D information from above. Complying with the consolidated trends in UAV image acquisition [45,46], the flights over the Rocca were planned with the aim of acquiring nadir and oblique images. According to a limited extension of the surveyed object; first, a double grid acquisition with the nadir camera axes was performed (Figure 6b). This first flight was made with a longitudinal overlapping of 80% and a lateral overlapping of 70%. Furthermore, a circular flight with the camera axes at 45° was carried out, in this case, the angle between the adjacent images was 5° with the camera being oriented to the centre of the circle. The Rocca was covered with 137 images with two different flights at an average flight height of 50 m (the height of the Rocca is about 32 m). The Pix4DCapture application was used for planning (Figure 6a) and performing the different flights to obtain a complete reconstruction of the Rocca (Figure 6b).

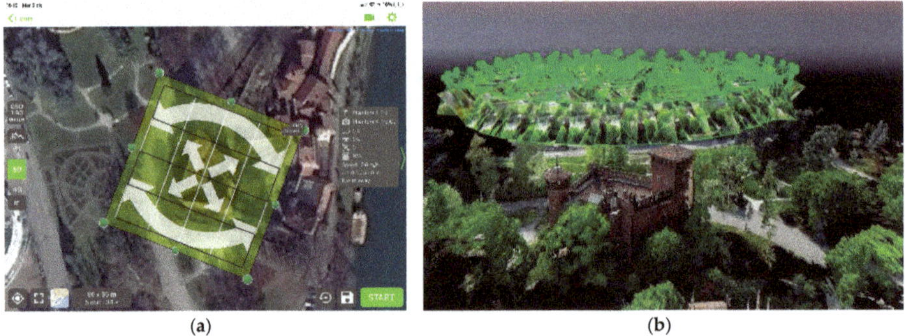

(a)   (b)

Figure 6. (a) The flights performed by Phantom 4 PRO (b) oriented UAV images block on the Rocca.

The advantage of using an ad hoc application for the image acquisition phase is surely connected to the full automation of the process, but the radiometric quality is not fully guaranteed, in fact, a recurrent drawback is the overexposure of images. During the different performed tests, especially in sunny conditions, the automatic procedure that was implemented in Pix4DCapture in most cases delivers moderately overexposed images. In the case of the Rocca, fortunately, the performed flights were carried out during a cloudy day with a perfect diffuse illumination, which allowed for us to acquire the images in the best light conditions. Another problem that was connected to the automatic acquisition using the aforementioned application is related to the fixed altitude of the flight. In other words, with Pix4DCapture, it is not possible to change the height of UAV during the flight according to the slope of the terrain. In difficult areas (with large height gaps), as a result, this limit could cause the acquisition of images with uneven Ground Sample Distance (GSD). In the present research, according to the shape of the area/object, the flights were set up to obtain a GSD value (~2.5 cm/px)—suitable for an architectural scale (1:100–1:200)—that measured 20 Ground Control Points (GCPs) and eight Check-Points (CPs), as reported in Table 2.

Table 2. UAV acquisition specifications.

Specifications	Phantom 4 Pro Obsidian
Height flight	50 m
Area covered (Area of Interest AoI)	0.106 km^2 (4000 m^2)
Strips	1 circular (oblique), 2 nadir
N° of images	137
Time	45 min
GSD (cm/px)	2.34 cm
N° of GCPs/CPs	20/8
Pt density (AoI)	mean 1800 pt/m^2 (17,500 pt/m^2)

Currently, the only way to solve the two reported drawbacks is to perform manual flight. However, that does not allow for following predetermined flight lines or circular acquisitions; therefore, according to the underlined problems, actually, where the environmental conditions allow for flight without any particular problem (high sunlight, obstacles, or other significant issues) the automatic strategy is preferred.

3.2.2. Close-Range Photogrammetric Acquisitions

To test other image-based sensors, in addition to the UAV data, different terrestrial acquisitions were carried out to document the analysed areas. The followed approach is the one that is well consolidated in the community that require large image overlapping using parallel and convergent images [47], when considering all the above internal/external light conditions.

In the Valentino Castle, low-cost platforms, such as DJI Osmo, GoPro Fusion 360, and high-resolution digital cameras (Sony ILCE-7RM2), were employed for recording the indoor and outdoor data. Using the Osmo and the GoPro, several videos were recorded, and moreover, the frames were extracted for performing the photogrammetric process, as reported in [48]. The acquisitions performed with Sony ILCE-7RM2, which, in our case, was equipped with a 12 mm focal length, were carried out according to the common practice. Normal and convergent images with large overlapping (more than 80%) were collected to perform the photogrammetric process that was based on SfM techniques. In the Rocca, the same low-cost sensors were employed, while the DSLR acquisition was performed using a Canon EOS 5DS R with a 24 mm lens. The strategies were the same used for the Valentino Castle. In the next sections, only the data that was obtained using the images that were acquired by the DSLR cameras will be analysed.

3.2.3. The Terrestrial LiDAR Scans Projects

According to the consolidated practice of traditional TLS acquisition [49,50] and scans registration, a huge number of scans data were obtained to cover all of the architectural contexts. As usual, the scans acquisition was supported by control points topographic measurements and distribution guaranteed the suitable surfaces coverage with adequate overlapping.

First of all, a large number of scans is connected to the complex shape of the Rocca, which is composed of narrow rooms that are connected to each other. Moreover, the furniture located in the surveyed areas was another important aspect that has increased the number of acquired scans. The scan resolution was set up at 1/5 (that means one point every 7.74 mm at 10 m), with a quality of 4× (that means that each point is measured four times). As usual, after the point clouds acquisition, the images were also captured to colour the point clouds. The high number of planned scans is also due to the prefiguration of accelerating the registration procedures, ensuring a high overlap between all of the scans that is necessary for assuring an easy registration of the different acquisitions in line with the cloud-to-cloud approach, based on the well-known Iterative Closest Point (ICP) algorithms. To speed up the scans' registration, the work was divided floor by floor, and subsequently all the data were connected, as shown in Figure 7. In the Medieval Rocca, the survey planning has gone along the complex distribution of the apartments, as summarized in Table 3 and the process is evaluated through values of mean for ICP phase and standard deviation (st. dev.) with Root Mean Square Error (RMSE) for the georeferencing one. A mean of seven scans per room was collected, with a sum of 93 scans, on a total surface of about 1000 m^2. Furthermore, as mentioned, the survey was also operated in the Castle, specifically, on the main floor of the building. This is composed of 11 main rooms with plaster decorations and frescoes, as reported before, and four ancillary rooms (two bathrooms and two corridors). The TLS acquisitions were carried out with the same Rocca's strategy. For each main room, at least five different scans were achieved (usually in the four corners and in the centre) to cover the complete shape of the area with the decorations as well. In total, 82 scans were recorded, a mean of five per room, on a total volume of decorated spaces of around 900 m^2.

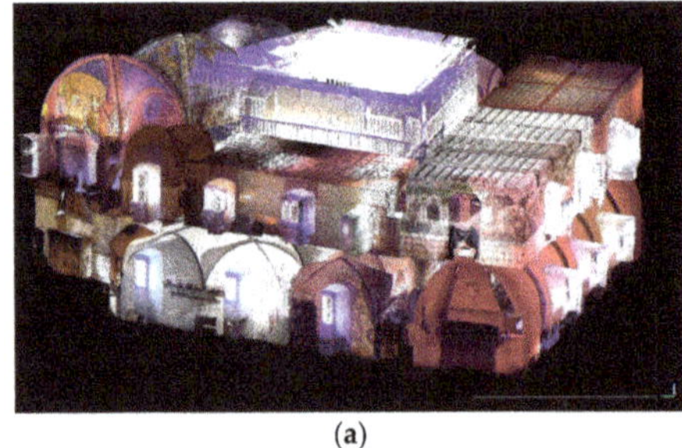

(a)

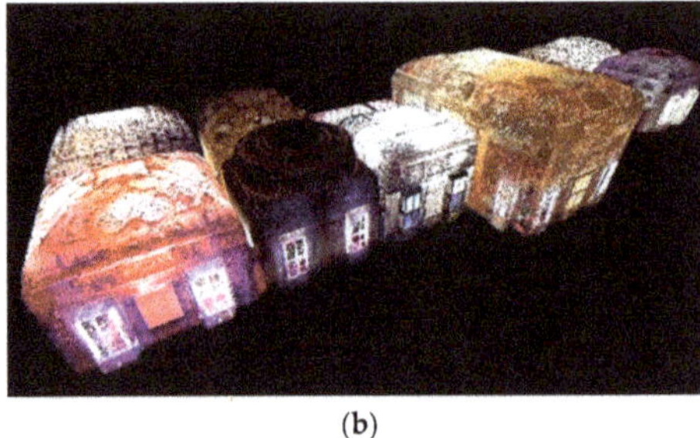

(b)

Figure 7. A view of the registered scans from (**a**) Rocca (n. 93) and (**b**) Castle (n. 82).

Table 3. Light Detection and Ranging (LiDAR) acquisition specifications.

| | *Environmental Features* | | | | *Scans* | | *Registration/Georeferencing* | | | | |
| | | | | | | | ICP | | Targets-Based | | |
	n° Rooms	Surface (m²)	Volume (m³)	Time (h)	n°	n°/Room	Mean Error (mm)	n°CPs	St.dev (mm)	RMSE (mm)
Castle	16	900	3600	18	82	~5	1.35	136	1.25	3.2
Rocca	13	1000	14,000	22	93	~7	1.05	116	2.69	4.65

3.2.4. Experimenting with 3D Mobile Mapping

According to the location characteristics, a test on the use of a portable mapping system was performed for evaluating the MMS strategy and the achievable products. As is commonly reported in several pieces of research, SLAM-based devices are almost the best solution for enclosed ambient rich in-features, because the SLAM algorithm also exploits the geometry to estimate 3D reconstruction of surveyed places. The optimal trajectories [2,19,41] require closed loops and roundtrips, rather than one-way mode (Figure 8a,b and Figure 9a) during the operator's rooms mapping.

One of the main matters affecting the ZEB Revo point cloud, in parallel to the accuracy of the points model, is the lack of positioning data. The origin of the reference system for each scan is the starting point and the orientation of axes XYZ is related to the rotating head: in fact, the X reference axis is normal to the scanner's back plate. In this way, it is possible, if initialized vertically, to rotate the X-axis to the East and to then perform the survey at least roughly following the North–South reference system, as investigated in [39]. Furthermore, the relative positions between multiple scans of the same survey set are also not assured to be aligned, due to the possible change of position of the rotating head at the initialization moment. Therefore, to manage all the scans in the same relative reference system, the simple single roto-translation and mutual ICP fitting applied on each ZEB scan can be now effectively avoided by the application of the "Merge" algorithm function (Figure 10), as implemented in the last desktop release of GeoSLAM Hub software for the scans processing [19].

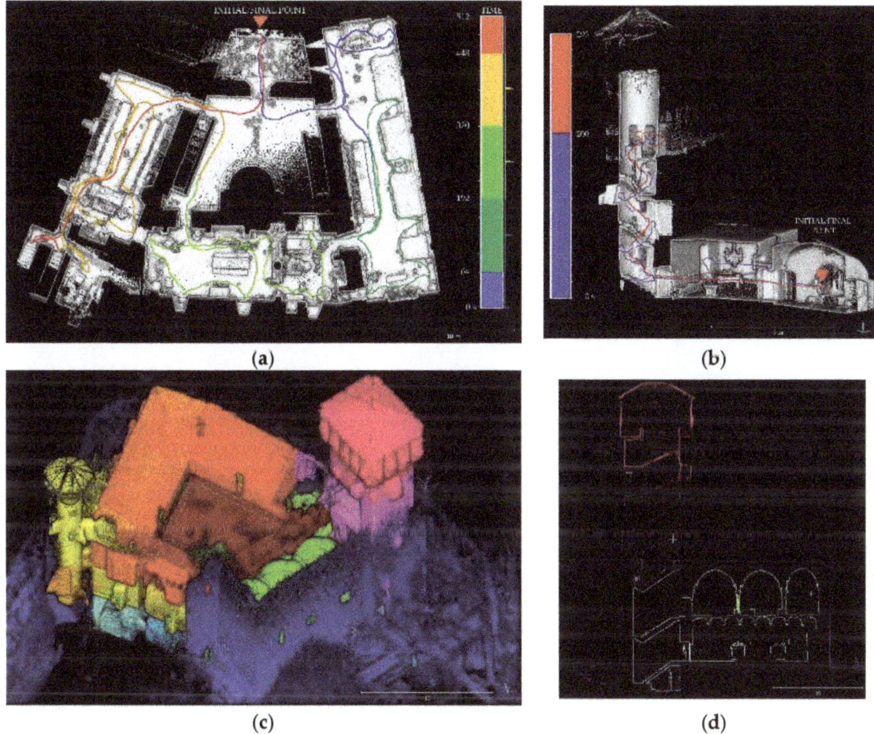

Figure 8. Example of SLAM-trajectory in the Rocca colourized by time: (**a**) Close loop with some roundtrip parts; (**b**) Roundtrip on the tower: going (blue) and return (red). (**c**) All of the reprocessed ZEB point clouds after the merge process, each colour corresponds to one scan, in blue the external one and (**d**) vertical section of the point cloud.

In the first step, the single scan is manually aligned one at a time to others to prepare the second step, which is the recomputation of the SLAM data; the implemented algorithm—based on joint SLAM+IMU data—combines 3D data derived from the overlapping parts between the scans and finally estimates a more accurate point cloud that is based on an improved trajectory.

In [2], the effectiveness of the "merge" and "reprocessing" functions are analysed and compared with the raw output data. Instead, to approach the solution of the problem of absolute positioning, this last software implementation plans to recognize and identify, along the trajectory, just those specific points in which the operator stopped for at least about five seconds. Concerning these so-called 'reference points', the processing algorithm extracts XYZ coordinates on each measured point, retrieving a set of points in a *.txt* output file. This is currently under research, but it is possible to say that this solution allows for obtaining a set of local coordinates of some vertices that are related to a topographic network, as measured by traditional TS and GNSS techniques. Subsequently, a roto-translation will enable conversion of the ZEB point cloud from the local system to the global positioning.

The collected ZEB-scans in the Castle and in the Rocca are thus processed in GeoSLAM Hub, where the reprocessing and the *merge* algorithms are applied to the two scans sets. Table 4 reports the two sets of acquisitions and an average value of the amount of data that is collected by the MMS.

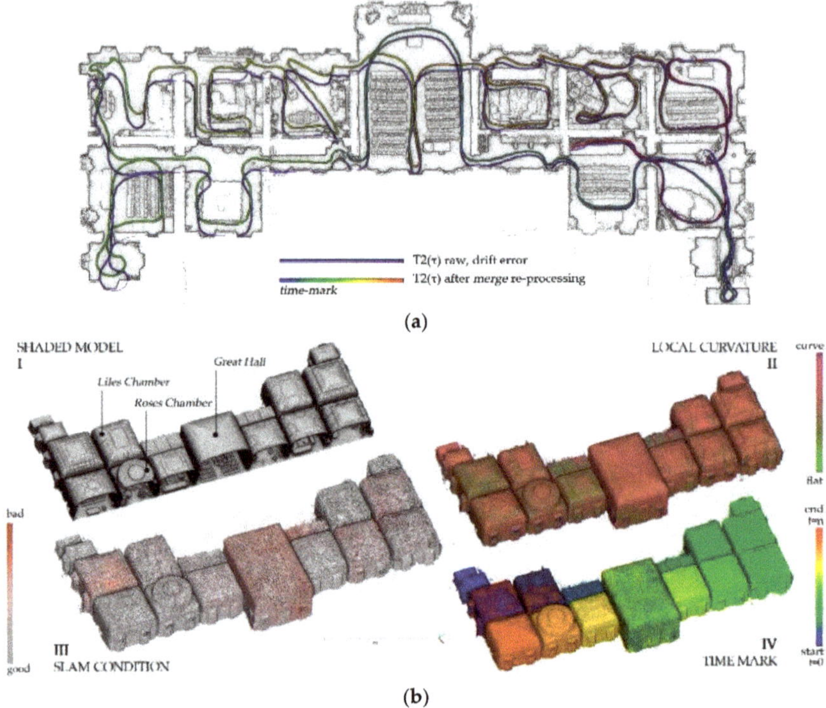

Figure 9. The ZEB survey in the Valentino Castle: (**a**) The plan view with the closed loop T2(τ), performed, before (blue) and after (range colours) the SLAM reprocessing by merge function correction; (**b**) the attributes related to ZEB-based point cloud processing, inside the GeoSLAM Hub platform. Along with an axonometric representation of the main floor in shaded view (**I**), there are normal mapping (**II**); SLAM quality condition during the scan (**III**); time-stamp information (**IV**). Each thematic result is associated with its own trajectory data as well.

Table 4. ZEB acquisition specifications.

Specifications	Rocca	Castle
N° of scans	11	2
Time	~3 h	~30 min
N° of points	~181,100,000	~35,410,000

The output point clouds consists of a set of different available formats (**.las; *.ply; *.E57; *.asc*) and very useful information can be calculated on the points surface, in the format of attributes that are associated with points (Figure 9b): i.e., normal, SLAM quality, time, elevation, etc. Section 4 contains discussion regarding these aspects. The selected area of the Rocca meets the ZEB requirements; in fact, it contains both indoor articulated places and open enclosed spaces, as previously described (Figure 8c,d). In the case of one-way blind tracks, as in the towers of the Rocca, the roundtrip trajectory was the only solution, to obtain good results. For this kind of environment, the employment of the

ZEB Revo represents a useful solution thanks to its manoeuvrability and time speed, which allows for the operator to move freely in narrow spaces.

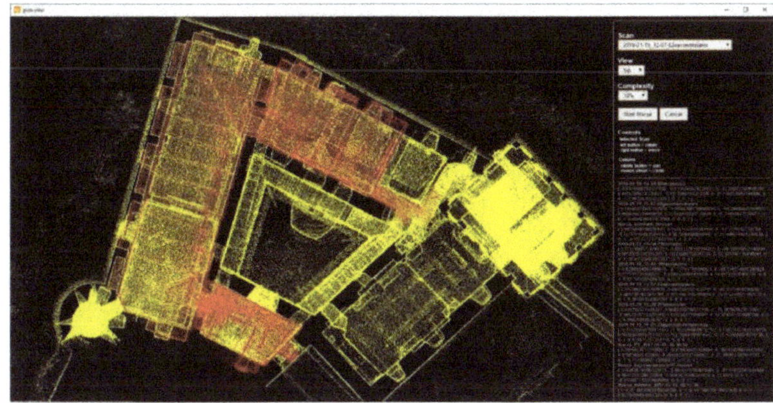

Figure 10. The desktop interface of the merge function in the GeoSLAM Hub software: in red, a single scan that is manually aligned, one at a time, to others, yellow. In the black box bottom right, the roto-translation matrix of each merged scan.

Contemporary SLAM-based acquisition was carried out in the Castle main floor as well the adopted strategies were similar to the ones that were performed in the Rocca area. The rooms of the main floor were surveyed testing two different strategies, with aimed to evaluate and validate the best approach for such large-range mapping in quite complex volumes, geometrically characterized objects, and scenes rich in furnishings. As reported in [2], the first T1(τ) (almost 6 min along 110 m, 0.3 m/s speed), was shorter and more simple, being performed with the best practices execution—a roundtrip—travelling along the corridor and mapping two rooms and the extreme sides of the scans. It has been processed with the "merge" function to help in the rigidity of the second T2(τ) (almost 12 min along 350 m, 0.5 m/s speed), the complete trajectory across all of the rooms on the floor coverage with a closing loop, to optimize the final alignment of the acquired data. In Figure 9a, the T2(τ) is shown in the original trend after the acquisition and in scale colours after using the merge tool, which is the reprocessing of the SLAM algorithm with the contribution of T1(τ).

4. Methods and Strategies for Complex Environments and Advanced 3D Modelling

The optimization-driven approach, which underlies the data integration strategy and the experimental process of fusion, tries to tackle, in the most all-encompassing way, but certainly not exhaustively, certain aspects typical of 3D digitization, downstream of the research experience conducted, and the methodological innovation tested in recent years.

The attempt, as introduced, is to balance a role for cutting-edge techniques, alongside those more consolidated approaches, to compare them and to obtain different deliverables that are useful to user-oriented scopes. Special attention is given, in this paragraph, to the 3D mapping via ZEB Revo MMS to investigate, in such kind of applications, how far it can be self-contained and where it needs supporting integration, or better, in which way its potential, in terms of speediness and manoeuvrability, can be a worthy basis for reasoning on potential data fusion processes.

Thus, the open issues, faced in these tests and clarified hereafter, mainly referred to the multisensor problem in the case of ZEB single data or a combined use within the mapping of complex scenarios, are tackled with experimental integration- or fusion-based approaches and are proposed for:

- the management of the reference system (Section 4.1): relative alignment and absolute (geo)positioning;

- the geometric reconstruction aptitude (Section 4.2), relating to decorative and morphological aspects; and,
- the examination of a fusion-based pipeline solution for the geometric/radiometric attributes enrichment in point clouds and surfaces (Section 4.3).

4.1. Positioning Issues for Complex and Extensive Indoor–Outdoor Environments

As introduced, the positioning question that is related to the ZEB point cloud could be currently solved by a roto-translation and ICP procedure with another positioned point cloud that was used as a reference (i.e., TLS model). This evidently requires a LiDAR survey (topographically based) to be performed in parallel to the MMS survey.

The research testing tries to split the problem matter and to lead a differential evaluation of ZEB performance in relation to a 3D model ground-truth: this is due for the possibility to consider the SLAM-based point clouds result in a "local" or "global" perspective. In fact, the operation of SLAM-based algorithms during the processing should simultaneously meet the needs of geometry (*locally*) and trajectory (globally), i.e., balancing of the influence of SLAM and IMU data in the trajectory estimation. This is the reason why ZEB mapping usually suffers from some drift errors in trajectory reconstruction (planimetric and altimetric drifts), even if the local reconstruction of a single space returns very positive values of deviation from the reference surface that employs Cloud-to-Cloud (C2C) analysis by CloudCompare software.

The configurations that are considered for the analysis are:

I. Local result, indoor, single enclosed area
II. Global result, indoor, single-level floor, evaluation of planimetric drift error
III. Global result, indoor, multilevel floor, evaluation of planimetric and Z drift error
IV. Global result, outdoor scenario with UAV data

I. Indeed, if we consider a single room, e.g., the Rocca's Throne Chamber (around 6 m × 13 m × 4 m dimensions) and the two point clouds, ZEB 2 mln points and TLS 16.6 mln points, their comparisons demonstrate very good results of ZEB technology in local 3D reconstruction. In particular, the dimensional comparison of main dimensions and areas estimation, as reported in Table 5, shows very accurate results, as modest residual deviation errors from the reference dimensions (LiDAR) exist. The computation of the C2C algorithm, in the second part of Table 5, considers the downsampling of the data by a filtering approach to the ZEB point cloud due to the noise errors that are typical of these SLAM-based acquisitions. In fact, the compared mean and the st. dev. results considered before and after the procedure, which are represented in Figure 11b,c, reach almost 1 centimetre (±2 st. dev.) and then less than 1 centimetre. It is also visible in the variation in the statistical distribution of points after the filtering process.

II. & III. If we consider now the global results of a ZEB set of mapping data in a typical indoor scene, it is possible to summarize that, in the case of both single floor and multilevel environments, a residual drift error in extensive and articulated trajectories is inevitable, but it is also possible to control and limit it. The test cases have been the honour floor in the Valentino Caste and the apartments of the Medieval Rocca. The execution of partially repeated and overlapping trajectories, for example, in the spaces that are designated for horizontal and vertical distribution (corridors, stairway blocks), together with the recomputation of SLAM algorithm with the "merge" reprocess, significantly improves the ZEB results according to a comparison with a LiDAR ground-truth surface, as is reported in [2].

IV. If we lastly consider the evaluation of the SLAM-based mapping into a global perspective in the outdoor scenario, another question that is connected with the ZEB data is the scale and the coverage of the point cloud. If the data is collected as usual by a walking operator in a close-range framework, the results will return a point cloud that is characterized by range distance proportional to the outdoor laser extension, declared as 15 m.

Table 5. Results of (I.) Dimensional and C2C comparison of ZEB and LiDAR point clouds on the Throne Chamber.

Cases	*n° of pts*		Dimensional Comparison (m)					
			Length	Width	Height	Area 1	Area 2	
ZEB	2,209,937		12.6289	6.130	4.2117	75.9021	22.7629	
TLS	16,661,778		12.6163	6.1231	4.2088	75.9006	22.7668	
		Δ	0.0126	0.0069	0.0029	0.0015	0.0038	
C2C comparison		No filter (Figure 11a)			Noise filter (Figure 11b)			
Mean		0.0131			0.0076			
St. dev.		0.0214			0.0058			

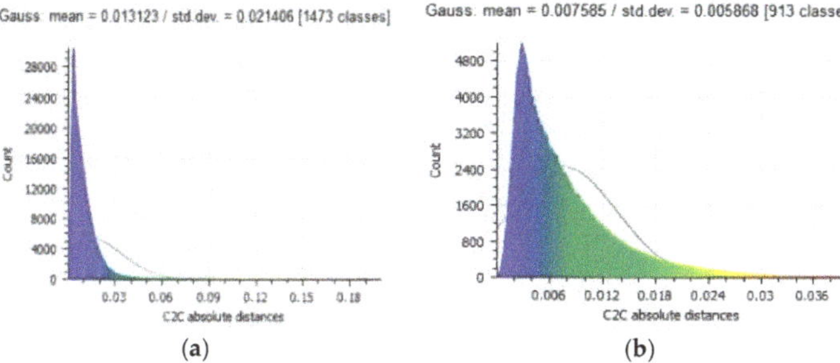

Figure 11. Statistical graphics of C2C comparison of ZEB and LiDAR point clouds on the *Throne Chamber*: mean and st. dev. distribution (**a**) before and (**b**) after noise filtering applied to ZEB data (C_1-C_2 cases).

In Figure 12, the research direction has considered, as recently investigated in different types of large-scale scenarios, the integration between these two assimilable strategies of rapid mapping and the expected comparable final scale: the terrestrial SLAM-based mapping with the contribution of high-scale aerial data coming from the UAV photogrammetric data.

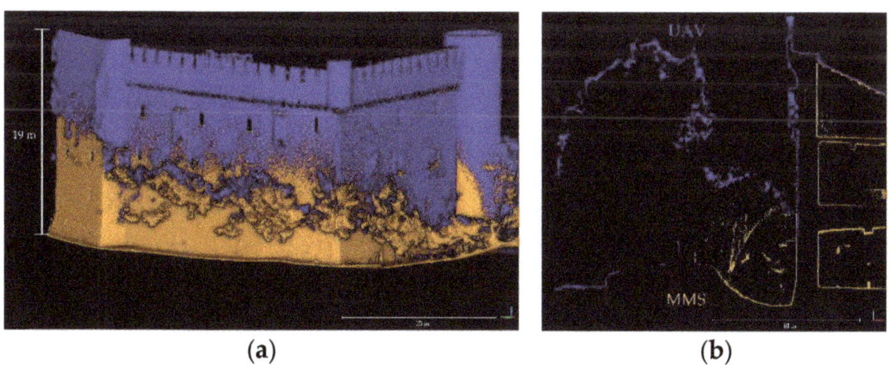

Figure 12. Extension of point clouds, UAV data coloured in blue, ZEB data in orange: (**a**) angular view of the external walls with the presence of vegetation and (**b**) vertical section of the Rocca.

As previously described, the Medieval Rocca is a great example of a complex architecture, where the performance of UAV, LiDAR, and MMS systems are proved, but the presence of vegetation hinders the ground level and covers a portion of the walls. The UAV point cloud Digital Surface Model (DSM) enables obtaining the external geometry of the surroundings and the Rocca, especially with its roofs and with the upper parts of its curtain walls and towers. Thanks to the easy manoeuvrability of ZEB Revo, it is possible to overcome this lack. Moreover, the employment of a handheld solution for the inner apartments also speeds up the acquisition and the processing phase. For these reasons, the integration of these two systems could be the best solution to be faced up. In fact, a surfaces' comparison that is proposed in Figure 12, and their C2C distance computation (Figure 13) show the potentialities of the enriched 3D descriptive capabilities of both integrated digitization of the Rocca. In Figure 12b, a noteworthy section of the integrated model and a critical point is the trees that lean against the curtain wall. Wherever the mapping operator managed to pass near the walls, the reconstruction was continuous, with a maximum displacement of distances, in the upper part of about 9 cm, and a minimum in the lower part of about 1.5 cm.

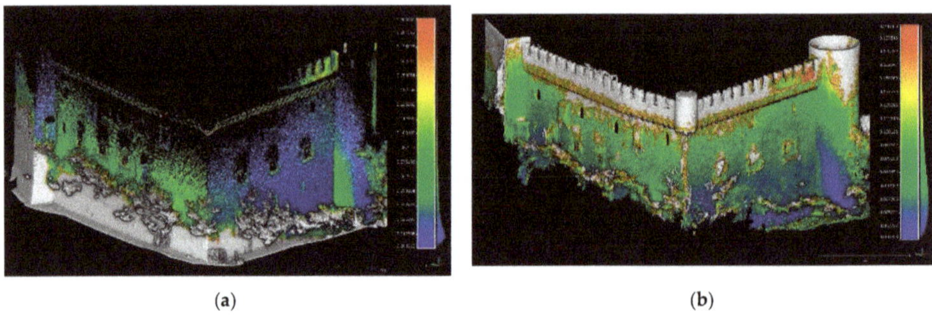

(a)   (b)

Figure 13. Complementary representation of C2C points distances between the two point clouds: (**a**) ZEB distances on UAV data (min. 1.33 cm, Max. 9.27 cm) and (**b**) vice-versa projection and values in Table 6.

Table 6. Results of C2C comparison ZEB point cloud and UAV Digital Surface Model (UAV DSM) on the *Rocca* volumes.

Cases	A	B	C	D
	Complete ZEB-Complete UAV	Outdoor ZEB-Complete UAV	Outdoor ZEB-Optimized UAV	n°12 Points-Based Alignment
RMSE (cm)	76.4	52.2	5.27	6.26 Min 1.33 Max 9.27

4.2. SLAM-Based 3D Modelling: Testing Descriptive Capabilities in Digitalizing Complex Surfaces

Hereafter, in this second analytical step, we will consider the problem that concerns the geometric and radiometric featuring of sensors' models in comparison with the ZEB surface, applying the validation strategies that are presented in [48]. In particular, a benchmarking analysis focuses on some examples of richly featured surface of frescoes and 3D details of decorative apparatus of the Valentino Castle, as previously described as a great example of complex architecture with completely decorated indoor spaces, that are, for distinctive reasons (Figure 9b above left): the *Great Salon* and the *Fleur-de-lis Chamber* with pavilion vaults and the particularly terraced *Roses Chamber* vault.

The surface models of typically decorated stucco's vaults, as extracted from the point cloud, are the ones in Figures 14 and 15. The problem of digitization of the emphasized 3D decorations is proposed in Figure 14, where the risk of vertical shadings is visible: the lack of information in the undercut areas corresponding to the stucco frames appeared as a drawback mostly in (b), caused by the fixed position of the scanner and moderately in (a) and (c). In Table 7, the quality values of the point clouds are only related to the vault parts: the close-range photogrammetric dense cloud can be compared with the massively rich LiDAR point cloud, only just after a 1:3 filtering ratio. Conversely, for the ZEB point cloud, the density ratio turns out to be 1:20. In Figure 15, the deviation distances map between LiDAR ground-truth and ZEB mapping is presented on the vaults of the Roses Chamber, the only great terraced vaulted structure of the honour floor. The values of C2C comparison imply average values of accuracy between 5 and 20 mm (blue to green).

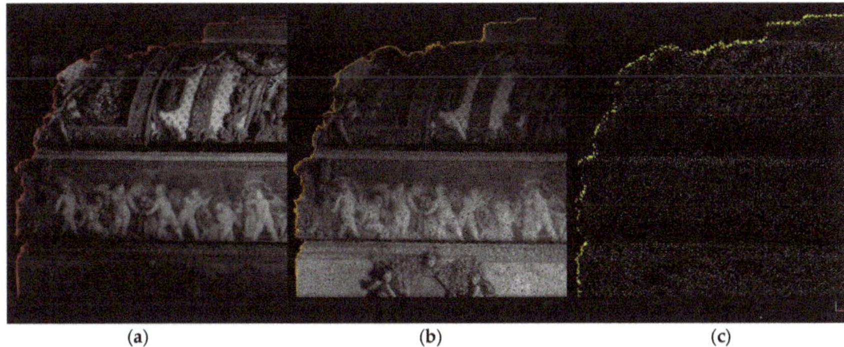

(a)   (b)   (c)

Figure 14. The benchmarking analysis on multisensors' data of Fleur-de-lis Chamber in the Valentino Castle. (**a**) SfM points cloud; (**b**) LiDAR points cloud; and (**c**) ZEB points cloud.

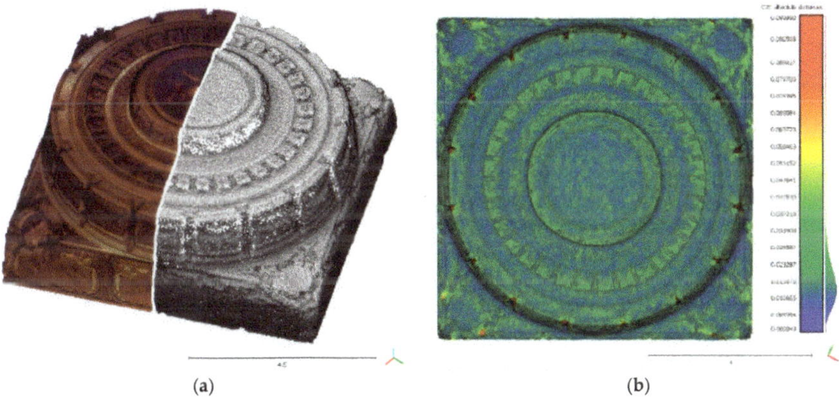

(a)   (b)

Figure 15. (**a**) The *Roses Chamber* vault and the (**b**) C2C distances analysis between LiDAR and ZEB.

Table 7. Comparison of point density related to the close-range photogrammetric model, LiDAR scan, and ZEB scan of the *Fleur-de-lis Chamber*. Similarities between (*) and between (#).

| | (a) Close-Range SfM | (b) LiDAR | | | | | (c) ZEB |
| | | Original | Filtering | | | | |
			3 mm	5 mm	10 mm	20 mm	
N° points	69,650,000 [*1]	213,270,000	67,420,000 [*2]	25,550,000	3,617,133	1,688,825 [#1]	1,440,000 [#2]
Density (pt/m^2)	197,800	556,000	185,000	60,000	15,000	3000	2400

In Figure 16, the density distribution of the *Roses Chamber* vault is presented in a colour range map, according to a density that is computed in the number of neighbours in a sphere with r = 0.05 m (CloudCompare density analysis implementation). The thematic map not only shows noticeably the difference of precision of the two sensors in the surface 3D digitalization, but it also makes the interpretation of statistical values of density in the left graphs possible.

In fact, from the distribution values and the curve trend, it is clearly visible that the ZEB data does not admit the accurate description of the detail's geometry in the recorded objects, as a TLS.

Specifically, when studying the ZEB descriptive aptitudes in surfaces' reconstruction, it should be considered that a single scan, travelling the whole honour floor in almost 30 min, and collecting almost 35 mln points, manages to model some interesting morphological aspects of the architectural surfaces, i.e., anomalies from the generative geometries behind the wall's course or vault's curvature.

In Figure 17b, the optimization of DSM of the Great Salon pavilion vault is thematized by height values, already investigated by the LiDAR-based approach finalized to the structural analysis of the restored wooden structures in [51]; in white, the extraction of isolines from LiDAR DSM and in black the ones that were extracted by ZEB-based DSM. The ZEB-based DSM scale detail certainly could not support high accuracy analysis for structural purposes, but these two examples support the hypothesis that ZEB Revo MMS allows, through a rapid mapping, a dense point cloud and the potentially advantageous collection of medium- to high-scale information, especially for the morphological analysis of internal settings. Figure 17a shows a displacement map of T_2–T_1, between two ZEB acquisitions at time T_1 and T_2 and identifies a localized anomaly on the central part of the *Fleur-de-lis Chamber* vault, which has undergone restoration works and bedding phase for the wooden structures: it has been detected by MMS acquisition and further confirmed by a LiDAR scan.

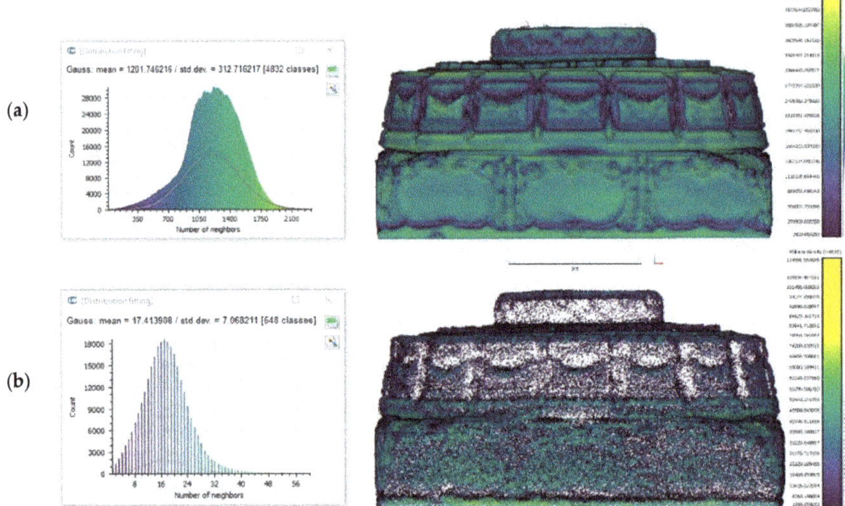

Figure 16. The two density analysis maps on the Roses Chamber in (**a**) LiDAR and (**b**) ZEB models.

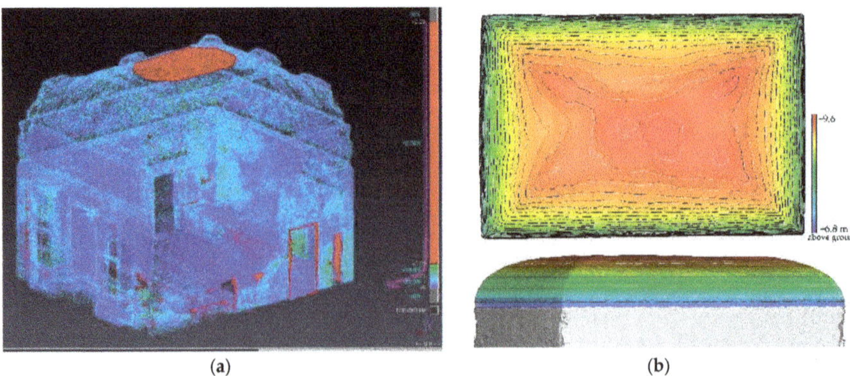

Figure 17. Morphological anomalies detected by the ZEB sensor: (**a**) the central part of the *Fleur-de-lis Chamber* vault in displacement map of T_2–T_1 and (**b**) in the Great Salon vault, with the comparison of isolines from the LiDAR DSM (white) and ZEB (black).

4.3. Fusion-Based Strategies Towards Geometric and Radiometric Enrichment and Self-Supporting

The fusion-based strategies are commonly developed where Cultural Heritage documentation projects are based on extensive and heterogeneous starting data, from passive and active sensors, for efficiency and optimization purposes in multiscale and multiresolution 3D models. As previously proposed in research [52,53], the concepts of fusion and integration methods are often the subject of crucial arguments regarding definitions and usages. First of all, the fusion needs arise from the purposes of 3D data features enrichment, i.e., the aim of overcoming the sensor's own limits (image-based and range-based). Moreover, the levels of processing steps on which the data are managed in the fusion can be the raw data or the processed ones or more post-processed ones. Lastly, the types of information behind the interchange and effective provision between approaches.

Hereafter, a set of these topics are discussed from the perspective of a fusion-driven approach based on data and methods fusion—mainly the photogrammetric approach with the range-based method being recognized as both the innovative MMSs point clouds and the contribution of more accurate terrestrial LiDAR data. The data fusion is aimed in a way oriented towards the optimization and validation of geometrically and radiometrically enriched MMS SLAM-based surfaces:

- Fusion in data processing
- Fusion in meshing model generation
- Fusion of geometric and radiometric data

Thus, this simplified workflow tries to answer some open-ended questions, such as the ones hereafter reported. Is it possible to constitute a schematic pipeline in which sensors' contributions are optimized? Is it conceivable to have a self-supporting use of the ZEB SLAM-based survey for a kind of extensive and building-scale mapping, even admitting as introduced, a forthcoming autonomy in the georeferencing issues? Hence, how accurate could be a photogrammetric blocks alignment, quickened by the non-involvement of topographic work, using alternatively geometric GCP that is extracted from georeferenced SLAM-based 3D models? Instead, what about the use of TLS-extracted GCPs? After a fusion-based Bundle Block Adjustment (BBA), what would happen if, simply, the block of oriented images was enough to interfacing with a ZEB point cloud properly georeferenced, evading the photogrammetric densification phase? In this case, is it promising to fuse the photogrammetric images, exploiting the oriented block tie points, with dense ZEB clouds in a typical Graphic User Interface (GUI) SfM platform, to perform the geometric and radiometric enrichment of point cloud and triangulated mesh surface in a unique space workflow?

4.3.1. ZEB-Based Photogrammetric Block-Orientation (o Images Block-Orientation): A Hybrid Model

The photogrammetric approach commonly requires a set of coordinates—involved as control points (geometric features as well as contrast markers) in the BBA and in the accuracy check phase—that are generally acquired during time-consuming topographical operations in a different survey moment than the images shooting phase. The availability of a geometrically rich 3D point cloud that is derived from an MMS as the ZEB Revo survey meets some remarkable requirements, together with time savings, practicality, and manageability.

For example, in this case, the ZEB point cloud would like to offer an experimental solution as a source of points coordinates to be exploited in the photogrammetric process of BBA, avoiding topographic measures. This method could be a useful solution in the case of a photogrammetric acquisition without measured targets or with data that is acquired at different times. This proposed data fusion solution to orient a set of photogrammetric images is hereby proven. The considered example is a dataset of 263 photos that were captured by Sony ILCE in the inner courtyard of the Rocca.

The images are aligned with coordinates that were obtained extracting a selection of n°20 points well distributed in the entire space. It has been chosen to compare the alignment with RMSE results, Table 8, using the manually extracted GCPs from the ZEB point cloud, with another two types of reference values:

- images BBA RMSE results using the same n°20 manually extracted points from the TLS point cloud, and
- images BBA RMSE results using n°20 topographic GCPs with typical contrast markers.

The main difficulties refer to the identification and manual extraction of the point in point clouds because of their lack of direct radiometric accurate data. In fact, RGB information is not currently available in ZEB point clouds, so it is not possible to exploit colour recognition (for instance, contrast markers, as in the case of LiDAR usual pipeline), but it is necessary to pinpoint the geometric features (for instance, architectural ones as corners, windows, edges, 3D decorative elements, or 3D targets), as visible in Figure 18. To reduce the operator imprecision in the manual collimation phase, each point has been selected and extracted n°5 times, by picking-points selection (as implemented in CloudCompare software), and then the arithmetic mean and *st. dev.* were calculated to estimate the precision of manual point extraction and the effectiveness of the proposed method. The *st. dev.* related to TLS-extracted coordinates varies from 0.11 cm to 1.47 cm; meanwhile, the ZEB ones vary from 0.44 cm to 9.42 cm, as reported in Table 8.

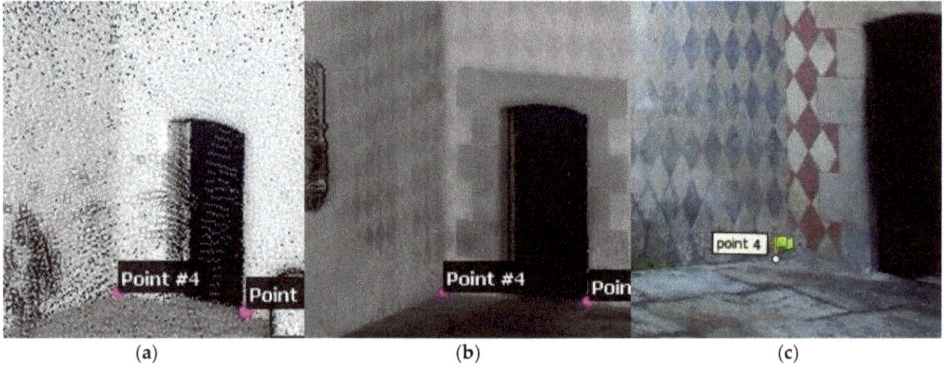

Figure 18. The n°4 points selected in (**a**) ZEB and in (**b**) TLS point cloud for coordinates' extraction and finally the point was detected and placed on the (**c**) Digital Single Lens Reflex (DSLR) photo, for the images block-orientation.

Finally, the mean X, Y, and Z values of each point selections were employed as GCPs coordinates to orient the photogrammetric block. To evaluate the effectiveness of these methods, the results of BBA are compared on CPs errors with the ones that were calculated with more accurate coordinates obtained when measuring targets by TS and GNSS surveys. Table 9 presents the results. It is tested that decorative corners are less recognizable in ZEB and LiDAR point clouds and affects the result.

Table 8. The precision of coordinates derived from the test of the manual pick-point selection, statistically repeated five times on 20 points and applied on TLS and ZEB point cloud.

5-Time Pick-Point	TLS-Extracted Coordinates				ZEB-Extracted Coordinates			
(m)	X	Y	Z	Total	X	Y	Z	Total
σ_{min}	0.0011	0.0024	0.0028	0.0021	0.0054	0.0044	0.0060	0.0053
σ_{MAX}	0.0120	0.0101	0.0147	0.0123	0.0599	0.0287	0.0942	0.0609
σ_{mean}	0.0047	0.0049	0.0061	**0.0053**	0.0163	0.0136	0.0249	**0.0182**

Table 9. Results of the photogrammetric block adjustment using differently accurate set points: the topographic reference measurements on target points, the Ground Control Points (GCPs) extracted from the LiDAR point surface and the ones from the ZEB point cloud.

BBA RMSE	Topographic Coordinates	TLS-Extracted Coordinates	MMS-Extracted Coordinates
15 GCPs error (cm)	0.43	0.95	7.72
5 CPs error (cm)	0.54	1.37	11.03

4.3.2. Fusion-Based Mesh Triangulation Oriented to High-Scale 3D Digitization and Optimization

The second experimental pipeline within the proposed workflow aims at exploiting the best of descriptive capabilities of each employed method in a kind of multisensor and multiscale survey approach. The chance of creating a multiresolution model, for example, in extensive and compound areas, as in [54], offers a solution towards data selection, segmentation, integration, and addressing to the appropriate level of detail wherever necessary to be increased and concretely approached.

It is based on a triangulated high-quality textured surface that is computed by an optimized multisensor points cloud. The starting data conveyed in the integration test process are:

- ZEB Revo point clouds based on portable rapid-mapping techniques;
- Photogrammetrically Oriented image block; and,
- LiDAR point cloud from TLS acquisition.

The efficient integration in the same pipeline of the LiDAR data and photogrammetric dense cloud has been partially investigated by low-cost 3DVEM–Register GEO tool in [23], but the processing was not yet in the same platform as the SfM workflow. Here, Photoscan Pro by Agisoft GUI (today, Metashape, https://www.agisoft.com) performed the photogrammetric based workflow for data fusion. The recently released interface can efficiently solve the problem of interoperability in point clouds data import/export from other sensors as well (it is possible to import in **.las, *.e57, *.ply, *.asc* formats). For example, the same image-based point cloud could be externally processed (filtered, segmented, or optimized), or a laser point cloud can be imported with or without its radiometric attributes. Moreover, in the GUI it is also possible to merge a photogrammetric image block (Figure 19a), oriented using the MMS-extracted GCPs coordinates, as reported before with ZEB data. As a result, the colourization phase is applied both to the MMS point cloud and to the fusion-driven built mesh, using as primary data the oriented photogrammetric image block. The test has focused on the *Fleur-de-lis Chamber* and it consisted of the use of different point cloud surfaces, as related to the Table 7 analysis: the photogrammetric-based one (almost 70 mln points); the LiDAR (original not considered, 5 mm filtering, 25.5 mln points, and 1 cm filtering, 3.6 mln points); ZEB SLAM-based point cloud (1.4 mln points). The normal vectors calculation of the input point clouds should be accurately controlled for a uniform distribution and orientation, if a finalization to the triangulated surface calculation (mesh), is foreseen. Although photogrammetric surface normals are correctly computed embedded in the SfM process and dense reconstruction, a range-based data should generally be subjected to normals data recomputation or finalizing.

A noticeable aspect in the GeoSLAM Hub processing platform is the managing of the shape attribute in output files, as previously reported (Figure 9b, III): this effectively concerns the computation of suitable embedded normal values data associated to the ZEB point cloud output, but the formats that are required for the export exclude **.las*, as incompatible with normal data.

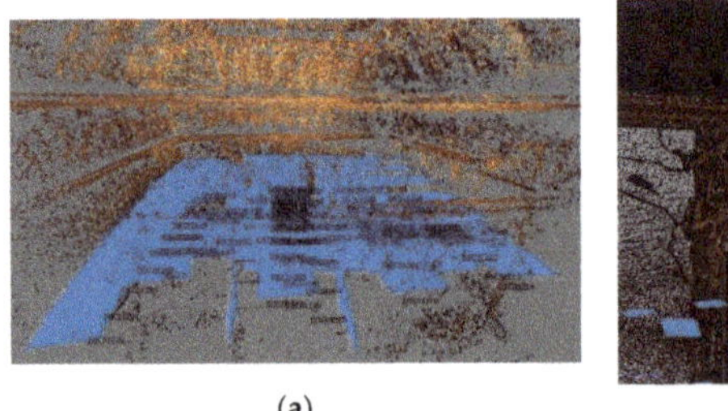
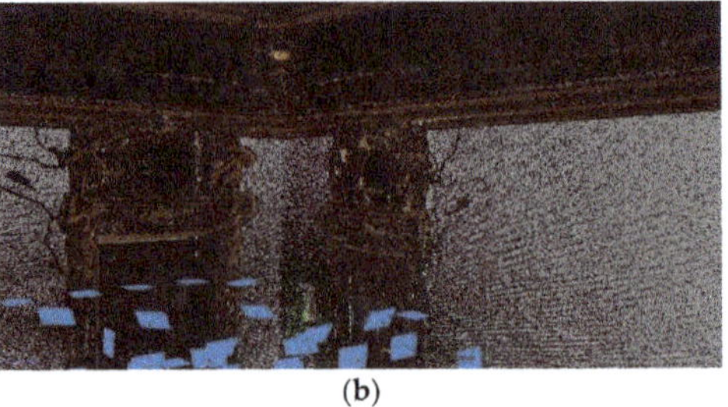

(a)   (b)

Figure 19. (a) The oriented images block in the *Fleur-de-lis Chamber*; (b) The Photoscan Pro Graphic User Interface (GUI) allows for importing external integrated point cloud for the fusion-based mesh triangulation.

The CloudCompare workspace offers an implemented Normal Computation algorithm, working on the octree strategy and using the setting of a point neighbours' radius area to simplify or improve the computing. Accordingly, the ZEB surface that was used for the meshing processing has been exploited in the original format, as exported from raw processing and with recomputed normals in CloudCompare, with a neighbour's radius of 5 mm and 15 mm. After the mesh triangulation process as compared in Table 10, simplified surfaces have been reconstructed, as shown in Figure 20a,b. Finally, as has been reported before, the oriented images that were acquired by Sony ILCE 7RM2 were used to associate the radiometric attribute with the 'colorize' function, as presented in Figure 20c.

Table 10. Results of mesh triangulation process, based on different sensors point clouds.

Mesh Model	Close-Range SfM	LiDAR (5 mm)	ZEB	Fusion-Based
N° triangles	13,937,030	5,110,790	318,232	3,391,239
.obj (.obj+ *.jpg)* file size (Mb)	1366 (1798+4)	215 (386+9)	24 (33+8)	186 (297+9)

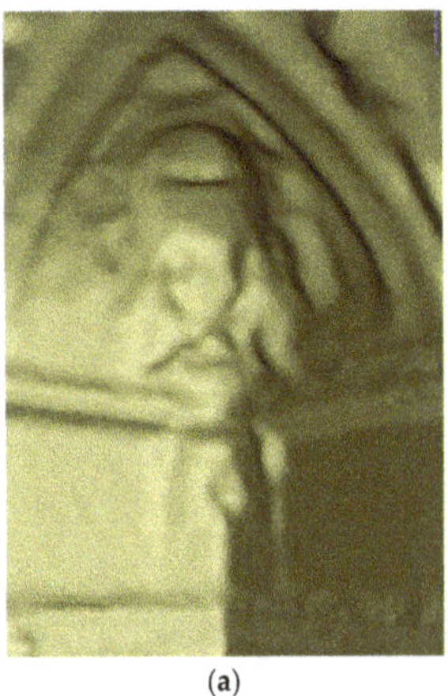
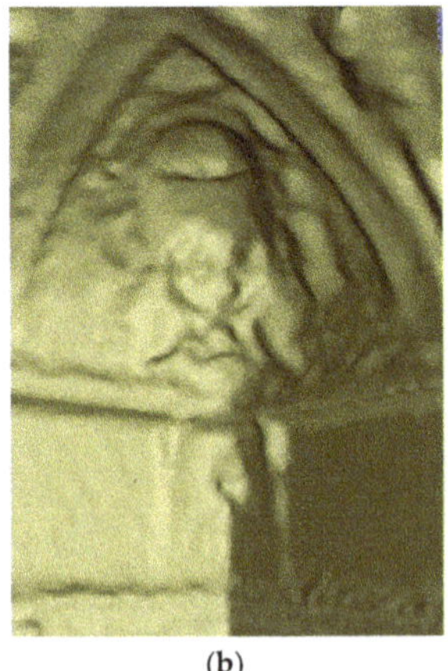

(a)   (b)   (c)

Figure 20. The ZEB-based surface reconstruction: (a) mesh processed with normal data computed with 15 mm radius and (b) computed with 5 mm; and, (c) with an HQ texture map applied on the (b) surface.

5. Discussion

When geomatics techniques encounter the difficulties in workflows that are arranged for surveying, modelling, organizing, and sharing large cultural heritage, as Valentino Castle and Borgo Medievale, some challenges must be considered, and various solutions must be critically analysed. Interesting solutions that are derived from sensor integration and data fusion are promising, such as sensors and software development to speed up recording and processing time, in operational workflows. This should be addressed to the best possible interoperability of multidimensional data with emerging Information and Communications Technology (ICT) innovations: digital devices and virtualization of models in the real scenarios are a useful implementation for the cultural heritage fields' users, allowing for digitally exploring 'immersive' 3D models and related information. It has been validated that geomatics increasingly provides strategies to deal with specific needs and favouring a multidisciplinary approach, also encouraging wide dissemination of digital culture. However, the 3D documentation phase of an extensive built complex still proves to be a difficult challenge nowadays. It implies various technical aspects concerning data acquisitions, their interoperable management of data, and their finalization to all user-oriented aims.

The present research tries to underline most of these aspects; in particular, advantages and disadvantages of the achieved results are summarized below, mostly focusing on the creation of hybrid 3D models (Section 5.1) and their usability and potentialities (Section 5.2).

5.1. Multisensor Data Complexity: Hybrid 3D Models

In recent decades, with the exponential increase of technological solutions based on sensor improvements and fusion, the Geomatic community proposed many ad hoc solutions of multiscale and multisensor methods, but little about the guidelines and practices to execute them, especially in the built heritage domain.

The generation of a hybridization-driven 3D model could surely return positive effects in the whole workflow, not only in data acquisition and processing phases, but also in terms of data modelling, editing, and optimization for user-oriented purposes. The main idea is constantly to change the perspective on the digitization approach and to preselect and pre-orient ex ante the data content to the request level of detail, instead of ex-post customary lightening, simplification, or drastic filtering. In this paper, the problems regarding the preservation of the descriptive abilities in 3D models are mainly addressed, both in terms of geometrical accuracy and in the integration of radiometric contents, which is such important in the projects that must satisfy future immersive enjoyments.

Thus, the combination of sensors is exploited according to their specific abilities in terms of resolutions, characteristics, and behaviour in certain operational fields, also in critical ones, both indoor and outdoor scenarios. A sensor that works effectively in all conditions does not exist; as well, a single-sensor approach represents a limited approach, although it is often the most used one. After the series of extensive testing, it is possible to summarize some consideration and evaluation, as reported in Figure 21. Obviously, the strategies for achieving multisensor and multiscale models are those chosen to face this challenge, maintaining high attention, as seen above in Sections 3 and 4, on models that were created by recent rapid-mapping systems, which offer leaner and handy points models, less heavy in terms of file weight as compared with those that were derived from the more traditional LiDAR technology.

The advantages and disadvantages of each sensor combined with a critical environment suggest undertaking a mixed approach of techniques to obtain a hybrid model. Integrated- and fusion-based approaches could be winning solutions, customizable according to specific needs, in terms of time, costs, and sensors that are finalized to obtain predetermined final products with specific levels of detail, as shown in Figure 21. Following general procedural steps (planning, survey, processing, analysis, and communication), different considerations can be stressed.

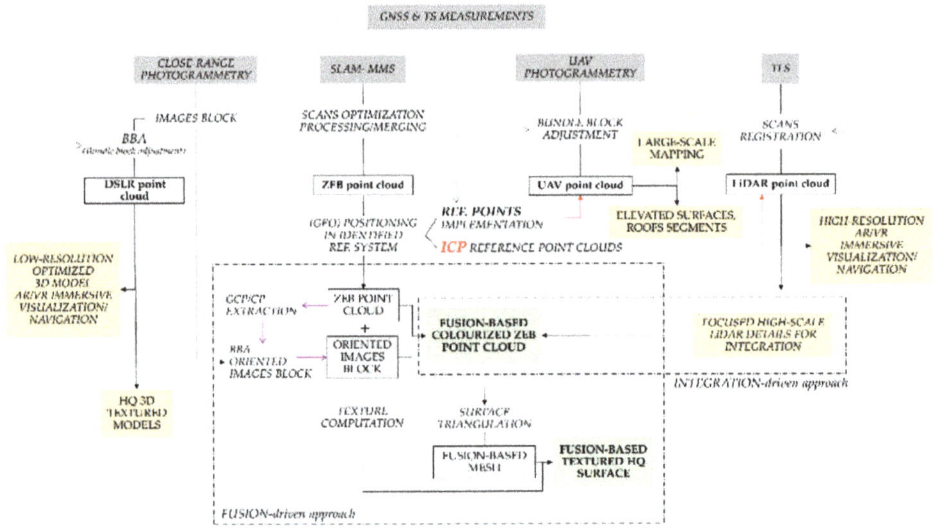

Figure 21. Graphic workflow of the hybridization of 3D models, deriving from integrated methods and fusion-based approaches. In grey, the acquisition methods, in green the achieved products with the possible connections. In the dashed box the fused and integrated results.

Furthermore, the possibility to integrate data that was acquired with different sensors and at different times underlines the ability of sensors to record geometrical differences and suggest further interesting analyses. In the planning phase, it is necessary to consider setting conditions. Spaces like the ones studied in this paper provide some challenging examples of where to perform a survey, because of building sizes, multilevel and articulated spaces, light conditions, presence of movable elements, little details, and fragile elements, as many cultural heritage assets show the occurrence of occlusions, as furnishing and artworks. In parallel, the presence of standing and leaning outdoor tall trees on the curtain walls influenced the survey very much: in most cases, they cover the architectural structure of the buildings and increase the non-continuity and noise errors of the dense clouds.

Sometimes—as demonstrated in 4.1(IV)—UAV photogrammetry can deliver affordable data comparable to MMS in terms of resolution, allowing for the georeferencing of MMS data and promoting their integration; in this way, a combined use also constitutes a rapid approach in architectural mapping in the case of extreme conditions, such as low overlapping areas. Furthermore, another fast approach can be pursued extracting 3D coordinates from MMS point cloud and using them to align a set of images that were captured with a DSLR. Referring to a deeper level of detail, a products fusion can be pursued exploiting LiDAR data for minute parts—as decorative elements—MMS data for a general overview of the context—such as walls or simple geometries—and close-range photogrammetry for radiometric quality that is derived by an oriented block of images. After the discussed experiences and the reported results, it is possible to assert that both the Rocca Medievale and the Valentino Castle justified such a kind of multisensor approach to obtain a complete 3D model. In Figure 21, a schema of the proposed approach for 3D model's hybridization is reported.

5.2. Usability and Flexibility Potential of 3D Models: Towards Navigation and Augmented Reality

Once the model is available, after the analysis and validation, an important issue is the use of the achieved products. Without addressing wide-ranging issues, such as the primary ones in the domain of communication of built heritage (four-dimensional (4D) recording, fruition of multidisciplinary information, 3D inventories, user friendly web platform), we intend to discuss the topics of usability potential related to 3D enriched data, with some examples of the most promising and cutting-edge applications that are related to the immersive use and navigability of the 3D data; thereafter, some examples.

It is very important to consider that many tools enabling model navigation (and video recording) are available; they permit overcoming, to a certain extent, the criticality of the difficult management of large amounts of data. The Recap Pro software by Autodesk (https://www.autodesk.com/products/recap), for example, offers navigation within points clouds that is particularly decimated, able to support navigation in real time, while also using smart devices that connect via the server for navigation of the model.

The application shown in Figure 22 was created, starting from the complete LiDAR points cloud, coloured by the automatic process of the FARO® system equipped with the coaxial camera. The total huge size of this registered cloud is equal to 66.4 Gb (in *.E57* standard format), as it is composed of 3268 mln points; once filtered, with 5 mm distance constant value, the point cloud size became around 325 mln (8.3 Gb *.las* format and 5.8 Gb *.e57* standard format). Despite the possibility of user interaction, with the 3D model being limited on this platform, the navigation with the subjective point of view that moves all around the space that is represented by the digital model is certainly immersive and suggestive, and it does not suffer from the amount of points data, which are only rendered with the level of detail to be zoomed in. It is undeniable, in fact, that this navigation makes it possible to observe and learn some of the main characteristics regarding the decorative apparatus of the rooms. In the *Fleur-de-lis Chamber*, a stucco frieze and triangular coves in the corners support the impressive composition of the vault, which is enriched with a dance of putti, ribbons, amphorae, and floral decorations. It is certainly not possible to recognize authors or learn other specialist information, but the communication purpose for general fruition is achieved.

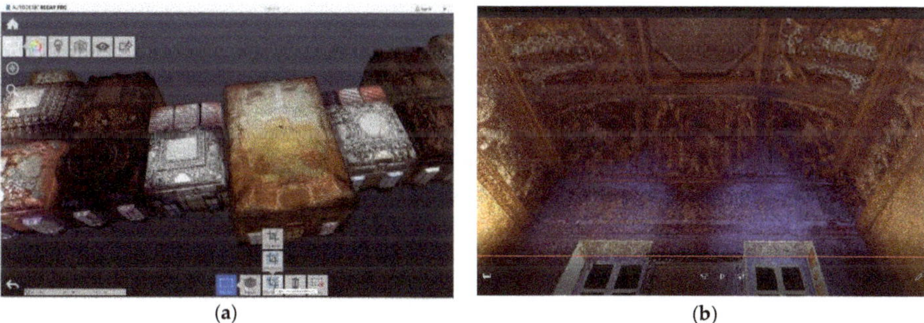

(a)	(b)

Figure 22. (a) A screenshot of the LiDAR points model managed in the Recap environment; (b) A screenshot from the video navigation achieved from LiDAR points cloud.

Using optimized models that are based on image or range-based approaches for fruition purposes and augmented reality (AR) or computer-mediated reality, that is a growing interest in the cultural heritage domain. They consist of enriching the human sensory perception through digital information, which would not be perceptible with the senses. Elements that 'increase' reality can be added through a mobile device, such as a smartphone, with the use of a PC that is equipped with a webcam or other sensors with vision devices. The new representation techniques (virtual reality (VR) and AR), provide immersive experiences in the virtual heritage field; the digital object can be queried by the user through the same dynamics that are present in the world of digital entertainment, a factor that surely is recognized by a relatively young audience, but intrigues a wider target audience, becoming a means of communication that meets today's expectations [55].

In the Honour Salon of the Valentino Castle, (Figure 23a), an AR test was performed [56] using the software solution by Metaio (no longer active, since being acquired by Apple). Here, it is possible to view the Honour Salon by simply moving the device, just as if you were inside the scene, and then, thanks to the framing by the camera of the simplified greyscale DSM of the dome, enable the visualization on it of the 3D model. It is a rather interesting shared use of technologies: the 6D Augmented Reality Holodeck technology (named 6D-AR) takes the advantages from SLAM algorithms and from the on-board IMUs motion sensors to attach the virtual environment to the real world, exploiting the camera of smart devices [57]. Moreover, in this case, it is obviously possible to learn the decorative themes of the frescoes, which consist in the exaltation of the Savoy family through the re-enactment of military enterprises. At the same time, it is possible to examine the expertise of the scenic installation that reproduces twisted columns and a balcony that was executed by Isidoro Bianchi who painted with the collaboration of his sons between 1633 and 1642 [31].

(a) (b)

Figure 23. (a) The SLAM for full six-dimensional virtual reality/augmented reality (6D VR/AR) application running on an iPad; (b) Interactive model of the hall showing typical points enabling to reach additional information.

From an operational point of view, here, a LiDAR model was preferred to the image-based one, because the large hall with openings on the short side is dark and difficult to illuminate, together with the criticality represented huge historic chandelier that has been masked in all of the images used for the photogrammetric matching and the SfM process. As already ascertained, the block of oriented images has been used to texturize the LiDAR model, which is superior and less noisy as far as metric accuracy is concerned. The triangulated continuous model achieved both with the open source MeshLab interface and with the commercial 3Dreshaper software, which had provided completely similar qualitative results, had been texturized and optimized with Blender software, being the most efficient and widespread in the field of simplification and optimization of texturized models: the initial filtered cloud of 10 mln points had been reduced to a polygonal model of about 180k vertices.

In the Hall of Feasts and Splendours, instead, with the richly decorated vault with light stuccos in high relief, the environmental condition allowed generation of the texturized model using DSLR images and SfM algorithms, obtaining a customary accuracy of around 1 cm [58]. The whole process of modelling the triangulated surface (about 5 mln polygons) and its texture were performed using the software Photoscan. The game engine that is selected for application development is Unity 3D, which is widespread and widely used in this sector, because it allows for high interoperability with different devices and platforms (the software is considered to be cross-platform, i.e., able to develop the WebGL project for PC, Android, iOS, and Windows Phone for mobile devices). Additionally, in this case, the final objective has been reached, that is, besides testing the interoperability of instruments and data between different application fields, it was to create a basic communication product composed of a navigable model that is characterized by an interactive system that is capable of providing informative details on the Hall, through the pressure of characteristic points arranged within the model (Figure 23b).

There are still many critical issues in integrating models derived from Geomatic techniques and game design software, mainly consisting in the criticality of using models derived from millions of points in systems designed to manage models of reality that are much more simplified.

6. Conclusions

Nowadays, 3D documentation of extensive built heritage complexes still proves to be a difficult challenge. It implies various aspects regarding acquisition, management of data, their interoperability, and their finalization to all user-oriented purposes (Figure 24).

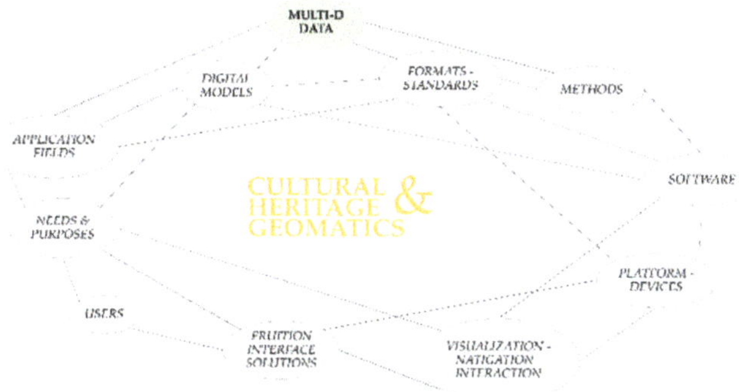

Figure 24. Relations between many deterministic factors turning around the Geomatics approach working in the cultural heritage domain.

The research proposed here mostly refers to an essential answer to the diffuse increasing spread of innovative input and traditional requests in the Cultural Heritage domain belonging to the digital era: it has to face many domains (disciplines, users, purposes, data, formats, software, supports, type, and mode of visualization) that are connected with each other.

With different depth, this paper has been dedicated to addressing both the evaluation of the effectiveness of the models and the current critical issues related to management tools. Moreover, the themes of data interoperability, operating times in acquisition and processing, and issues that are related to the management of large amounts of data have been discussed. The achieved results connected to technologies' progress are above presented; nonetheless, to pursue it, some issues remain and must be solved.

It has been demonstrated that geomatics provide strategies to deal with specific needs and purposes, favouring a multidisciplinary approach, while also encouraging the wide dissemination of digital technologies. Geomatics for heritage digitization addresses, in these cases, the entire workflow to manage the concept of complexity behind the building challenges (Figure 21). Once derived from data acquisition and processing, reality-based models offer a wide range of applications, responding to specific needs of costs and purposes, such as documentation, analysis, and sharing. The possibility to integrate other sources—textual and/or graphics—lead to the creation and management of digital georeferenced databases.

Thanks to different supports—PCs, smartphones, tablets, visors—various users can explore these models in online or offline modes; in these visualizations, immersive, interactive, augmented, or virtual realities offer a more and more expanding way to understand and discover.

These aspects require the capability of handling huge amounts of multidimensional data—among others, geometry, time, and radiometry—characterized by digital formats and standards created, managed, and visualized in various software, applications, and platforms. If the future idea is to work in the direction of allowing the user to move freely in the reconstructed scenario, then it is obviously essential to consider the completeness constraints of the model, which constitutes, in addition to the previous, one of the quality parameters for the evaluation of the models of the reality.

Author Contributions: Conceptualization, F.C., An.S., G.S. Al.S.; Data curation, Al.S.; Formal analysis, G.S. and Al.S; Investigation, Al.S; Methodology, G.S. and An.S.; Resources, F.C.; Supervision, F.C. and An.S.; Validation, An.S.; Writing – original draft, G.S. and Al.S; Writing – review & editing, F.C., G.S. and An.S.

Funding: the research presented here is supported by the joint research project *'Turin 1911'* between Politecnico di Torino and University of California San Diego.

Acknowledgments: a special thanks to the University of California San Diego for the joint research project *'Turin 1911'*: Cristina Della Coletta. The authors thank also L. Teppati Losè, G. Patrucco, E. Colucci, D. Einaudi from the Laboratory of Geomatics for Cultural Heritage (G4CH Lab) of Politecnico di Torino for data acquisition and C. Bonfanti from MESA srl for ZEBRevo RT by GeoSLAM for suggestions and help.

Conflicts of Interest: The authors declare no conflict of interest.

References

1. Donadio, E. 3D Photogrammetric Data Modeling and Optimization for Multipurpose Analysis and Representation of Cultural Heritage Assets. Ph.D. Thesis, Politecnico di Torino, Torino, Italy, 2018.
2. Chiabrando, F.; Della Coletta, C.; Sammartano, G.; Spanò, A.; Spreafico, A. "TORINO 1911" Project: A Contribution of a SLAM-based Survey to Extensive 3D Heritage Modeling. *ISPRS Int. Arch. Photogramm. Remote Sens. Spat. Inf. Sci.* **2018**, *42*, 225–234. [CrossRef]
3. Ramos, M.M.; Remondino, F. Data fusion in Cultural Heritage—A Review. *ISPRS Int. Arch. Photogramm. Remote Sens. Spat. Inf. Sci.* **2015**, *40*, 359–363. [CrossRef]
4. Pierrot-Deseilligny, M.; De Luca, L.; Remondino, F. Automated Image-Based Procedures for Accurate Artifacts 3D Modeling and Orthoimage Generation. *Geoinform. FCE CTU* **2011**, *6*, 291–299. [CrossRef]
5. Guidi, G.; Russo, M.; Ercoli, S.; Remondino, F.; Rizzi, A.; Menna, F. A Multi-Resolution Methodology for the 3D Modeling of Large and Complex Archeological Areas. *Int. J. Arch. Comput.* **2009**, *7*, 39–55. [CrossRef]
6. Remondino, F.; Girardi, S.; Rizzi, A.; Gonzo, L. 3D modeling of complex and detailed cultural heritage using multi-resolution data. *J. Comput. Cult. Herit.* **2009**, *2*, 1–20. [CrossRef]
7. Murtiyoso, A.; Grussenmeyer, P.; Suwardhi, D.; Awalludin, R. Multi-Scale and Multi-Sensor 3D Documentation of Heritage Complexes in Urban Areas. *ISPRS Int. J. Geo-Inf.* **2018**, *7*, 483. [CrossRef]
8. Tucci, G.; Bonora, V.; Conti, A.; Fiorini, L. Digital Workflow for the Acquisition and Elaboration of 3D Data in a Monumental Complex: The Fortress of Saint John the Babtist in Florence. *ISPRS Int. Arch. Photogramm. Remote Sens. Spat. Inf. Sci.* **2017**, *42*, 679–686. [CrossRef]
9. Farella, E.; Menna, F.; Nocerino, E.; Morabito, D.; Remondino, F.; Campi, M. Knowledge and valorization of historical sites through 3D documentation and modeling. *Int. Arch. Photogramm. Remote Sens. Spat. Inf. Sci. ISPRS Arch.* **2016**, *41*, 255–262. [CrossRef]
10. Bitelli, G.; Dellapasqua, M.; Girelli, V.A.; Sanchini, E.; Tini, M.A. 3D Geomatics Techniques for an Integrated Approach to Cultural Heritage Knowledge: The Case of San Michele in Acerboli's Church in Santarcangelo di Romagna. *ISPRS Int. Arch. Photogramm. Remote Sens. Spat. Inf. Sci.* **2017**, *42*, 291–296. [CrossRef]
11. Kim, J.; Lee, S.; Ahn, H.; Choi, C. Feasibility of employing a smartphone as the payload in a photogrammetric UAV system. *ISPRS J. Photogramm. Remote Sens.* **2013**, *79*, 1–18. [CrossRef]
12. Masiero, A.; Fissore, F.; Piragnolo, M.; Guarnieri, A.; Pirotti, F.; Vettore, A. Initial Evaluation of 3D Reconstruction of Close Objects with Smartphone Stereo Vision. *ISPRS Int. Arch. Photogramm. Remote Sens. Spat. Inf. Sci.* **2018**, *42*, 289–293. [CrossRef]
13. Fangi, G.; Nardinocchi, C. Photogrammetric Processing of Spherical Panoramas. *Photogramm. Rec.* **2013**, *28*, 293–311. [CrossRef]
14. Pérez Ramos, A.; Robleda Prieto, G. Only Image Based for the 3D Metric Survey of Gothic Structures by Using Frame Cameras and Panoramic Cameras. *ISPRS Int. Arch. Photogramm. Remote Sens. Spat. Inf. Sci.* **2016**, *41*, 363–370. [CrossRef]
15. Grussenmeyer, P.; Landes, T.; Voegtle, T.; Ringle, K. Comparison Methods of Terrestrial Laser Scanning, Photogrammetry and Tacheometry Data for Recording of Cultural Heritage Buildings. *ISPRS Int. Arch. Photogramm. Remote Sens. Spat. Inf. Sci.* **2008**, *XXXVI*, 213–218.

16. Lerma, J.L.; Navarro, S.; Cabrelles, M.; Villaverde, V. Terrestrial laser scanning and close range photogrammetry for 3D archaeological documentation: The Upper Palaeolithic Cave of Parpalló as a case study. *J. Archaeol. Sci.* **2010**, *37*, 499–507. [CrossRef]
17. Mandelli, A.; Fassi, F.; Perfetti, L.; Polari, C. Testing Different Survey Techniques to Model Architectonic Narrow Spaces. *ISPRS Int. Arch. Photogramm. Remote Sens. Spat. Inf. Sci.* **2017**, *XLII-2/W5*, 505–511. [CrossRef]
18. Eyre, M.; Wetherelt, A.; Coggan, J. Evaluation of automated underground mapping solutions for mining and civil engineering applications. *J. Appl. Remote Sens.* **2016**, *10*, 046011. [CrossRef]
19. Sammartano, G.; Spanò, A. Point clouds by SLAM-based mobile mapping systems: Accuracy and geometric content validation in multisensor survey and stand-alone acquisition. *Appl. Geomatics* **2018**, *10*, 317–339. [CrossRef]
20. di Filippo, A.; Sánchez-Aparicio, L.; Barba, S.; Martín-Jiménez, J.; Mora, R.; González Aguilera, D. Use of a Wearable Mobile Laser System in Seamless Indoor 3D Mapping of a Complex Historical Site. *Remote Sens.* **2018**, *10*, 1897. [CrossRef]
21. Rottensteiner, F.; Trinder, J.; Clode, S.; Kubik, K. Building detection by fusion of airborne laser scanner data and multi-spectral images: Performance evaluation and sensitivity analysis. *ISPRS J. Photogramm. Remote Sens.* **2007**, *62*, 135–149. [CrossRef]
22. Fassi, F.; Achille, C.; Fregonese, L. Surveying and modelling the main spire of Milan Cathedral using multiple data sources. *Photogramm. Rec.* **2011**, *26*, 462–487. [CrossRef]
23. Munumer, E.; Lerma, J.L. Fusion of 3D data from different image-based and range-based sources for efficient heritage recording. In Proceedings of the 2015 Digital Heritage, Granada, Spain, 28 September–2 October 2015; pp. 83–86.
24. Della Coletta, C. *World's Fairs Italian-Style: The Great Exhibitions in Turin and Their Narratives, 1860–1915*; University of Toronto Press: Toronto, ON, Canada, 2016; ISBN 9781487520564.
25. Best, S.; Kellner, D. *Postmodern Theory Critical Interrogations*; Macmillian: Basingstoke, UK, 1991.
26. Bianchi, C. *Il Valentino (Storia di un parco)*; Il piccolo editore: Torino, Italy, 1984; ISBN 88-7654-023-7
27. AA.VV. *Il Borgo Medievale. Nuovi Studi*; Pagella, E., Ed.; Edizioni Fondazione Torino Musei: Torino, Italy, 2011; ISBN 978-88-88103-83-9.
28. Un Borgo Colla Dominante Rocca. *Studi Per la Conservazione del Borgo Medievale di Torino*; Bartolozzi, C., Ed.; Celid: Torino, Italy, 1995; ISBN 88-7661-212-2.
29. Donato, G. *Omaggio al Quattrocento: Dai Fondi D'Andrade, Brayda, Vacchetta*; Borgo Medievale: Torino, Italy, 2006.
30. Roggero, C.; Dameri, A. *The Castello del Valentino*; Allemandi & C.: Torino, Italy, 2008; ISBN 978-88-422-1664-3.
31. Roggero, C.; Scotti, A. *Il Castello del Valentino/The Valentino Castle*; Politecnico di Torino: Torino, Italy, 1994.
32. Bertolini-Cestari, C.; Chiabrando, F.; Invernizzi, S.; Marzi, T.; Spanò, A. Terrestrial Laser Scanning and Settled Techniques: A Support to Detect Pathologies and Safety Conditions of Timber Structures. *Adv. Mater. Res.* **2013**, *778*, 350–357. [CrossRef]
33. Nocerino, E.; Menna, F.; Remondino, F.; Toschi, I.; Rodríguez-González, P. Investigation of indoor and outdoor performance of two portable mobile mapping systems. In Proceedings of the SPIE Videometrics, Range Imaging, and Applications XIV, Munich, Germany, 25–29 June 2017; Remondino, F., Shortis, M.R., Eds.; 2017.
34. Rodríguez-Gonzálvez, P.; Fernández-Palacios, B.J.; Muñoz-Nieto, Á.L.; Arias-Sanchez, P.; Gonzalez-Aguilera, D. Mobile LiDAR system: New possibilities for the documentation and dissemination of large cultural heritage sites. *Remote Sens.* **2017**. [CrossRef]
35. Lagüela, S.; Dorado, I.; Gesto, M.; Arias, P.; González-Aguilera, D.; Lorenzo, H. Behavior analysis of novel wearable indoor mapping system based on 3d-slam. *Sensors* **2018**, *18*, 766. [CrossRef] [PubMed]
36. Cabo, C.; Del Pozo, S.; Rodríguez-Gonzálvez, P.; Ordóñez, C.; González-Aguilera, D. Comparing Terrestrial Laser Scanning (TLS) and Wearable Laser Scanning (WLS) for Individual Tree Modeling at Plot Level. *Remote Sens.* **2018**, *10*, 540. [CrossRef]
37. Thomson, C.; Apostolopoulos, G.; Backes, D.; Boehm, J. Mobile Laser Scanning for Indoor Modelling. *ISPRS Ann. Photogramm. Remote Sens. Spat. Inf. Sci.* **2013**, *5*, 289–293. [CrossRef]
38. Zlot, R.; Bosse, M.; Greenop, K.; Jarzab, Z.; Juckes, E.; Roberts, J. Efficiently capturing large, complex cultural heritage sites with a handheld mobile 3D laser mapping system. *J. Cult. Herit.* **2014**, *15*, 670–678. [CrossRef]

39. Lehtola, V.V.; Kaartinen, H.; Nüchter, A.; Kaijaluoto, R.; Kukko, A.; Litkey, P.; Honkavaara, E.; Rosnell, T.; Vaaja, M.T.; Virtanen, J.P.; et al. Comparison of the selected state-of-the-art 3D indoor scanning and point cloud generation methods. *Remote Sens.* **2017**, *9*, 796. [CrossRef]
40. Dewez, T.J.B.; Yart, S.; Thuon, Y.; Pannet, P.; Plat, E. Towards cavity-collapse hazard maps with Zeb-Revo handheld laser scanner point clouds. *Photogramm. Rec.* **2017**, *32*, 354–376. [CrossRef]
41. Dewez, T.J.B.; Plat, E.; Degas, M.; Richard, T.; Pannet, P.; Thuon, Y.; Meire, B.; Watelet, J.-M.; Cauvin, L.; Lucas, J. Handheld Mobile Laser Scanners Zeb-1 and Zeb-Revo to map an underground quarry and its above-ground surroundings. In Proceedings of the 2nd Virtual Geosciences Conference, Bergen, Norway, 21–23 September 2016.
42. Tucci, G.; Visintini, D.; Bonora, V.; Parisi, E. Examination of Indoor Mobile Mapping Systems in a Diversified Internal/External Test Field. *Appl. Sci.* **2018**, *8*, 401. [CrossRef]
43. Bosse, M.; Zlot, R.; Flick, P. Zebedee: Design of a Spring-Mounted 3-D Range Sensor with Application to Mobile Mapping. *IEEE Trans. Robot.* **2012**, *28*, 1104–1119. [CrossRef]
44. Cadge, S. Welcome to the ZEB REVOlution. *GEOmedia* **2016**, *20*, 22–26.
45. Rossi, P.; Mancini, F.; Dubbini, M.; Mazzone, F.; Capra, A. Combining nadir and oblique uav imagery to reconstruct quarry topography: Methodology and feasibility analysis. *Eur. J. Remote Sens.* **2017**. [CrossRef]
46. Xiao, X.; Guo, B.; Shi, Y.; Gong, W.; Li, J.; Zhang, C. Robust and rapid matching of oblique UAV images of urban area. In Proceedings of the MIPPR 2013: Pattern Recognition and Computer Vision, Wuhan, China, 26–27 October 2013; Cao, Z., Ed.; p. 89190Y.
47. Aicardi, I.; Chiabrando, F.; Grasso, N.; Lingua, A.; Noardo, F.; Spanò, A.T. UAV Photogrammetry with Oblique Images: First Analysis on Data Acquisition and Processing. *ISPRS Int. Arch. Photogramm. Remote Sens. Spat. Inf. Sci.* **2016**, *41*, 835–842. [CrossRef]
48. Calantropio, A.; Patrucco, G.; Sammartano, G.; Teppati Losè, L. Low-cost sensors for rapid mapping of cultural heritage: First tests using a COTS Steadicamera. *Appl. Geomat.* **2018**, *10*, 31–45. [CrossRef]
49. Luis, J.; Navarro, S.; Cabrelles, M.; Elena, A.; Haddad, N.; Akasheh, T. Integration of Laser Scanning and Imagery for Photorealistic 3D Architectural Documentation. In *Laser Scanning, Theory and Applications*; InTech: London, UK, 2011; ISBN 978-953-307-205-0.
50. Zalama, E.; Gómez-García-Bermejo, J.; Llamas, J.; Medina, R. An Effective Texture Mapping Approach for 3D Models Obtained from Laser Scanner Data to Building Documentation. *Comput. Civ. Infrastruct. Eng.* **2011**, *26*, 381–392. [CrossRef]
51. Bertolini-Cestari, C.; Spanò, A.T.; Invernizzi, S.; Donadio, E.; Marzi, T.; Sammartano, G. *Historical Earthquake-Resistant Timber Framing in the Mediterranean Area*; Cruz, H., Saporiti Machado, J., Campos Costa, A., Xavier Candeias, P., Ruggieri, N., Manuel Catarino, J., Eds.; Lecture Notes in Civil Engineering; Springer International Publishing: Cham, Switzerland, 2016; Volume 1, ISBN 978-3-319-39491-6.
52. Vosselman, G. Fusion of laser scanning data, maps, and aerial photographs for building reconstruction. In Proceedings of the IEEE International Geoscience and Remote Sensing Symposium, Toronto, ON, Canada, 24–28 June 2002; Volume 1, pp. 85–88.
53. Dasarathy, B.V. *Decision Fusion*; IEEE Computer Society Press: Los Alamitos, CA, USA, 1994; ISBN 0818644524.
54. Guidi, G.; Remondino, F.; Russo, M.; Spinetti, A. Range sensors on marble surfaces: Quantitative evaluation of artifacts. In Proceedings of the SPIE on Videometrics, Range Imaging, and Applications X, San Diego, CA, USA, 2–6 August 2009; Remondino, F., Shortis, M.R., El-Hakim, S.F., Eds.; 2009; Volume 7447, p. 744703.
55. Martina, A. *Virtual Heritage: New Technologies for Edutainment*; Politecnico di Torino: Torino, Italy, 2014.
56. Dalpozzi, M. *Modelli 3D e Realtà Aumentata: Un'applicazione sul Salone d'Onore al Castello del Valentino*; Politecnico di Torino: Torino, Italy, 2014.
57. SLAM For Full 6D VR/AR. Available online: https://my.metaio.com/dev/sdk/tutorials/slam-for-full-6d-vrar/index.html (accessed on 1 January 2015).
58. Crivellari, G. Nuove Frontiere Delle Tecnologie per la Fruizione dei Beni Culturali: Un'app per la Sala Delle Feste e dei Fasti. Master's Thesis, Politecnico di Torino, Torino, Italy, 2016.

 International Journal of
Geo-Information

Article

Multi-Scale and Multi-Sensor 3D Documentation of Heritage Complexes in Urban Areas [†]

Arnadi Murtiyoso [1],*, Pierre Grussenmeyer [1], Deni Suwardhi [2] and Rabby Awalludin [2]

[1] ICube Laboratory UMR 7357, Photogrammetry and Geomatics Group, INSA Strasbourg,
24 Boulevard de la Victoire, 67084 Strasbourg, France; pierre.grussenmeyer@insa-strasbourg.fr

[2] Remote Sensing and GIS Group, Bandung Institute of Technology (ITB), Jalan Ganesha No. 10,
Bandung 40132, Indonesia; deni@gd.itb.ac.id (D.S.); rabbyawalludin@students.itb.ac.id (R.A.)

* Correspondence: arnadi.murtiyoso@insa-strasbourg.fr; Tel.: +33-3-88-14-47-33

† This article is an extended version of a conference paper previously published under the title of "Digital Documentation Workflow and Challenges for Tropical Vernacular Architecture in the Case of the Kasepuhan Palace in Cirebon, Indonesia", presented at the Digital Heritage 2018 Congress & Expo, 26–30 October 2018, San Francisco, USA.

Received: 07 December 2018; Accepted: 14 December 2018; Published: 17 December 2018

Abstract: The 3D documentation of heritage complexes or quarters often requires more than one scale due to its extended area. While the documentation of individual buildings requires a technique with finer resolution, that of the complex itself may not need the same degree of detail. This has led to the use of a multi-scale approach in such situations, which in itself implies the integration of multi-sensor techniques. The challenges and constraints of the multi-sensor approach are further added when working in urban areas, as some sensors may be suitable only for certain conditions. This paper describes the integration of heterogeneous sensors as a logical solution in addressing this problem. The royal palace complex of Kasepuhan Cirebon, Indonesia, was taken as a case study. The site dates to the 13th Century and has survived to this day as a cultural heritage site, preserving within itself a prime example of vernacular Cirebonese architecture. This type of architecture is influenced by the tropical climate, with distinct features designed to adapt to the hot and humid year-long weather. In terms of 3D documentation, this presents specific challenges that need to be addressed both during the acquisition and processing stages. Terrestrial laser scanners, DSLR cameras, as well as UAVs were utilized to record the site. The implemented workflow, some geometrical analysis of the results, as well as some derivative products will be discussed in this paper. Results have shown that although the proposed multi-scale and multi-sensor workflow has been successfully employed, it needs to be adapted and the related challenges addressed in a particular manner.

Keywords: heritage; photogrammetry; laser scanning; tropical; vernacular; multi-scale; multi-sensor

---

1. Introduction

Heritage documentation is an important aspect in any conservation effort. Apart from the traditional 2D drawings and photographs, digital 3D documentation of historical sites currently presents a useful tool in the analysis and interpretation of historic buildings, as well as its eventual reconstruction in the event of damage [1]. This is more important in the presence of important threats, both natural [2,3] and anthropological [4]. Digital documentation uses various sensors in capturing the reality. This may be done either by image-based or range-based techniques [5]. Each technique has its own advantages and disadvantages; as such, it is not rare to see the combination of both techniques in a thorough documentation project [6,7]. Indeed, the integration of heterogeneous data is an important research issue [8].

The Terrestrial Laser Scanner (TLS) is a range-based system that enables the acquisition of many points in a short period of time [9] and has seen much use in the field of heritage documentation [10,11]. As regards to image-based techniques, photogrammetry has in recent times seen important leaps, and even more so with the advent of Unmanned Aerial Vehicles (UAVs) or drones [12,13]. Both range-based and image-based techniques generate dense point clouds from which derivative geospatial products may be generated.

In the context of the documentation of heritage complexes, the extent of the site makes it logical to use a multi-scalar approach. Larger areas do not need fine resolution data, as opposed to smaller buildings or even architectural elements. In this approach, the site is digitized in several scale steps, according to the required resolution for each level [14]. The use of the multi-scalar approach also means that more than one sensor could be employed in order to cover each scale step. Furthermore, for the case of urban areas, many constraints such as the geography and urban density mean that the use of one single sensor may not be sufficient. It is in this regard that multi-sensor and multi-scalar documentation became a logical solution to the problem of documenting historical complexes.

Even though the use of 3D documentation methods has seen a significant rise in the past few years, vernacular architecture usually presents different challenges depending on its architectural style. This may also be influenced by the climate and geography of the site. The Indonesian archipelago has a lot of examples of traditional vernacular architecture, characterized by distinctive building design and layout, which takes into account its tropical climate (e.g., open-air pavilions, rooftops, etc.). Some of these sites also possess historical values, and therefore merit a 3D digitization and conservation.

Within the context of the Franco-Indonesian PHC NUSANTARA research program, a digital documentation mission for a representative Indonesian vernacular and historical site was made possible in May 2018. In this regard, the Kasepuhan Palace, located in the city of Cirebon, West Java, was chosen as a prime example of such a site. The mission was conducted by a multidisciplinary team of surveyors and architects. The Kasepuhan Palace is an urban palatial complex located at the center of Cirebon city, on the northern coast of Java island. It was the center of the Kasepuhan Sultanate, one of the main successors to the old Sultanate of Cirebon. Currently, the palace no longer commands any political influence, but still plays a role as a center of Cirebonese culture. The site is also interesting due to the fact that it had been previously drawn by architects using manual measurements, enabling a comparison and eventually updating of the existing drawings by means of the 3D data acquisition.

The oldest parts of the palace date from the 13th Century [15], with many additions and embellishments throughout the centuries. Many of its buildings demonstrate the local architecture suited for the tropical climate, with influences from Islamic, Chinese, and European architecture. Following these constraints and influences, the palace proper does not consist of a single contiguous building, but rather a complex of smaller buildings located within the same premises. While some enclosed buildings do exist, including the main building, many parts of the palace complex consist of small pavilions, often with few walls in order to increase air flow [16]. The whole palatial complex was modeled in a lower resolution in order to get a general idea of the site and its surroundings within the context of the historical Cirebon city center. One area of particular interest within the complex was identified to be digitized at a higher resolution, namely the Siti Inggil ("Elevated Ground") area near the entrance (see Figure 1). The Siti Inggil area dates from the 15tth Century [17] and served as a viewing platform for the Sultan, as it directly faces the city's main square ("alun-alun").

The objective of this paper is to address the challenges in implementing a multi-scalar and multi-sensor 3D documentation workflow for tropical vernacular cases. The paper will report these challenges and how we addressed them during the project. The paper will also describe some geometrical analysis and quality control, in keeping with the required geometric precision and accuracy of digital heritage documentation. Both image-based and range-based techniques were employed in this regard. Several sensors were also used, including a UAV, which enables both a bird's eye view of the palace's surroundings, and also close-range aerial photogrammetry, which complements the classical terrestrial photogrammetry and laser scanning.

Figure 1. The area of Siti Inggil within the palace compound, which was documented in higher detail. To the left, an aerial view of the area with several main structures overlaid. To the right, the Royal Pavilion of the Siti Inggil compound. Note the characteristic tropical architecture in the form of double-tiered roofs and the absence of walls.

2. Related Work

The 3D documentation of heritage buildings and architectural elements has been addressed many times in the literature. The use of photogrammetry, laser scanning, and indeed the combination of both is seen as a winning solution for digital documentation [1]. Laser scanning has the advantage of being a fast system, able to produce millions of points in a very short amount of time [18]. It is therefore a practical solution for most digitizing projects. Nevertheless, different types of laser scanners are required for different ranges and accuracies; one of its disadvantages is therefore its price. This is more so when the site to be digitized requires different scales.

Photogrammetry has seen many improvements and renewed interest in the last decade, partly due to significant developments in terms of sensor manufacturing [19]. This is also helped by breakthroughs in the field of dense matching [20], as well as the democratization of UAVs [21]. As regards heritage documentation, photogrammetry is often employed in its close-range configuration. Using this technique, a flexible ground sampling distance (GSD) can be designed according to the requirements. It is therefore capable of capturing very intricate details. Added with the possibility to recover real-life texture from the images, this gives photogrammetry an advantage over other methods. UAVs give another edge to photogrammetry by adding aerial points of view. However, photogrammetric data processing requires extensive time, experience, and resources. Historically, photogrammetry requires a well-calibrated and even metric sensors. Today, however, faster computing capabilities and algorithms have widened the scope of sensors that may be processed according to photogrammetrical principles. The DSLR camera is typically used in a standard heritage documentation project due to its relative stability and possibility to choose acquisition parameters [5,22]. However, other lower end sensors have also seen a rise in quality and therefore usability in recording heritage objects, e.g., smart-phone cameras [23,24] or spherical panoramic cameras [25,26].

Laser scanning, on the other hand, has been around since the 1980s and was a revolutionary technology in the domain of 3D mapping. Contrary to traditional total stations, the TLS technology enables the recording of the environment in a fast and relatively accurate manner while being fairly easy to use. This abundance of data, however, may also be a disadvantage as it generates large files of point clouds. This is especially true when the object in question is a complex building, as is often the case in heritage documentation [1]. Occlusions may also occur, which render the final point cloud incomplete [27]. The TLS also has limitations when it comes to point cloud colors and textures, even with the addition of cameras attached to the device [28]. Meanwhile, registration between

TLS stations has been studied in many literature works, mainly based on a coarse 3D conformal transformation followed by a fine closest neighbor-based registration [7,8,29].

Due to the advantages and disadvantages of each recording method, both can be complementary, and their integration is often performed [30,31]. Various integration workflows have been proposed, which mainly depend on the particular case. Early attempts include the possibility to integrate the 3D data by means of a Geographic Information System (GIS) [32]. Another approach performs independent georeferencing on each dataset in the same coordinate system, with direct integration at the end of each georeferencing [33,34]. In terms of photogrammetric workflow, several protocols stress the need for a good camera network for the bundle adjustment part and enough overlaps for the dense matching parts [35,36]. UAV acquisition strategies can also be used in order to extract as much information as possible from aerial flights, for example by performing oblique image acquisitions [37].

Architectural documentation has been performed using these methods in many research works [38]. It is often used in extracting orthophotos [39], vector models of facades [40], 3D models [41], as well as inputs for heritage building information models (HBIM) [42]. Both photogrammetry and laser scanning have been employed in the documentation of vernacular buildings [43,44]. They have also been employed in the case of sites located in a tropical climate [32], as well as for the distinctive Asian architecture [45,46].

However, few research works have been conducted on vernacular Indonesian architecture documentation. One research paper discussed the digitizing of the Borobudur temple [47] using a similar multi-sensor and multi-scale approach. However, the site in question is of a very different nature than the one encountered in the Kasepuhan Palace; therefore, different problems and challenges arise. Indeed, the Kasepuhan Palace presents a different layout and architecture from other monolithic monuments in that it is spread out in a larger area. Another research work that was more similar to this project was conducted in a traditional village in Sumatra; however, the paper reported only the use of close-range photogrammetry [48].

3. Proposed Workflow

The proposed workflow involves a multi-sensor and multi-scale digitization of the Kasepuhan Palace site, with the intention to create a hybrid photogrammetric-laser scanning 3D model. This hybrid model can then be used to generate various derivative products that support the work of architects and archaeologists, while also providing a medium for the dissemination of information to the public. Photogrammetry was mainly used to capture close-range objects with intricate details, which require higher resolution, as well as an aerial point of view of the site. Laser scanning, on the other hand, was used to capture the buildings in general and thus acts as a bridge between the resolution generated by aerial photogrammetry and that by close-range photogrammetry. In terms of the sensors used, the main types of sensors employed include the following:

- Aerial photogrammetry: Aerial photogrammetry was performed in order to obtain a global view of the site and its surroundings. The DJI Phantom 4 (Normal) UAV was used in this regard, equipped with a 12-megapixel camera and a small 3.6-mm lens. The flight was conducted at an average altitude of 80 m.
- Close-range photogrammetry: The DSLR camera Canon EOS 5DSR was used to capture close-range images. This camera generates images with a 50-megapixel resolution. Two types of lenses were used during the acquisition: a 24-mm one was used to capture larger architectural elements and buildings (e.g., columns, ceilings, walls, etc.), while a 40-mm lens was used for elements requiring finer details (e.g., carvings, fine woodwork, etc.). Additional close-range images were acquired using the UAV from varying flying heights from 5–20 m.
- Laser scanning: A TLS was used to obtain the general environment and 3D model of the buildings within the palace complex. A Faro Focus 3D was used to this end. The Faro Focus 3D is a phase-based TLS, which works well within a medium range of distance.

- Topographical surveying: In parallel with the 3D data acquisition, a topographical survey was conducted to measure several aerial premarks, as well as the 3D coordinates of some control points. These control points were then integrated into the aerial and close-range photogrammetric project, as well as the laser scanning project in order to georeference them to the same system. The total station Topcon GTS-212 was used for the tacheometric measurements, which in turn was attached to the Indonesian national UTM system by means of GNSS measurements.

The multi-scale aspect of the project refers to the different scale steps that were acquired (see Table 1). UAV flights were conducted to take images of the Kasepuhan Palace and its surroundings; this constitutes Step 1 of the assigned scale steps. Step 2 involves a digitizing process at a larger scale, this time encompassing a particular area of the complex: the Siti Inggil area near the entrance. This area was prioritized during the documentation mission due to its historical value. Close-range UAV data were combined with a middle-range (10–25 m) TLS point cloud in order to achieve this level. The TLS acquisition was aided by artificial spheres, which facilitate the registration process. The third scale step was comprised of monuments, pavilions, or other single buildings in general. Close-range photogrammetry using the 24-mm lens, as well as the TLS at a closer distance (5–10 m) were employed in this case. The final step consisted of fine details such as carvings and intricate decorative woodwork on some of the architectural elements. The 40-mm lens was used in this case.

The method of 3D data integration follows the existing workflow as described in [33]. Instead of a block integration of all the data, which would have taken an immense amount of time and resources, integration was performed by means of independent georeferencing for each dataset. This means that photogrammetric projects were georeferenced using the absolute orientation method, while laser scanning projects used the measured coordinates of some of the artificial spheres to perform a rigid-body transformation. All control coordinates for Steps 1 and 2 were measured via total station and GNSS survey and attached to the same projection system (UTM 49S). Since both TLS and photogrammetry data were georeferenced to the same system, the two data were automatically merged at the end of the georeferencing process. The fact that all data were georeferenced to the same absolute system also means that future missions can be superposed easily. The control points were each measured twice from two stations, in order to enable spatial intersection. Note that for Steps 3 and 4, georeferencing by means of surveyed control points would mean a large amount of control points. In order to economize the available time and resources, manual control points were measured from the georeferenced TLS point cloud. The downside of this method, however, would be the propagation of error from the TLS georeferencing process.

Architects have also previously created drawings and plans of the Kasepuhan Palace, mostly using measuring tapes and distance meters (Figure 2). These preliminary data were useful for the mission planning and may also serve as a comparison to the obtained 3D data. These drawings, however, are somehow simplified with only measurements for one facade given. The 3D documentation therefore also permits a more detailed measurement on the dimensions of various objects, which may be very useful for architects and archaeologists alike.

Table 1. The different scale steps in this project, with information on the methods used, sensors employed, and approximate GSD (for photogrammetry) and resolution (for TLS) for each step.

Step	Methods	Sensors	GSD (mm)	Resolution (mm)
1	Aerial photogrammetry	UAV	34.2	-
2	Close-range photogrammetry	UAV	8.6	-
	Laser scanning	TLS (10–25 m)	-	ca. 0.9–2.2
3	Close-range photogrammetry	DSLR (24 mm)	0.9	-
	Laser scanning	TLS (5–10 m)	-	ca. 0.6–1.2
4	Close-range photogrammetry	DSLR (40 mm)	0.1	-

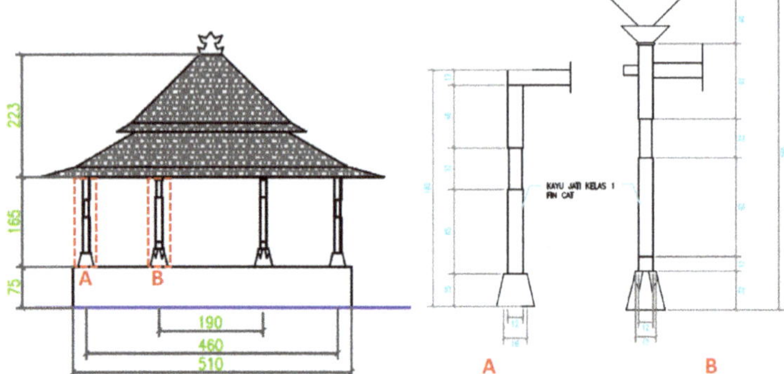

Figure 2. An excerpt of the architectural drawings for the Central Pavilion of Siti Inggil commissioned by the Tourism and Cultural Service of the Province of West Java. Note the two-dimensional nature of the drawings, which shows only measurements for one facade, while assuming that all of the object's sides are symmetrical.

4. Challenges and Constraints

Various challenges specific to the conditions encountered in the Kasepuhan site were mostly related to the local climate and culture. As has been previously mentioned, the Kasepuhan Palace compound consists of a number of separate pavilions. These pavilions serve as both shelters against the elements and as a cultural statement of the palace as a historic center of Cirebonese culture. Due to the tropical and coastal climate of Cirebon, the weather is usually hot and humid. These conditions dictated the design of the vernacular architecture present in Kasepuhan. The use of walls was scarce on the majority of buildings, which provides a free flow of air and natural cooling [16]. Walls are limited to some boundaries between the different parts of the compound. This lack of walls presents a particular challenge to TLS data acquisition, since this means that there are more hidden angles to cover and more masking problems. This translates into more stations and sometimes higher resolution scans in order to cover the hidden parts. This problem was addressed by taking additional photogrammetric images of some of the parts where the TLS is most difficult to station. The resulting point cloud can then be integrated into the TLS one to cover some of the most problematic parts.

The climate of the site also poses certain challenges with regards to the acquisition mission. Cirebon possesses a tropical and coastal climate, being located at the northern coast of the island of Java. The temperature during the mission averaged between 30 °C and 34 °C, with very high levels of humidity. These conditions play a role in the performance of some surveying tools, which rely on infrared and/or lasers. The total station encountered small problems in measuring distances due to this factor, and at some points, the measurement had to be performed several times. This may also be due to the strong solar exposure. In this regard, the use of the spatial intersection method for the computation of control points is useful, since it may be based only on angle measurements, which are less influenced by the temperature and humidity. This method has the added benefit of giving a more robust result due to the fact that each point is measured from at least two stations. Solar intensity was also an important influencing factor, as it renders some passive sensors problematic due to overexposure. Problems with overexposure in photos may be rectified in manual mode; however, the panoramic images taken by the TLS were more difficult to compensate. In some processing steps where it was required to select points, intensity rather than RGB images were therefore used (Figure 3).

Figure 3. Example of a section of a panoramic image captured by the TLS. The figure to the left shows the RGB image, with strong overexposure due to sunray intensity. The figure to the right shows the intensity image, which was preferred in the tie point identification process.

The dense vegetation in and around the Kasepuhan complex was also a challenge, particularly for UAV acquisition. Ground control points (GCPs) were to be placed around the complex following standard photogrammetric convention, but this was hindered by the amount of tree canopy. Premark placement for GCPs was therefore very limited to small open areas, which do not necessarily correspond to the ideal photogrammetric ground control network. It also posed a problem for GNSS measurements, necessitating a longer measurement time in order to get to the required centimetric precision. In regards to TLS and close-range terrestrial photogrammetry, the dense vegetation also generated noises, which must then be cleaned from the resulting point cloud.

Another problem with the site is that residential housing is virtually mixed with palatial buildings in some areas, sometimes with no walls or fences to delineate them. This means that residents were free to enter the palace compounds, rendering the site non-sterile. Particular care in the handling of surveying tools and TLS spheres was therefore of the utmost importance. Members of the team had to stand by on several spots with artificial TLS spheres in order to avoid them being moved by passers-by. Several control points were therefore also measured on immobile detail points (e.g., roof edges, brick intersections, etc.) to mitigate problems that may arise from moved targets. Tourists also came in large groups due to the palace's guided tour system. Complete authorization for the sterilization of the site was difficult due to the amount of tourists who visited Kasepuhan, even during normal work days. This added complexity during the acquisition and noises in the point cloud.

5. Results and Discussions

Overall, 310 aerial images, over 430 close-range UAV images, and 1060 terrestrial close-range images were taken during the mission. In terms of TLS stations, 23 stations were acquired throughout the palace compounds. Photogrammetric processing was performed using Agisoft PhotoScan, while TLS registration and georeferencing were conducted using the Faro Scene software.

5.1. Preliminary Processing

The Step 1 images oriented in PhotoScan are shown by Figure 4. Overall, eight control points spread over a 23 hectares area were measured on the field and used during the image orientation process. Five points were used as GCPs, yielding an overall RMS value of 3.5 cm. The remaining three points were used as check-points, giving an RMS value of 9.6 cm. Since the main objective of the Step 1 processing was only to give a general overview of the site, not enough time was available to measure too many control points for the whole area (the main area of interest for the project being the Siti Inggil area). While far from ideal, this configuration was largely enough for the Step 1 requirements. This first step is useful in assessing the Kasepuhan site and its environment and may be used to generate maps and orthophotos with centimetric resolution.

Meanwhile, from the 23 TLS stations that were acquired, a complete registration of all stations was performed using Faro Scene, with a 4.5-mm average distance between tie points ("tensions"). These values rest within the tolerance when considering the theoretical GSD as described in Table 1.

The TLS point cloud was georeferenced using the coordinates of six artificial spheres measured during the topographical survey using a total station and is a separate step from the GNSS measurement used for the aerial control points.

Figure 4. Images of the Step 1 scale level oriented in PhotoScan.

As regards to the other scale steps, the georeferenced TLS point cloud was used as a reference for the absolute orientation of the photogrammetric projects. Manual control points were selected on the TLS point cloud on parts where the photogrammetric result overlapped. Even though this process was performed on the full resolution of the TLS data, further errors in terms of accuracy were still possible due to the error propagation. The TLS data were then subsampled into 5-mm, 1-cm, and 2-cm point clouds with the aim to reduce the file size.

Figure 5 shows some results with regards to the multi-scale approach used in this project. Aerial photogrammetry was used in the creation of the Step 1 3D model; this scale step can be used as an overview of the Kasepuhan site and its surroundings. In total, 370 images were taken for this Step 1. Orthophotos can also be generated from this 3D model, from which an update of existing maps can also be derived. Some interesting preliminary remarks based on this result include the fact that some of the areas of the palace are not aligned to the cardinal directions. The Siti Inggil area is quite notable as it demonstrates a parallelogram shape, in addition to the non-alignment to the cardinal directions.

Step 2 shows results from the TLS dataset, from which a higher resolution can be observed in the case of certain areas (Siti Inggil in Figure 5). Advancing to even higher resolution, Step 3 shows the Royal Pavilion within Siti Inggil, which was modeled using close-range photogrammetry images. In this resulting point cloud, characteristic architectural elements can already be identified individually. A total of 640 images were taken for various details of the palace, averaging 150 images for each object of interest. Step 4 shows the finest resolution, that of decorative elements such as carvings. In Figure 5, a decorative plinth of a column generated from 30 images is showcased.

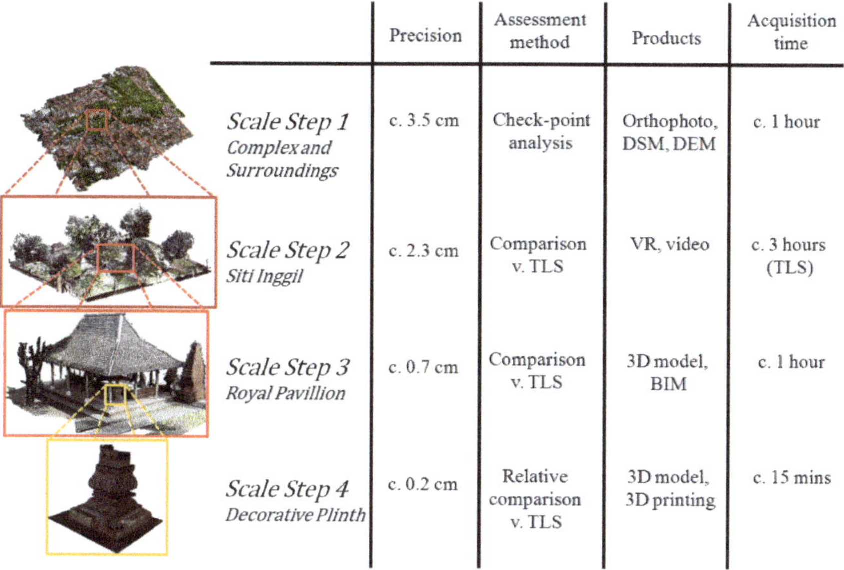

Figure 5. Example of the result of the multi-scale approach. In Step 1, the whole Kasepuhan complex and its surroundings are modeled by aerial photogrammetry. Step 2 shows the Siti Inggil area within the palace compounds; here is shown the registered TLS point cloud. Step 3 shows a building, the Royal Pavilion, within Siti Inggil, which was modeled using close-range photogrammetry. Finally, Step 4 shows an architectural detail, in this case a column's plinth, also modeled using close-range photogrammetry.

5.2. Quality Assessment

Assessment of the quality of the results was performed on several aspects, namely comparison against the pre-existing architectural drawings, analysis of the quality of the close-range photogrammetry results in Steps 2 and 3, and an evaluation of the TLS dataset in the scope of Step 4. This will hopefully give an overall, as well as more detailed views of the quality of the 3D documentation project.

5.2.1. Comparison against Architectural Drawings

One of the first analyses that was conducted on the result was a comparison against the pre-existing architectural drawings. Table 2 shows a comparison of the dimensions of the elevated platform sides and the pillar heights of the Central Pavilion of Siti Inggil. For this purpose, measurements were taken from the Step 3 model. One of the disadvantages of the existing drawings is the fact that they assume that all facades of the object are symmetrical, and therefore only give measurements on one side. Similarly, pillars were categorized into taller interior and lower exterior columns, with singular measurements for each. 3D documentation uses the reality-based approach, and therefore captures the object as it is. Step 3 of the 3D model gives a centimetric resolution, and may therefore detect variations of this order.

When observing Table 2, the platform sides had differences (Δ) ranging from 0.4–2.7 cm. The average difference was 1.4 cm, which is acceptable considering that the measurements on the drawings were made using measuring tapes. An interesting observation can however be seen in the Δ for the pillar height. A seemingly systematic error of around 2 cm can be observed in this case, which may also be due to errors that occurred during the manual measurement.

Table 2. Comparison of some dimensions of the Central Pavilion of Siti Inggil between the resulting 3D model and the architectural drawings. Shown here are comparisons for the pavilion's platforms' sides and pillar heights.

No.	Platform Sides		Δ (cm)	No.	Pillar Heights		Δ (cm)
	3D Model (cm)	Drawings (cm)			3D Model (cm)	Drawings (cm)	
1	508.7		1.3	1	159.6		2.6
2	509.2		0.8	2	159.4	157.0	2.4
3	507.3		2.7	3	159.2		2.2
4	509.0	510.0	1.0	4	159.2		2.2
5	511.3		1.3	5	185.8		2.2
6	508.4		1.6	6	186.6	188.0	1.4
7	511.9		1.9	7	187.2		0.8
8	509.6		0.4	8	185.4		2.6
	Mean		1.4		*Mean*		2.1

The objective of this part of the assessment is to confirm the manual measurements that were made for the architectural drawings. Since the drawings serve as a reference for architects, archaeologists, and conservators working on the site, it was deemed important to compare them to the 3D documentation method. It has also shown that the 3D methods were able to give more precise measurements in three dimensions, instead of simplifying the dimensions into the constraints of a 2D drawing.

5.2.2. Close-Range Photogrammetry Assessment (Steps 2 and 3)

In the case of Steps 2 and 3, the point cloud resulting from the photogrammetry method was georeferenced using common point coordinates obtained from the TLS point cloud. In this regard, the TLS provides a reference to the close-range photogrammetry result. This assumption was made from the fact that the TLS data were georeferenced and did not require a scaling factor, making it more reliable for the georeferencing of the close-range photogrammetry results. In terms of acquisition time, this method reduces much of the field time as it means that only TLS and aerial photogrammetry control points are required to be measured during the topographical survey. However, this may mean that a propagation of error might occur in the results.

Figure 6a shows the result of the combination of point clouds generated by TLS and UAV photogrammetry for the Siti Inggil area in Step 2. As can be seen, the UAV complements the TLS data by filling some holes otherwise impossible to acquire from a terrestrial point of view. This includes buildings roofs, but also tree crowns. Although visually, the composite point cloud seems to have been combined correctly, a further quantitative analysis was performed in order to investigate the quality of the data. In this regard, the cloud-to-cloud distance analysis was performed using the CloudCompare software. This analysis computes the absolute distance between nearest points belonging to two point clouds. Owing to the absence of a precise mesh model that may act as an absolute reference in this step, local surface modeling was conducted using the TLS data as a reference. The results are shown in Figure 6b, which gave a mean distance between the TLS and UAV photogrammetry point clouds of 2.3 cm with a standard deviation of 1.8 cm. Despite the millimetric theoretical GSD for the TLS data, this centimetric result was generated by the UAV data's GSD, which was indeed in the order of 1 cm (see Table 1). Furthermore, in Figure 6b, it can also be observed that the largest distances are located either in the periphery of the Siti Inggil or on parts that are considered movable (e.g., tree crowns and grasses). However, by observing the paved pathway, as well as most of the rooftops of the pavilions, the absolute distance is very small, in the order of 0–7 mm. A similar analysis can be observed in the comparison between TLS and terrestrial photogrammetry data, with similar results.

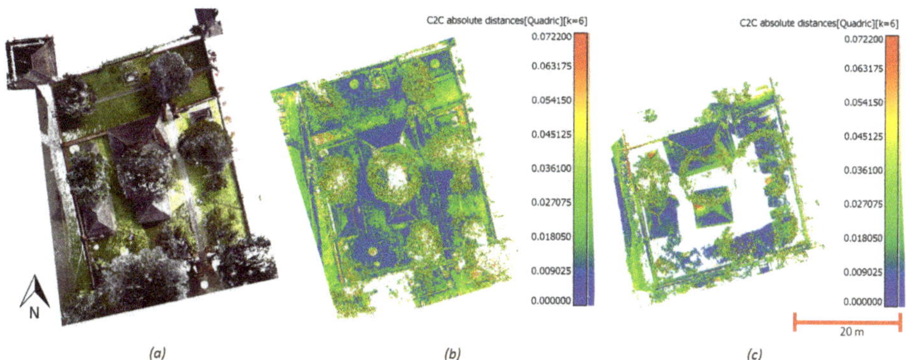

Figure 6. Results for the Siti Inggil area in Step 2. (**a**) shows the TLS point cloud. (**b**) shows the absolute distance between points from the TLS and UAV point clouds, with an average distance of 2.3 cm and a standard deviation value of 1.8 cm. (**c**) shows a similar analysis between TLS and terrestrial photogrammetry, yielding an average distance of 2.0 cm and a standard deviation of 1.6 cm.

In terms of Step 3, the Central Pavilion was chosen as a sample to be analyzed. This building was chosen due to its central position with open space around it, which permitted the generation of a relatively complete point cloud. Figure 7a shows the composite point cloud from combined TLS and close-range photogrammetry results. Pure TLS data did not manage to capture the entirety of the building, as it did not manage to scan the building roofs and difficult angles in the interior. Photogrammetry was useful in adding details of the interior ceilings, while additional UAV images were used in completing the roofs. However, the terrestrial photogrammetry encountered a problem in this case when trying to reconstruct the pavilion's floors. The floors of the pavilion are made of ceramic tiles, which when photographed from a low angle (as is the case with the terrestrial close-range acquisitions) became reflective surfaces and rendered the dense matching algorithm problematic in generating points. Here, the complementarity of TLS and photogrammetry is showcased, as both techniques completed each other and generated a complete model of the pavilion. This complementarity, however, requires a good registration between the two data. In this regard, another cloud-to-cloud distance analysis was performed in order to quantify the quality of the registration.

Figure 7b shows the result of such analysis in CloudCompare, generating a mean distance between TLS and photogrammetry point clouds of 7 mm with a standard deviation of 5 mm. These values were also higher when considering the theoretical GSD as expressed in Table 1. By observing Figure 7b, a slight systematic error can be seen to the right part of the model. This may be due to the imperfect iterative closest point (ICP) registration process, caused by the noises in both datasets. One main problem of the ICP process was the presence of a tree behind the pavilion, rendering some parts of the roofs very noisy in a different way for each dataset. This may have contributed to the slight rotation between the two datasets, as evidenced in Figure 7. Nevertheless, the resulting average distance and its standard deviation were still within the centimetric level, which is enough in this case.

Figure 7. Results for the Central Pavilion in Step 3; (**a**) shows a composite of TLS and close-range photogrammetry point clouds. In (**b**), the absolute distance between points from each point cloud is shown with an average distance of 7 mm and a standard deviation value of 5 mm.

5.2.3. TLS Assessment (Step 4)

Georeferencing of Step 4 results used common points identified on the TLS data. However, in terms of point cloud resolution, the close-range photogrammetry data provided much finer details. The close-range data also enabled the creation of a detailed mesh, which could then serve as a reference surface. Another analysis based on cloud-to-mesh signed Euclidean distance was therefore performed in order to assess the quality of the TLS data in recording these details. It should be noted that the Faro Focus 3D was designed as a medium-range TLS.

Figure 8 shows the 3D reconstruction of a plinth located within the Royal Pavilion. The close-range photogrammetry point cloud resulted in a very detailed point cloud due to the acquisition setup. This point cloud was meshed using the Poisson method in order to serve as a reference for the TLS. Figure 8d shows the signed distance between the TLS point cloud and the photogrammetric mesh. A mean error of 1.5 mm was acquired, as well as a standard deviation value of 2 mm. The mean error was quite high for this level of resolution and indicates the presence of a systematic error. This was probably caused by errors during the registration of both point clouds. The standard deviation value, however, roughly corresponds to the theoretical resolution of the TLS acquisition.

A similar analysis was performed on the woodwork on the Central Pavilion's ceilings (Figure 9). A mean error of 0.4 mm and a standard deviation value of 5.6 mm were obtained, as shown by Figure 9d. The mean error was near zero, which indicates that the registration between the two datasets was correct, and only a negligible systematic error was detected. That being said, in Figure 9d, we can observe a systematic tilt at the upper right and lower left corners of the ceilings. This fact may not be noticed when the errors are not plotted as such in Figure 9d, as the tilt may have compensated the value of the mean error. Moreover, the value of the standard deviation was higher in this case, almost thrice that of the plinth. This might be caused by the sensor to object distance, as both photogrammetry and TLS data were acquired from a slightly farther distance than that of the plinth, which reduces the resolution of both TLS and photogrammetry data.

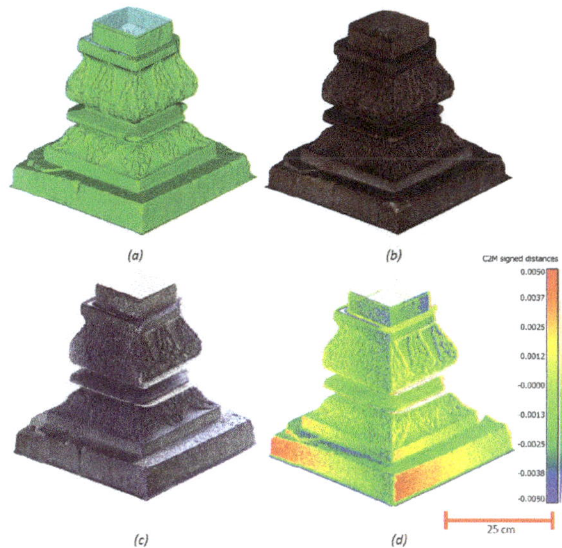

Figure 8. Results for the eastern upper plinth of the Royal Pavilion; the 3D mesh (**a**) and textured (**b**) models generated by photogrammetry were used as a reference against the TLS point cloud (**c**). In (**d**), the cloud-to-mesh distance analysis yielded a mean error of 1.5 mm and a standard deviation value of 2 mm.

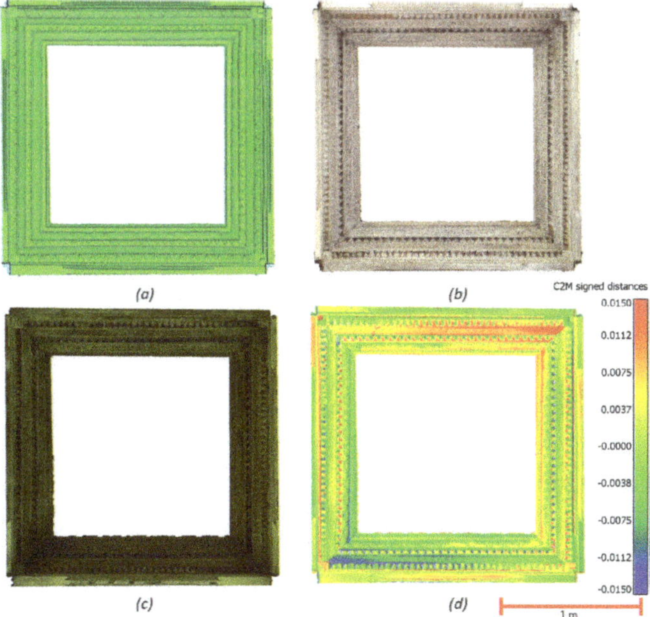

Figure 9. Results for the woodwork on the Central Pavilion's ceilings; the 3D mesh (**a**) and textured (**b**) models used as a reference against TLS point cloud (**c**). The cloud-to-mesh analysis (**d**) yielded a mean error of 0.4 mm with a standard deviation of 5.6 mm.

5.3. Derived Products

Once the preliminary processing was finished, several derivative products were generated from the resulting hybrid photogrammetric and TLS point cloud. This section will briefly describe some of the products that were created in this project, which were mainly done in order to increase the management and dissemination of the Kasepuhan site and Siti Inggil in particular. This includes an orthophoto map, a photorealistic 3D model, a BIM model, and a virtual reality environment.

5.3.1. Orthophotography of Kasepuhan and Its Surroundings

The orthophoto map (Figure 10) was one of the first derived products to be generated, as it requires only the finishing of the Step 1 scale level. The map was generated by aerial images taken from a flying height of around 80 m from the ground, yielding a pixel resolution of 3 cm. This map was not by any means meant to be a true topographic map, but rather an overview of the Kasepuhan Palace and its surroundings. In this sense, it followed the (semi-) automatic workflow in PhotoScan, and the DTM (digital terrain model) from which the orthographic projection was based did not undergo a thorough and rigorous manual clean up.

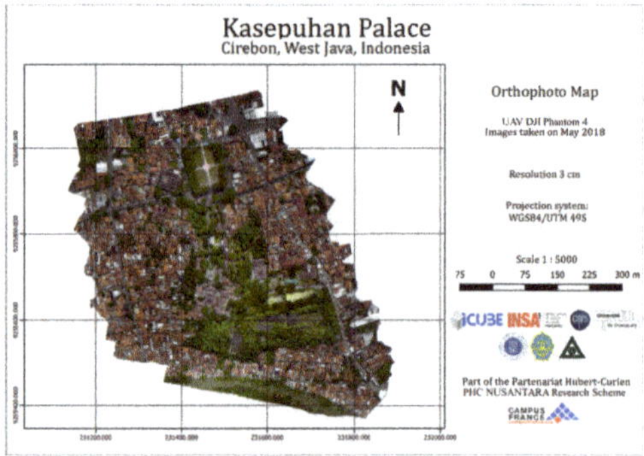

Figure 10. Orthophoto map of Kasepuhan and its surroundings, produced from aerial UAV images. This type of product is useful to assess the overall condition of the site.

Indeed, the purpose of the map was to serve as a support during the acquisition mission itself, by showing the general layout of the site and enabling a more detailed planning for the acquisitions of the next scale steps. Although a CAD drawing of the palace layout was available, the high-resolution orthophoto texture helped in performing qualitative interpretation of some ground features. Processing speed was therefore of the essence, and a first version of the orthophoto map was already available at the end of the first day of the mission. Regardless, the map still provides valuable information to other stakeholders such as architects and palace managers. The up-to-date nature of the data from which it was created made it a useful tool in revising the existing maps of the palace complex.

5.3.2. Photo-Textured 3D Meshed Model

A surface reconstruction of Siti Inggil was made possible by meshing the point cloud into 3D meshed models. 3D meshes have smaller file sizes and may be superimposed by textures, therefore both easier for data management and better for visualization purposes. Mesh triangles can be textured using an interpolation of point colors, but this results in a blurred texture, which is not photorealistic. Since photogrammetric data were also available, the superimposition of photorealistic texture was also feasible. The proposed workflow in the creation of these photorealistic textured 3D model will be elaborated in this section. The results from several main steps are shown in Figure 11.

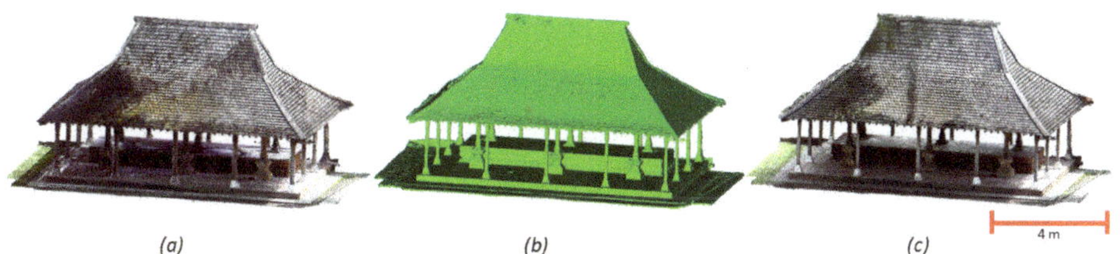

(a)   (b)   (c) 4 m

Figure 11. 3D models of the Royal Pavilion; (**a**) the combined point cloud from TLS and photogrammetry, (**b**) 3D meshed model of the combined point cloud, and (**c**) photorealistic textured 3D mesh model. Note the homogenization of textures in (**c**).

Starting from the registered and combined TLS and photogrammetry point cloud (Figure 11a), a 3D mesh was generated using the Poisson method (Figure 11b). Since both photogrammetric and TLS point cloud were already georeferenced to the same absolute coordinate system, the integration of these two data sources were immediate. The meshing was performed using the CloudCompare software. The combination of TLS and photogrammetry was important, since using only photogrammetric point cloud results in some missing parts, and vice versa with the TLS point cloud. In the photogrammetric point cloud, this problem is exacerbated by the ceramic tiles used on the pavilions' flooring, often rendering the floors missing. For the TLS point cloud, texturing is important since textures from the inherent panoramic images are not good enough and are often hindered by occlusions from the station's point of view.

The resulting mesh was then re-imported into PhotoScan, bearing in mind that the coordinate systems of the PhotoScan project and the mesh were identical. This is crucial in order to keep the texture mapping properties already set in the PhotoScan project and avoid shifted textures. This would bring the mesh into the PhotoScan project's network of oriented images. Photorealistic textures can then be superimposed on this mesh using PhotoScan's texturing function (Figure 11c). The resulting 3D model has a reduced file size compared to the original point cloud, while possessing a photorealistic texture. This procedure was performed for all Step 3 objects, including all the pavilions and the 15th Century brick wall surrounding Siti Inggil. The photorealistic 3D models provide a better visualization compared to the point cloud or simple meshes. These models can then be integrated into the virtual visit environment, which was also developed in this project.

5.3.3. BIM 3D Model

In terms of site management, the 3D mesh is not sufficient as it could not store semantic information. One way to address this problem is the creation of a building information system (BIM), which permits the storing of semantic data within the 3D model. The BIM 3D model was created using the Archicad software, based on a wireframe CAD model, which were created using photogrammetry using the software PhotoModeler (EosSystems). The 3D model was drawn as a test, and in this regard, a repeating pattern texture image was used instead of the photo textures. The resulting model of the Central Pavilion can be seen in Figure 12. Note that the BIM model still consists solely of geometric data, with semantics and ontology to be added in a later phase. The representation of complex geometries in a BIM environment is an interesting subject and may be addressed in several ways, e.g., representation as mesh rather than geometric primitives.

5.3.4. Virtual Visit Environment

In order to create an immersive medium for visualizing the results, a virtual visit environment was developed. Both the point cloud and photorealistic 3D meshed models were used in this regard. A combination of texturing methods was opted, with the Step 3 objects (i.e., pavilions and brick walls) displayed in photorealistic textures, while the less interesting parts of Siti Inggil (e.g., lamp post, ground, trees) only used interpolated color originating from the point cloud color. This was done in order to reduce processing time and focus more on the objects of interest.

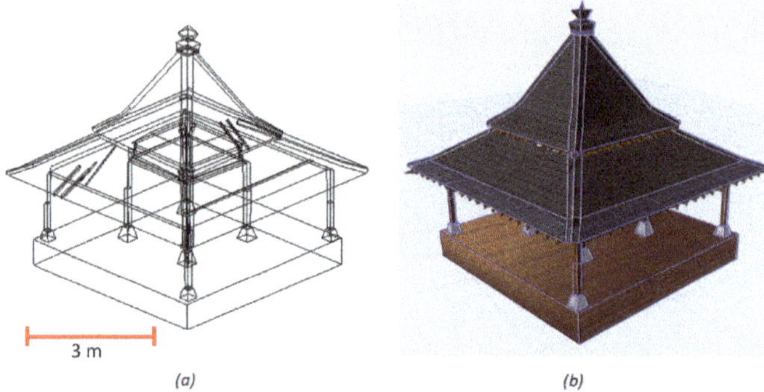

Figure 12. 3D models of the Central Pavilion in the BIM environment in (**a**) wireframe and (**b**) solid models.

Two products were generated from this virtual visit environment, displayed here in Figure 13. The first one is a virtual visit video of Siti Inggil, which showcases the photorealistic models of the main objects of interest. The second product is a Virtual Reality (VR) system, which was developed using the HTC VIVE VR goggles. These derivative products is very useful in helping the dissemination of information about the site to the public. The immersive environment enables people to visit the site remotely and appreciate more the Kasepuhan Palace in its empty state. It is also an interesting tool for architects in order to examine various details of the area without having to go to the site.

Figure 13. The virtual visit environment developed for the Siti Inggil area. To the left, the combined 3D mesh models from various scale steps. Notice that the 3D model of the pavilion is at a higher resolution than the ground or tree beside it. To the right, a virtual visit of the same area using VR gear.

6. Conclusions and Outlook

This paper described the workflow used in the digitizing of a tropical vernacular architecture in the context of cultural heritage documentation. The workflow included a multi-scale approach, in which the varying components of 3D data were linked together using measured 3D coordinates to create a georeferenced hybrid 3D point cloud. The multi-scale aspect divided the project into four scale steps, going from Step 1 (aerial photogrammetry for the site and its surroundings), Step 2 (TLS and UAV close-range images for smaller specific areas of interest), Step 3 (terrestrial close-range photos for architectural objects), up to Step 4 (terrestrial close-range images for fine decorative elements). The results showed that the approach is suitable for this type of site, where the area is spread out in a large complex with smaller buildings within. Furthermore, some challenges during the acquisition have been discussed in this paper. Some adaptive measures to answer these challenges were also briefly mentioned. Indeed, the type of architecture encountered in Kasepuhan poses a special case adapted for the surrounding climate, and therefore must be addressed differently. Challenges regarding the conditions during the acquisition were also described in the paper.

The analysis conducted in this paper showed that for Step 1, UAV photogrammetry is a very suitable solution. It provides a fast result in the form of the orthophoto map and a rough DTM, which permits a more detailed planning within the period of the acquisition mission. TLS still proves to be the most practical to document a site for a Step 2 scale level such as Siti Inggil. It is much faster in terms of both acquisition and processing and provides a rather complete overview of the site. However, as has been seen in this paper, some missing data are still possible due to difficult angles and occlusions (e.g., roofs and pillars). It is in this regard that photogrammetry can be a complementary technique in providing the missing data. Naturally, this can also be performed by other types of TLS (e.g., handheld laser scanner), but the cost constraints may give photogrammetry the advantage in this particular case.

In terms of Step 3, the TLS is still a pertinent solution. However, within this scale step, the problem for TLS comes in regard to its texturing options, which is very limited. Again, photogrammetry may complement this technique in providing not only missing parts, but also photorealistic textures. However, for Step 4, photogrammetry is by far the best option for the recording process. The TLS used in the project simply is not designed for such close-range application. Close-range photogrammetry can be employed in this case rather quickly and simply, while also providing photorealistic textures. Again, this may be substituted with a more precise laser scanner such as triangulation-based scanners (e.g., FARO FreeStyle, Konica Minolta Range 7, etc.). However, the downside of this solution would be the acquisition of multiple laser scanners for such a project where the multi-scale aspect is important. The analysis in this paper also showed that certain metrics and statistics (e.g., standard deviation with regards to a reference) are sometimes required in order to better represent the quality of the results (case in point as seen in Figure 9).

This type of documentation project is very important for many stakeholders. Architects and conservators may find it useful in conservation efforts, while policy makers may use it to decide on necessary measures to preserve the site. The 3D data may also be used for other, more sophisticated purposes. For example, the creation of a heritage building information model (HBIM) will be very useful in the management of the heritage site. Virtual reality solutions also enable a democratization of 3D technology and the diffusion of historical information to the public. Finally, with the existing 3D documentation acting as digital archives, a physical reconstruction in the case of damage can be performed via, for example, 3D printing technology. Further work will focus on several points, including formalization of a standard procedure for heritage mapping for tropical vernacular cases, elaboration of VR systems to include semantic information, and the automation of some of the time-consuming steps during the data processing.

Author Contributions: Conceptualization, A.M., P.G. and D.S.; data curation, R.A.; formal analysis, A.M. and R.A.; funding acquisition, P.G. and D.S.; investigation, A.M., P.G., D.S. and R.A.; methodology, A.M., P.G. and D.S.; project administration, P.G. and D.S.; resources, D.S.; supervision, P.G. and D.S.; validation, A.M. and R.A.; visualization, R.A.; writing, original draft, A.M.; writing, review and editing, A.M., P.G. and D.S.

Funding: This research is part of the Franco-Indonesian Partenariat Hubert-Curien (PHC) NUSANTARA program under the aegis of the Indonesian Ministry of Research and Higher Education (KEMENRISTEKDIKTI), the French Ministry for Europe and Foreign Affairs (MEAE), and the French Ministry of Higher Education, Research, and Innovation (MESRI).

Acknowledgments: The authors would like to express their gratitude to His Majesty the Sultan Sepuh XIV PRA. Arief Natadiningrat for authorizing the digital documentation project to be conducted within the premises of the Kasepuhan Palace. The authors would also like to thank PT. Exol Innovindo for providing the TLS used in this project. The following individuals also provided valuable and crucial help during the project: Iwan Purnama of the Cirebon Technological College, Sutrisno Murtiyoso of the Indonesian Institute for History of Architecture, as well as Dirga Sumantri, Shafarina Wahyu, and Gina Andryana from the ITB Remote Sensing and GIS Group. The research also benefits from the Indonesian Endowment Fund for Education (LPDP), Republic of Indonesia.

Conflicts of Interest: The authors declare no conflict of interest. The founding sponsors had no role in the design of the study; in the collection, analyses, or interpretation of data; in the writing of the manuscript; nor in the decision to publish the results.

References

1. Barsanti, S.G.; Remondino, F.; Fenández-Palacios, B.J.; Visintini, D. Critical factors and guidelines for 3D surveying and modeling in Cultural Heritage. *Int. J. Herit. Dig. Era* **2014**, *3*, 141–158. [CrossRef]
2. Achille, C.; Adami, A.; Chiarini, S.; Cremonesi, S.; Fassi, F.; Fregonese, L.; Taffurelli, L. UAV-based photogrammetry and integrated technologies for architectural applications—Methodological strategies for the after-quake survey of vertical structures in Mantua (Italy). *Sensors* **2015**, *15*, 15520–15539. [CrossRef] [PubMed]
3. Baiocchi, V.; Dominici, D.; Mormile, M. UAV application in post-seismic environment. *Int. Arch. Photogramm. Remote Sens. Spat. Inf. Sci.* **2013**, *XL-1/W2*, 21–25. [CrossRef]
4. Grussenmeyer, P.; Al Khalil, O. From metric image archives to point cloud reconstruction: Case study of the Great Mosque of Aleppo in Syria. *Int. Arch. Photogramm. Remote Sens. Spat. Inf. Sci.* **2017**, *XLII-2/W5*, 295–301. [CrossRef]
5. Remondino, F.; Del Pizzo, S.; Kersten, T.P.; Troisi, S. Low-cost and open-source solutions for automated image orientation—A critical overview. *Prog. Cult. Herit. Preserv.* **2012**, *7616 LNCS*, 40–54.
6. Grenzdörffer, G.J.; Naumann, M.; Niemeyer, F.; Frank, A. Symbiosis of UAS Photogrammetry and TLS for Surveying and 3D Modeling of Cultural Heritage Monuments—A Case Study About the Cathedral of St. Nicholas in the City of Greifswald. *Int. Arch. Photogramm. Remote Sens. Spat. Inf. Sci.* **2015**, *XL-1/W4*, 91–96.
7. Murtiyoso, A.; Grussenmeyer, P. Comparison and assessment of 3D registration and georeferencing approaches of point clouds in the case of exterior and interior heritage building recording. *Int. Arch. Photogramm. Remote Sens. Spat. Inf. Sci.* **2018**, *XLII-2*, 745–751. [CrossRef]
8. Lachat, E.; Landes, T.; Grussenmeyer, P. Comparison of Point Cloud Registration Algorithms for Better Result Assessment—Towards an Open-Source Solution. *Int. Arch. Photogramm. Remote Sens. Spat. Inf. Sci.* **2018**, *XLII-2*, 551–558. [CrossRef]
9. Grussenmeyer, P.; Alby, E.; Landes, T.; Koehl, M.; Guillemin, S.; Hullo, J.F.; Assali, P.; Smigiel, E. Recording approach of heritage sites based on merging point clouds from high resolution photogrammetry and Terrestrial Laser Scanning. *Int. Arch. Photogramm. Remote Sens. Spat. Inf. Sci.* **2012**, *39*, 553–558. [CrossRef]
10. Lachat, E.; Landes, T.; Grussenmeyer, P. First Experiences with the Trimble SX10 Scanning Total Station for Building Facade Survey. *Int. Arch. Photogramm. Remote Sens. Spat. Inf. Sci.* **2017**, *42*, 405. [CrossRef]
11. Lerma, J.L.; Navarro, S.; Cabrelles, M.; Villaverde, V. Terrestrial laser scanning and close-range photogrammetry for 3D archaeological documentation: the Upper Palaeolithic Cave of Parpallo as a case study. *J. Archaeol. Sci.* **2010**, *37*, 499–507. [CrossRef]
12. Murtiyoso, A.; Grussenmeyer, P.; Koehl, M.; Freville, T. Acquisition and Processing Experiences of Close Range UAV Images for the 3D Modeling of Heritage Buildings. In *EuroMed 2016: Digital Heritage. Progress in Cultural Heritage: Documentation, Preservation, and Protection, Part I, Proceedings of the 6th International Conference, Nicosia, Cyprus, 31 October–5 November 2016*; Ioannides, M., Fink, E., Moropoulou, A., Hagedorn-Saupe, M., Fresa, A., Liestøl, G., Rajcic, V., Grussenmeyer, P., Eds.; Springer International Publishing: Cham, Switzerland, 2016; pp. 420–431.
13. Remondino, F.; Spera, M.G.; Nocerino, E.; Menna, F.; Nex, F.; Gonizzi-Barsanti, S. Dense image matching: Comparisons and analyses. In Proceedings of the 2013 Digital Heritage International Congress (DigitalHeritage), Marseille, France, 28 October–1 November 2013; Volume 1, pp. 47–54.
14. Fiorillo, F.; Jiménez Fernández-Palacios, B.; Remondino, F.; Barba, S. 3d Surveying and modeling of the Archaeological Area of Paestum, Italy. *Virtual Archaeol. Rev.* **2013**, *4*, 55–60. [CrossRef]
15. Djunaedi, A.; Sudaryono; Suryo, D. Spatial constructs of spiritual consciousness: The case of Keraton Kasepuhan in Cirebon, Indonesia. *ISVS e-J.* **2016**, *4*, 16–28.
16. Prianto, E.; Bonneaud, F.; Depecker, P.; Peneau, J. Tropical-Humid Architecture in Natural Ventilation Efficient Point of View: A Reference of Traditional Architecture in Indonesia. *Int. J. Archit. Sci.* **2000**, *1*, 80–95.
17. Flood, F.; Necipoglu, G. *A Companion to Islamic Art and Architecture*; Blackwell Companions to Art History; Wiley: Hoboken, NJ, USA, 2017.

18. Dore, C.; Murphy, M. Current State of the Art Historic Building Information Modelling. *Int. Arch. Photogramm. Remote Sens. Spat. Inf. Sci.* **2017**, *XLII-2/W5*, 185–192. [CrossRef]

19. Murtiyoso, A.; Koehl, M.; Grussenmeyer, P.; Freville, T. Acquisition and Processing Protocols for UAV Images: 3D Modeling of Historical Buildings using Photogrammetry. *ISPRS Ann. Photogramm. Remote Sens. Spat. Inf. Sci.* **2017**, *4*, 163–170. [CrossRef]

20. Remondino, F.; Spera, M.G.; Nocerino, E.; Menna, F.; Nex, F. State of the art in high density image matching. *Photogramm. Rec.* **2014**, *29*, 144–166. [CrossRef]

21. Nex, F.; Remondino, F. UAV: Platforms, regulations, data acquisition and processing. In *3D Recording and Modelling in Archaeology and Cultural Heritage: Theory and Best Practices*; Remondino, F., Campana, S., Eds.; Archaeopress: Oxford, UK, 2014; pp. 73–86.

22. Bedford, J. *Photogrammetric Applications for Cultural Heritage*; Historic England: Swindon, UK, 2017; p. 128.

23. Nocerino, E.; Lago, F.; Morabito, D.; Remondino, F.; Porzi, L.; Poiesi, F.; Rota Bulo, S.; Chippendale, P.; Locher, A.; Havlena, M.; et al. A smartphone-based 3D pipeline for the creative industry—The replicate eu project. *Int. Arch. Photogramm. Remote Sens. Spat. Inf. Sci.* **2017**, *42*, 535–541. [CrossRef]

24. Kim, J.; Lee, S.; Ahn, H.; Seo, D.; Park, S.; Choi, C. Feasibility of employing a smartphone as the payload in a photogrammetric UAV system. *ISPRS J. Photogramm. Remote Sens.* **2013**, *79*, 1–18. [CrossRef]

25. Ramos, A.P.; Prieto, G.R. Only image based for the 3D metric survey of gothic structures by using frame cameras and panoramic cameras. *Int. Arch. Photogramm. Remote Sens. Spat. Inf. Sci. ISPRS Arch.* **2016**, *41*, 363–370. [CrossRef]

26. Fassi, F.; Achille, C.; Fregonese, L. Surveying and modeling the main spire of Milan Cathedral using multiple data sources. *Photogramm. Rec.* **2011**, *26*, 462–487. [CrossRef]

27. Lachat, E.; Landes, T.; Grussenmeyer, P. Combination of TLS Point Clouds and 3D Data from Kinect V2 Sensor to Complete Indoor Models. *Int. Arch. Photogramm. Remote Sens. Spat. Inf. Sci.* **2016**, *XLI-B5*, 12–19. [CrossRef]

28. Hassani, F. Documentation of cultural heritage techniques, potentials and constraints. *Int. Arch. Photogramm. Remote Sens. Spat. Inf. Sci.* **2015**, *XL-5/W7*, 207–214. [CrossRef]

29. Fabado, S.; Seguí, A.E.; Cabrelles, M.; Navarro, S.; García-De-San-Miguel, D.; Lerma, J.L. 3DVEM Software Modules for Efficient Management of Point Clouds and Photorealisticic 3D Models. *Int. Arch. Photogramm. Remote Sens. Spat. Inf. Sci.* **2013**, *XL-5/W2*, 255–260. [CrossRef]

30. Munumer, E.; Lerma, J.L. Fusion of 3D data from different image-based and range-based sources for efficient heritage recording. In Proceedings of the 2015 Digital Heritage, Granada, Spain, 28 September–2 October 2015; Volume 304, pp. 83–86.

31. Grussenmeyer, P.; Landes, T.; Voegtle, T.; Ringle, K. Comparison methods of terrestrial laser scanning, photogrammetry and tachometry data for recording of cultural heritage buildings. *Int. Arch. Photogramm. Remote Sens. Spat. Inf. Sci.* **2008**, *XXXVI*, 213–218.

32. Von Schwerin, J.; Richards-Rissetto, H.; Remondino, F.; Agugiaro, G.; Girardi, G. The MayaArch3D project: A 3D webGIS for analyzing ancient architecture and landscapes. *Lit. Linguist. Comput.* **2013**, *28*, 736–753. [CrossRef]

33. Murtiyoso, A.; Grussenmeyer, P.; Guillemin, S.; Prilaux, G. Centenary of the Battle of Vimy (France, 1917): Preserving the Memory of the Great War through 3D recording of the Maison Blanche souterraine. *ISPRS Ann. Photogramm. Remote Sens. Spat. Inf. Sci.* **2017**, *IV-2/W2*, 171–177. [CrossRef]

34. Grussenmeyer, P.; Landes, T.; Alby, E.; Carozza, L. High Resolution 3D Recording and Modelling of the Bronze Age Cave "Les Fraux" in Perigord (France). *Int. Arch. Photogramm. Remote Sens. Spat. Inf. Sci.* **2010**, *XXXVIII*, 262–267.

35. IGN—Laboratoire MAP/GAMSAU. *TAPENADE Protocole: Elements D'architecture*; Laboratoire MAP/ GAMSAU: Marseille, France, 2011.

36. Wenzel, K.; Rothermel, M.; Fritsch, D.; Haala, N. Image acquisition and model selection for multi-view stereo. *Int. Arch. Photogramm. Remote Sens. Spat. Inf. Sci.* **2013**, *XL*, 251–258. [CrossRef]

37. Murtiyoso, A.; Grussenmeyer, P. Documentation of heritage buildings using close-range UAV images: Dense matching issues, comparison and case studies. *Photogramm. Rec.* **2017**, *32*, 206–229. [CrossRef]

38. Grussenmeyer, P.; Hanke, K.; Streilein, A. Architectural Photogrammety. In *Digital Photogrammetry*; Kasser, M., Egels, Y., Eds.; Taylor & Francis: Milton Park, UK, 2002; pp. 300–339.

39. Chiabrando, F.; Donadio, E.; Rinaudo, F. SfM for orthophoto generation: A winning approach for cultural heritage knowledge. *Int. Arch. Photogramm. Remote Sens. Spat. Inf. Sci.* **2015**, *XL-5/W7*, 91–98. [CrossRef]

40. Cefalu, A.; Abdel-wahab, M.; Peter, M.; Wenzel, K.; Fritsch, D. Image based 3D Reconstruction in Cultural Heritage Preservation. In Proceedings of the 10th International Conference on Informatics in Control, Automation and Robotics, Reykjavík, Iceland, 29–31 July 2013; pp. 201–205.

41. Verhoeven, G.J. Mesh Is More-Using All Geometric Dimensions for the Archaeological Analysis and Interpretative Mapping of 3D Surfaces. *J. Archaeol. Method Theory* **2016**, *35*, 1–35. [CrossRef]

42. Quattrini, R.; Malinverni, E.S.; Clini, P.; Nespeca, R.; Orlietti, E. From TLS to HBIM. High quality semantically-aware 3d modeling of complex architecture. *Int. Arch. Photogramm. Remote Sens. Spat. Inf. Sci.* **2015**, *XL-5/W4*, 367–374. [CrossRef]

43. Akbaylar, I.; Hamamcioglu-Turan, M. Documentation of a Vernacular House With Close-Range Digital Photogrammetry. In Proceedings of the XXI International CIPA Symposium, Athens, Greece, 1–6 October 2007.

44. Brown, N.; Laing, R.; Scott, J. The doocots of Aberdeenshire: An application of 3D scanning technology in the built heritage. *J. Build. Apprais.* **2009**, *4*, 247–254. [CrossRef]

45. Lu, Y.C.; Shih, T.Y.; Yen, Y.N. Research on Historic BIM of Built Heritage in Taiwan -a Case Study of Huangxi Academy. *Int. Arch. Photogramm. Remote Sens. Spat. Inf. Sci.* **2018**, *XLII-2*, 615–622. [CrossRef]

46. Tsai, F.; Chang, H.; Lin, Y.W. Combining 3D Volume and Mesh Models for Representing Complicated Heritage Buildings. *Int. Arch. Photogramm. Remote Sens. Spat. Inf. Sci.* **2017**, *XLII-2/W5*, 673–677. [CrossRef]

47. Suwardhi, D.; Menna, F.; Remondino, F.; Hanke, K.; Akmalia, R. Digital 3D Borobudur—Integration of 3D Surveying and Modeling Techniques. *Int. Arch. Photogramm. Remote Sens. Spat. Inf. Sci.* **2015**, *XL-5/W7*, 417–423. [CrossRef]

48. Hanan, H.; Suwardhi, D.; Nurhasanah, T.; Bukit, E.S. Batak Toba Cultural Heritage and Close-range Photogrammetry. *Procedia Soc. Behav. Sci.* **2015**, *184*, 187–195. [CrossRef]

 International Journal of *Geo-Information*

Article

Out of Plumb Assessment for Cylindrical-Like Minaret Structures Using Geometric Primitives Fitting

Bashar Alsadik [1,2,*], Nagham Amer Abdulateef [2] and Yousif Husain Khalaf [3]

[1] International Committee of Architectural Photogrammetry CIPA, 7500–7549 Enschede, The Netherlands
[2] Department of Surveying, College of Engineering, University of Baghdad, Baghdad 10071, Iraq; nagham.amer1984@gmail.com
[3] Department of Water Resources, College of Engineering, University of Baghdad, Baghdad 10071, Iraq; yousif.hussain1976@gmail.com
* Correspondence: bsalsadik@gmail.com

Received: 21 December 2018; Accepted: 27 January 2019; Published: 29 January 2019

Abstract: Cultural heritage documentation and monitoring represents one of the major tasks for experts in the field of surveying, photogrammetry and geospatial engineering. Cultural heritage objects in countries like Iraq and Syria have suffered from intentional destruction or demolition during the last few years. Furthermore, many heritage sites in the mentioned places have an added religious value, and were either destroyed or are still in danger. Mosques, churches and shrines typically include one or multiple tower structures, and these towers or minarets are in many cases cylindrical-like objects. Because of their tall and relatively thin body, and adding in their age of construction, observing their inclination or out of plumb is of high importance. Accordingly, it is highly necessary for the continuous monitoring and assessment of their preservation and restoration. In this paper, we suggest an out of plumb assessment procedure using a geometric primitives least squares fitting technique, namely, cylinders, cones, and 3D circles. The approach is based on reconstructing a dense point cloud of the minaret tower which is scaled to reality by control points. Accordingly, the out of plumb is computed by fitting one of the mentioned 3D primitives to the minaret point cloud where its major axis orientation is computed. Two experimental tests of heritage objects in Iraq are presented: the lost heritage of the minaret al Hadbaa in the city of Mosul (1173 AD) and an existing inclined minaret of the religious shrine of Imam Musa AlKadhim in Baghdad (1058 AD). The results show the efficiency of the suggested methodology where the out of plumb is computed as 0.45m±1cm for the shrine minaret and 1.90m±10cm for the model of the minaret al Hadbaa.

Keywords: out of plumb; close range photogrammetry; cultural heritage preservation; cylinder fitting; point cloud; lost heritage

---

1. Introduction

The preservation of cultural heritage objects requires regular monitoring and rehabilitation work. Normally, experts in the field of surveying, photogrammetry and geospatial engineering take the lead in such monitoring because of the high precision and reliability needed to conduct the measurements. Valuable cultural heritage sites in countries like Iraq and Syria have suffered from intentional destruction or demolition during the last few years. Mosques, churches and shrines are at the top of the list of heritage objects that need attention because of their added religious and spiritual value to the people living there [1]. These religious cultural heritage objects typically include one or more tower structures, and these towers or minarets are in many cases cylindrical-like objects.

Minarets are vertical elongated structures with a relatively small base diameter. This, in addition to their construction material types, non-robust foundations, and their age of construction, makes observing their inclination or out of plumb of high importance for their preservation and maintenance.

Technically, out of plumb measuring and monitoring techniques can be categorized as:

-   Geodetic techniques: these techniques are the most accurate, where several control targets are placed on specific preselected locations of the study object. Total station instruments are used to fix the coordinates of the control points and baselines with millimetric accuracy in a predesigned network [2]. Frequently, GPS is also used for absolute positioning of the control network. This technique is based on measuring sparse target control points on the structure surface body (Figure 1a). The cylindrical axis is then defined at multiple heights from the best fit circles of the points. Other old techniques included using theodolite instruments, whereby angles are halved to define the cylindrical axis (Figure 1b) at multiple heights.

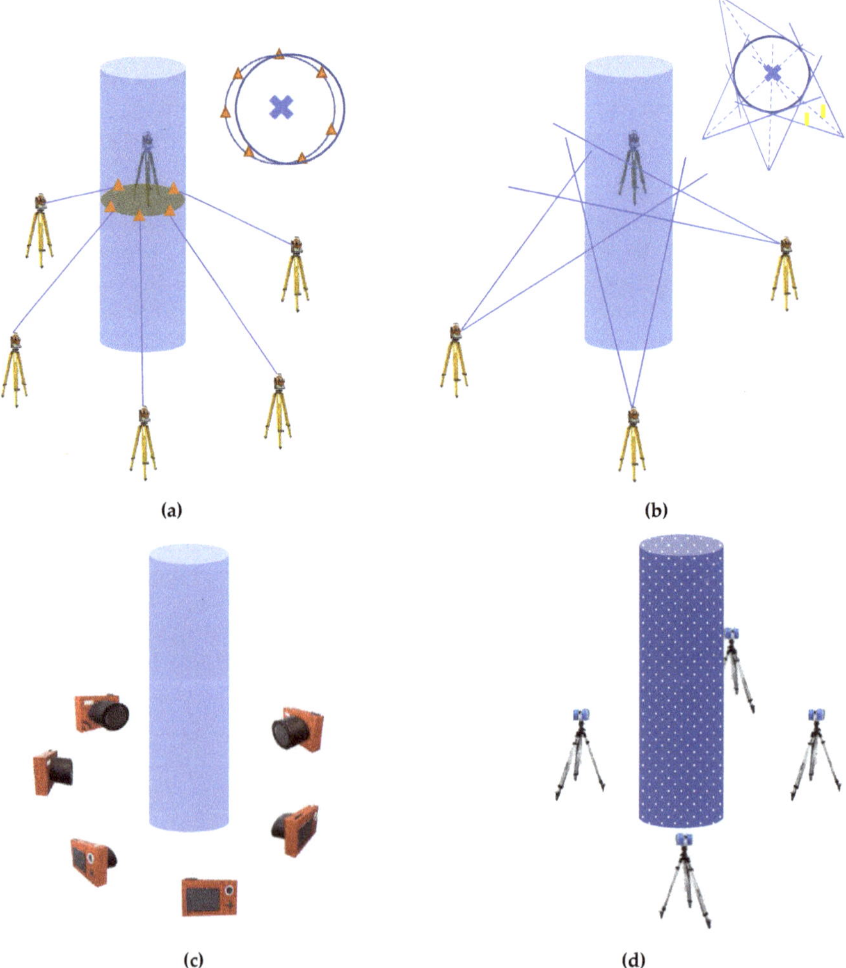

Figure 1. Different out of plumb monitoring techniques. (**a**) Geodetic approach using target point measurements. (**b**) Geodetic approach using angular measurements. (**c**) Close range photogrammetric approach. (**d**) TLS approach.

In recent years, geodetic techniques for tall structures like towers and skyscrapers have started using a combined sensor-assisted approach [3]. As an example, Leica Geosystems has developed the Core Wall Control Survey System (CWCS), which uses GPS sensors combined with high-precision inclination sensors and total stations to deliver precise and reliable coordinates [4].

- Laser scanning techniques TLS: these techniques are applied normally from terrestrial-based time of flight (TOF) ranging scanners or phase-based laser scanners (Figure 1d). The TLS approach [5–10] has the advantage of high positioning accuracy, while it depends on the intrinsic noise of the instrument and how good the coregistration of the multiple scanning point clouds is. Compared to image-based approaches, the TLS is higher in cost, heavier, and requires a level of proficiency.
- Photogrammetric techniques: these techniques are applied using image-based methods (Figure 1c). These techniques [11,12] range from centimetric to sub-centimetric accuracy depending on the type of camera, image network preplanning, accessibility around the object, illumination, etc.

In this paper, we use photogrammetric image-based methods to create the minaret point clouds, because they have the following advantages:

- They have lower cost compared to the geodetic and laser scanning approaches.
- They are portable and can be applied in situ without disturbing site visitors.
- The available state-of-the-art software tools for automated image-based 3D modeling [13–15].
- The possibility to reconstruct lost heritage models using archived and crowdsourced images and videos.

Accordingly, this ease of use, highly detailed reconstruction of the objects, and the permitted accuracy mentioned above comprise the advantages of using this photogrammetric technique for the out of plumb assessment of heritage minarets. In Section 2, below, we introduce three possible techniques of the geometric primitive fitting for the out of plumb calculations. Then, two experimental tests will be discussed in Section 3 for the out of plumb assessment of cultural heritage minaret structures.

2. Out of Plumb Computation Method

To calculate the out of plumb inclination of a cylindrical-like minaret body, we suggest a geometric primitive fitting, namely a cylinder, cone or a 3D circle fitting.

2.1. Cylinder Fitting

There are different methods of least squares fitting of a cylindrical shape [16–18]; either linear- or nonlinear-based solutions. Normally, seven parameters are used to define a cylinder (Figure 2):

- Point of origin lying on the cylinder axis, defined as xo, yo, zo.
- Cosine direction normal a, b, c of the cylinder axis.
- Radius of the cylinder R.

In this subsection, we will introduce a modified approach in computing the unknown parameters xo, yo, zo, R and normal cosine direction [a, b, c] of the cylinder axis using nonlinear least squares adjustment as follows:

- The first derivation step is to apply the cross product between the vector derived from the coordinate differences and the cosine direction as:

$$E = \begin{bmatrix} a \\ b \\ c \end{bmatrix} \times \begin{bmatrix} x - xo \\ y - yo \\ z - zo \end{bmatrix} \rightarrow d = \| E \| \tag{1}$$

where $\times$ refers to cross product and $\|\ \|$ refers to norm.

Therefore, the objective function is to minimize the difference between the distances from the fitting points to the cylinder axis and its radius R as illustrated in Figure 2.

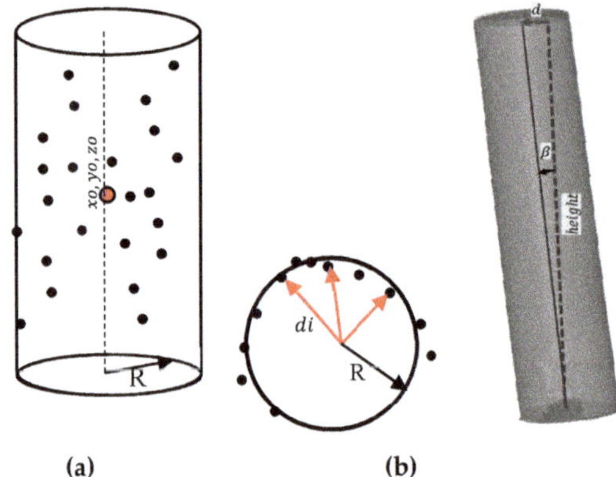

(a)

(b)

Figure 2. Fitting points to a cylinder primitive and the out of plumb illustration. (**a**) Fitting of a cylinder defined by seven parameters. (**b**) Out of plumb illustration.

Or

$$EF = \min. \sum (di - R)^2 \tag{2}$$

The observation equation $V = B\Delta - F$ can be formed for n points as follows:

$$\underbrace{\begin{bmatrix} v_1 \\ v_2 \\ \vdots \\ v_n \end{bmatrix}}_{V_{n*1}} = \underbrace{\begin{bmatrix} \frac{\partial F_1}{\partial a} & \frac{\partial F_1}{\partial b} & \frac{\partial F_1}{\partial c} & \frac{\partial F_1}{\partial R} & \frac{\partial F_1}{\partial xo} & \frac{\partial F_1}{\partial yo} & \frac{\partial F_1}{\partial zo} \\ \vdots & \vdots & \vdots & \vdots & \vdots & \vdots & \vdots \\ \frac{\partial F_n}{\partial a} & \frac{\partial F_n}{\partial b} & \frac{\partial F_n}{\partial c} & \frac{\partial F_n}{\partial R} & \frac{\partial F_n}{\partial xo} & \frac{\partial F_n}{\partial yo} & \frac{\partial F_n}{\partial zo} \end{bmatrix}}_{B_{n*7}} \underbrace{\begin{bmatrix} \Delta a \\ \Delta b \\ \Delta c \\ \Delta R \\ \Delta xo \\ \Delta yo \\ \Delta zo \end{bmatrix}}_{\Delta_{7*1}} - \underbrace{\begin{bmatrix} F_1 \\ F_2 \\ \vdots \\ F_n \end{bmatrix}}_{F_{n*1}} \tag{3}$$

where

Δ: vector of corrections to fitting parameters.
V: vector of residual errors.

Furthermore, to avoid the rank deficiency in the partial derivatives B matrix, two constraint equations C_1 and C_2 are added to the system as follows:

1st constraint: the cosine direction vector a, b, and c should be normalized as $(a^2 + b^2 + c^2 = 1)$, Therefore, the constraint equation matrix C_1 will be:

$$C_1 = \begin{bmatrix} 2a & 2b & 2c & 0 & 0 & 0 & 0 \end{bmatrix}, \ g_1 = \begin{bmatrix} 1 - a^2 - b^2 - c^2 \end{bmatrix} \tag{4}$$

2nd constraint: the second constraint equation that should be added is based on the dot product between the cosine direction a, b, and c of the cylinder axis and the origin center point 0, or:

$$a.xo + b.yo + c.zo = 0 \tag{5}$$

Therefore:

$$C_2 = \begin{bmatrix} xo & yo & zo & a & b & c & 0 \end{bmatrix}, \quad g_2 = -[a.xo + b.yo + c.zo] \tag{6}$$

Accordingly, Helemert's method is used to solve the constrained problem as follows [19]:

$$\begin{bmatrix} N & C^t \\ C & 0 \end{bmatrix} \begin{bmatrix} \Delta \\ k_c \end{bmatrix} = \begin{bmatrix} t \\ g \end{bmatrix} \text{ where } C = \begin{bmatrix} C_1 \\ C_2 \end{bmatrix}, g = \begin{bmatrix} g_1 \\ g_2 \end{bmatrix}, N = B^t B, \text{ and } t = B^t F \tag{7}$$

Because of the large number of fitting points, it is expected to have some amount of noisy points, and as a result, a possibly error-prone estimation of the parameters. Therefore, the detection of blundered points using a voting method is applied in the form of the well-known "RANdom SAmple Consensus" (RANSAC) [20,21]. In the RANSAC algorithm, we assume the observational data to include both: inliers and outliers. Inliers can be described as observations without blunders with respect to a specific model, while outliers do not fit that model in any condition. It should be noted that the RANSAC algorithm can find adjustment models even if 50% of the input data set is affected by noise points. On the other hand, there is no guarantee of reaching the optimal solution if we limit the number of iterations.

2.2. Cone Fitting

A similar approach of least squares fitting can be applied to the cone shape [16]. Normally, eight parameters are used to define a cone (Figure 3):

- Point of origin lying on the cone axis and defined as xo, yo, zo.
- Cosine direction numbers a, b, c of the cone axis (pointing to the apex).
- Orthogonal distance from the point of origin to cone s.
- The cone's apex semi-angle θ.

The least squares adjustment technique will be applied to minimize the computed distance d between the cone axis and the fitting points as:

$$d = f_i \cos\theta + g_i \sin\theta - s \tag{8}$$

where

$$f_i = \sqrt{u^2 + v^2 + w^2} \tag{9}$$

$$\begin{aligned} u &= c(y_i - y) - b(z_i - z) \\ v &= a(z_i - z) - c(x_i - x) \\ w &= b(x_i - x) - a(y_i - y) \end{aligned} \tag{10}$$

$$g_i = a(x_i - x) + b(y_i - y) + c(z_i - z) \tag{11}$$

Therefore, the objective function is to minimize the distances from the fitting points to the cone surface, as illustrated in Figure 3.

Or

$$F = \min. \sum (di)^2 \tag{12}$$

The observation equation $V = B\Delta - F$ can be formed for n points as follows:

$$
\begin{bmatrix} v_1 \\ v_2 \\ \vdots \\ v_n \end{bmatrix}_{V_{n*1}} = \underbrace{\begin{bmatrix} \frac{\partial F_1}{\partial a} & \frac{\partial F_1}{\partial b} & \frac{\partial F_1}{\partial c} & \frac{\partial F_1}{\partial s} & \frac{\partial F_1}{\partial xo} & \frac{\partial F_1}{\partial yo} & \frac{\partial F_1}{\partial zo} & \frac{\partial F_1}{\partial \theta} \\ \vdots & \vdots & \vdots & \vdots & \vdots & \vdots & \vdots & \vdots \\ \frac{\partial F_n}{\partial a} & \frac{\partial F_n}{\partial b} & \frac{\partial F_n}{\partial c} & \frac{\partial F_n}{\partial s} & \frac{\partial F_n}{\partial xo} & \frac{\partial F_n}{\partial yo} & \frac{\partial F_n}{\partial zo} & \frac{\partial F_n}{\partial \theta} \end{bmatrix}}_{B_{n*8}} \underbrace{\begin{bmatrix} \Delta a \\ \Delta b \\ \Delta c \\ \Delta s \\ \Delta xo \\ \Delta yo \\ \Delta zo \\ \Delta \theta \end{bmatrix}}_{\Delta_{8*1}} - \underbrace{\begin{bmatrix} F_1 \\ F_2 \\ \vdots \\ F_n \end{bmatrix}}_{F_{n*1}}
\tag{13}
$$

Similar geometric constraints mentioned in the cylinder fitting problem are used to run a full rank least squares adjustment (Equation (7)) of the best cone fitting.

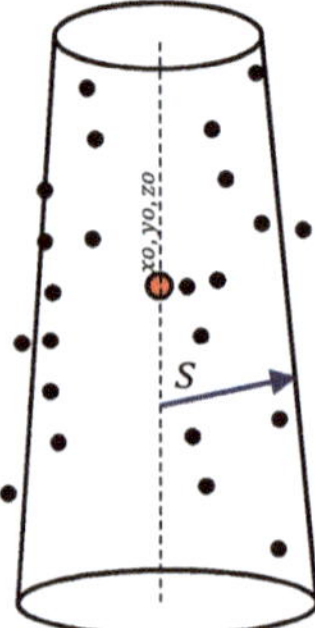

Figure 3. Fitting points to a cone defined by eight parameters.

After the calculation of the fitting parameters either for the cylinder primitive or the cone, the angular orientation of the cylinder axis is computed as follows:

$$
\begin{aligned}
\alpha &= \cos^{-1}(a/\sqrt{a^2 + b^2 + c^2} \\
\beta &= \cos^{-1}(b/\sqrt{a^2 + b^2 + c^2} \\
\gamma &= \cos^{-1}(c/\sqrt{a^2 + b^2 + c^2}
\end{aligned}
\tag{14}
$$

However, for such vertical primitive objects, we only need the γ inclination angle to compute the out of plumb as follows (Figure 2b):

$$
\text{Out of plumb} = \gamma \times \text{height}
\tag{15}
$$

Whereas α and β are used to determine the out of plumb direction if needed. The standard deviation of the out of plumb in Equation (15) can be computed using the propagation of errors technique. However, tests showed that the computed standard deviations of the cosine directions are of small magnitude, and the out of plumb accuracy is mainly related to the estimated height accuracy. Accordingly, the estimated accuracy of the out of plumb in the following experimental tests are assumed equivalent to the accuracy of the primitives fitting of the point clouds.

2.3. Circle Fitting in 3D space

Fitting a circle in 3D space (Figure 4) needs a multistep solution approach, as follows [16,22]:

1. Compute the best fit least squares plane of the data as explained previously.

2.   Rotate the points such that the least squares plane is the XY plane.
3.   Rotate data points onto the XY plane.
4.   Compute the 2D circle fit in the XY plane.
5.   Rotate back to the original orientation in 3D space.

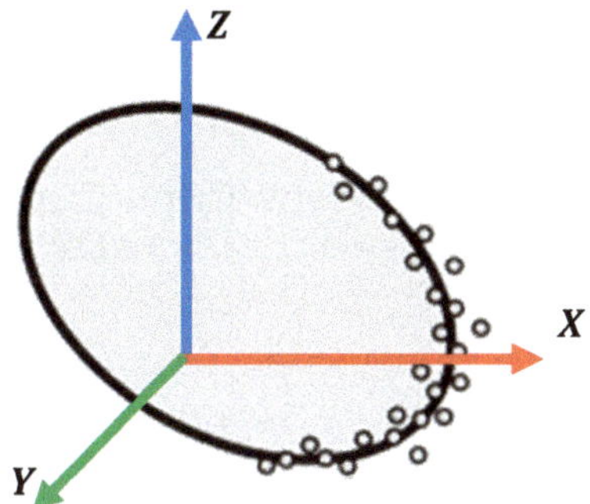

Figure 4. 3D circle fit.

2.3.1. Fitting a 2D Circle

The least squares adjustment of a circle best fit in a 2D space is shown in Figure 5 as follows:

Given: XY points, i = 1:n
Required: the adjusted radius r, and circle center x_c, y_c

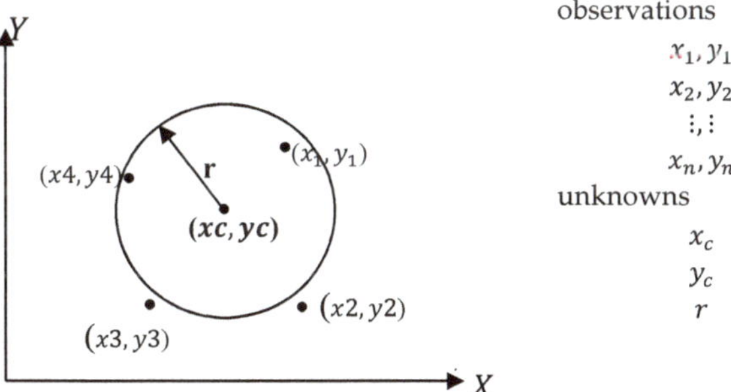

Figure 5. 2D circle fit.

Circle equation can be formulated as:

$$(x - x_c)^2 + (y - y_c)^2 - r^2 = 0 \tag{16}$$

The observation equation will be as follows in the form $V = B\Delta - F$

$$
\begin{bmatrix} v_{x1} \\ v_{y1} \\ v_{x2} \\ \vdots \\ v_{yn} \\ v_{zn} \end{bmatrix}
=
\begin{bmatrix} \frac{\partial F_1}{\partial x_c} & \frac{\partial F_1}{\partial y_c} & \frac{\partial F_1}{\partial r} \\ \vdots & \vdots & \vdots \\ \frac{\partial F_n}{\partial x_c} & \frac{\partial F_n}{\partial y_c} & \frac{\partial F_n}{\partial r} \end{bmatrix}
\begin{bmatrix} \delta x_c \\ \delta y_c \\ \delta r \end{bmatrix}
-
\begin{bmatrix} r^2 - (x_1 - x_c^o)^2 - (x_1 - y_c^o)^2 \\ \vdots \\ r^2 - (x_n - x_c^o)^2 - (x_n - y_c^o)^2 \end{bmatrix}
\tag{17}
$$

The partial derivatives of the function to the unknowns will be as follows:

$$\begin{aligned} \frac{\partial F_1}{\partial x_c} &= -2(x - x_c^o) \\ \frac{\partial F_1}{\partial y_c} &= -2(y - y_c^o) \\ \frac{\partial F_1}{\partial R} &= -2r^o \end{aligned} \tag{18}$$

Then a least squares adjustment is performed as $\Delta = \left(B^t B\right)^{-1}\left(B^t F\right)$.

2.3.2. Rodrigues Rotation Formula

To apply the 2nd step of rotating the points to the best fit plane of points, we can utilize the Rodrigues rotation formula to rotate 3D points and get their XY coordinates in the coordinate system of the plane. This is simply a problem of rotating from normal vector n1 to normal n2, as shown in Figure 6.

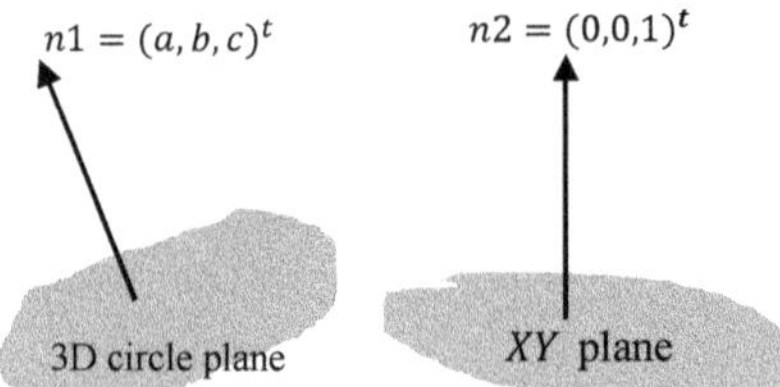

Figure 6. Rotating from normal vector n1 to normal n2.

To compute the rotation matrix, we will apply the following steps:

1. Find the axis and angle of rotation using cross product and dot product respectively. The axis of rotation k is a cross product ($\times$) between plane normal n1 and the normal of the new XY coordinates. Thus, $n2 = (0,0,1)^t$ and $k = n1 \times n2$. Furthermore, $\theta = \cos^{-1}\left(\frac{\vec{n1}.\vec{n2}}{\|\vec{n1}\|.\|\vec{n2}\|}\right)$

2. Find the rotation matrix M using exponential map:

$$M = I_{3*3} + \hat{k}\sin(\theta) + \hat{k}^2(1 - \cos\theta) \tag{19}$$

where

-   Skew—symmetric matrix $\hat{k} = \begin{bmatrix} 0 & k(3) & k(2) \\ k(3) & 0 & k(1) \\ k(2) & k(1) & 0 \end{bmatrix}$

-   $k = (n1 \times n2)$, then normalized.

The 3rd step after rotation of points is to find the best fit circle Cpoints in 2D space (xc, yc, r). Finally, rotate back to 3D space by taking n2 as the plane normal and $n1 = (0,0,1)^t$

$$C = Cpoints - [xc, yc, 0] \tag{20}$$

$$C = C * M' + \text{Points\_mean} \tag{21}$$

3. Results and discussion

Two experiments were conducted at Iraqi cultural heritage sites using the suggested geometric primitive fitting. The first experiment was carried out on the inclined minaret of the Imam AlKadhim shrine in Baghdad, and the second experiment was carried out on the lost al Hadbaa minaret in Mosul.

3.1. First Experiment

The Imam Musa AlKadhim shrine was first constructed in 1058 AD (450 Hijri), and included a mosque and one minaret; in 1098 AD, two further minaret structures were added. Then, during the rule of caliph Al Mustansir Billah in 1226 AD, the whole shrine structure was completed [23,24]. In 1508 AD, Shah Ismail Safavi added two more minarets to become four, as is still the case in our current time, and as shown in Figure 7 with two domes beside the four minaret structures.

Figure 7. Shrine of Imam Musa AlKadhim in Baghdad, Iraq with its four minaret structures.

Each minaret has a foundation 11 m in the ground and an elevation 28 m above the ground, which means a total of 39 m from the base to top. In 2013, an out of plumb increase was observed in the minaret structures, in addition to some cracks in the walls, both inside and outside. This indicated a problem with the site foundations and the soil due to the groundwater and the renovations previously applied at the shrine site.

Therefore, there was a need to ensure the visitors' safety and to find suitable construction engineering solutions to resolve the problems with the foundations and the groundwater. Furthermore, monitoring the minarets' stability, calculating the out of plumb, and fixing the wall cracks in the shrine structures was necessary. This was achieved by an extensive project which included TLS scanning of the whole shrine site in addition to geodetic monitoring of the minaret inclinations [24].

Conventionally, geodetic monitoring techniques are applied through multiple epoch measurements of the questionable minarets of the heritage site. Total station measurements are applied to selected critical points on the cylindrical body of each minaret, which comprise circular planes at multiple selected elevations. Then for each circular plane/section, the center coordinates are determined by applying the least squares adjustment for fitting a 2D circle.

The monitoring of the shrine site was carried out from May 2014 to August 2015 using the described geodetic methods, and with photogrammetric techniques, as well. The applied photogrammetric techniques were based on manually measuring the interest points of the minaret using an SLR camera at individual strips around the four minarets [24].

For both measuring techniques, 2500 monitoring object points were measured in the designed multi planes to compute their center points, defining the vertical axis of the minaret structures. It was observed that two minaret structures were stable, while the inclination was critical for the two other minarets [24].

In this paper, the inclination of one of the suspected renovated minaret structures was determined after three years of treatment to check its stability and whether there is still a critical out of plumb in its structure.

3.1.1. Close-Range Photogrammetry

First, a pre-analysis of the camera network around the minaret structure was carried out by reconnaissance and checking accessibility spaces. Unfortunately, it was only possible to capture the image from the ground, and no drone photogrammetry was permitted at the shrine site during the daytime. Therefore, it was decided to capture the images in a highly overlapped strip of images. Images were captured using a DSLR Canon EOS Mark III camera with a fixed focal length of 24 mm and an image frame size of 5760 × 3840 pixels. An average capture distance of 6 m was designated in order to guarantee a depth accuracy of 5 mm with an average ground sample distance (GSD) of 2 mm, as shown in Equation (22) [25].

$$Z = \sqrt{3fB\frac{\sigma_Z}{p}} \tag{22}$$

where

σ_Z = precision at object space [mm]
f = focal length [mm]
B = base line (step size) [mm]
p = pixel pitch (width) [mm]

However, higher parts of the minaret were expected to have larger depth accuracy and GSD.

Practically, a total of 450 images were captured in two rings around the minaret structure in different portrait and landscape orientations. To avoid the long processing time required to create the dense point cloud in our laptop machine, and to perform an efficient imaging network configuration for future image capture, we applied the minimal camera network approach presented in [26].

The minimal camera network filtering technique is based on the concept of having at least three images (cameras) viewing the object point simultaneously. The approach starts by labeling the derived sparse point cloud of the object as either over-covered or fair-covered. Over-covered points are the points that appear in more than three images, while fair-covered points, refer to the points that typically appear in three cameras. Therefore, the images are considered redundant and are filtered out if they exclusively contribute points that are covered by more than three cameras, as shown in Figure 8d. The camera filtering is allowed if it maintains the desired small Base/Distance ratio. The filtering is repeated until no more redundant cameras are involved in imaging over-covered points. Accordingly, the number of images around the shrine minarets is reduced from 450 to 169 images (Figure 8a,b) and oriented using the Metashape software tool [13].

The quality of the reconstructed minaret object is evaluated by computing the variance-covariance matrix of the tie points using a separate code. The quality computations are applied through the bundle adjustment of the images and the calculation of the mean of the error ellipsoid axes for each tie point. Then every point is colored with respect to the computed mean error, as shown in Figure 8c. Table 1 summarizes the camera specifications and the imaging network orientation quality.

Table 1. Camera specifications and image orientation quality of the first experiment.

Image Orientation Quality		Adjusted Camera Parameters	
GSD [mm]	1.5	Focal length [mm]	23.88
Pixel size [µm]	6.0	p.p in x [mm]	0.113261
Maximum reprojection error [pixel]	1.36	p.p in y [mm]	0.131599
Median reprojection error [pixel]	0.23	Radial lens distortion k1	−0.133579
Mean reprojection error [pixel]	0.37	Radial lens distortion k2	0.092059

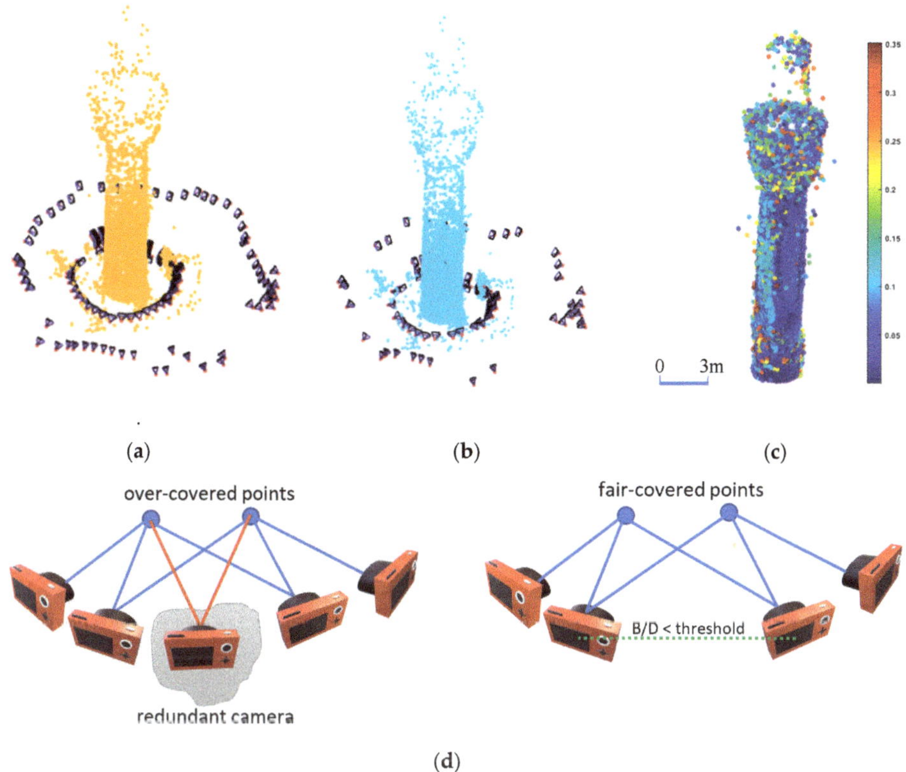

Figure 8. Imaging network around the shrine minaret. (**a**) Full imaging network with 450 images. (**b**) Filtered imaging network with 169 images. (**c**) Positional quality estimation represented in colors in cm. (**d**) The concept of filtering a redundant camera. (left) before filtering, (right) after filtering.

Mathematically, a minimum of three control points is sufficient to geo-reference the imaging network and, subsequently, the derived point cloud. However, adding more control points is necessary to reliably check the quality of the orientation and the produced point cloud. In this experiment, it was difficult to fix more than four control points because of the limitations regulated by the shrine authorities during the field work.

The root mean squared error (RMSE) of the four control points we used is computed as: $\sigma x = 1.2$ mm, $\sigma y = 2.4$ mm and $\sigma z = 1.6$ mm, respectively.

Afterward, the highly dense 3D point cloud of ≈ 3 million points was reconstructed, as shown in Figure 9a. A density analysis showed that the average number of points was $15,300 \pm 4150$ points/m^2, which is a high level of density that will certainly give a reliable 3D representation of the minaret (Figure 9b).

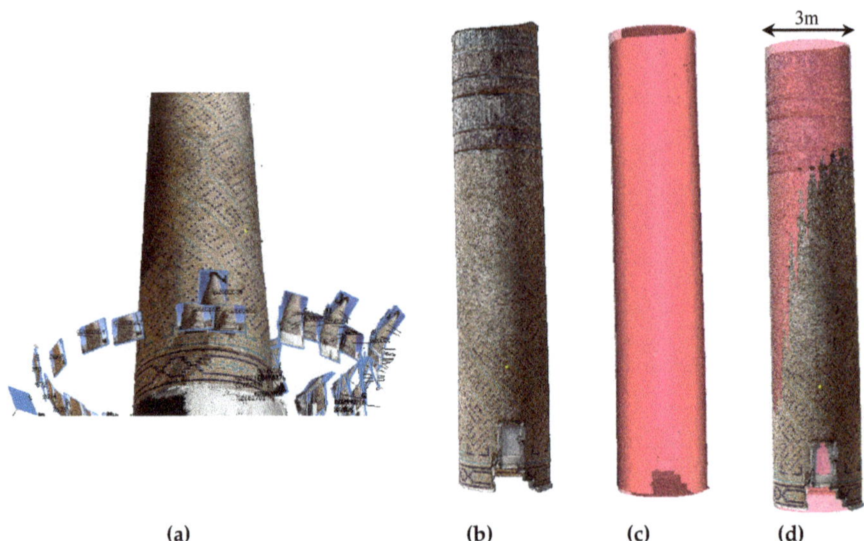

(a)   (b)   (c)   (d)

Figure 9. The produced image-based point cloud of the shrine minaret. (**a**) Minaret body with the oriented images around it. (**b**) Dense point cloud of the minaret. (**c**) Best fitting cylinder. (**d**) Fitting cylinder fused with the point cloud.

3.1.2. Geometric Primitive Fitting of the Shrine Minaret

As mentioned, a geometric cylinder and cone fitting is applied to the minaret point cloud to estimate the out of plumb of the minaret axis. Figure 9c shows the best fitting cylinder, and in Figure 9d, the best fitting cylinder is fused with the minaret point cloud. Using the formulas in Section 2.1, the least squares solution is converged to optimal values (Figure 10).

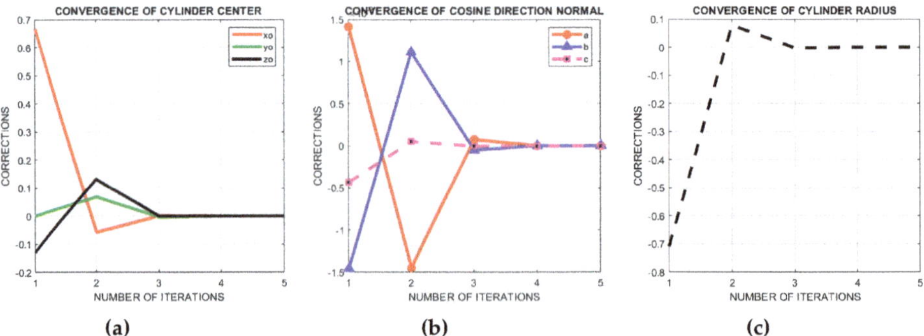

(a)   (b)   (c)

Figure 10. Nonlinear least squares cylinder fitting parameters converged to optimal.

Geometrically, it is clear that the minaret body is not a perfect cylinder, because whenever the points are higher, the diameter is slightly decreased. Therefore, a more efficient fitting is applied either by assuming a cone minaret body for the whole point cloud or by fitting multiple cylinders in elevated adjacent sections.

Originally, the out of plumb computations using Equation (15) are applied using a cylinder and a cone fitting for the whole point cloud, and the results are listed as follows in Table 2, where the computed out of plumb is shown:

Table 2. Fitting the whole point cloud with a cylinder and a cone primitive.

	xo [m]	yo [m]	zo [m]	a	b	c	Out of Plumb [m]
Cylinder fitting	136.94	114.89	31.37	0.019216	−0.019097	0.99963	0.43
Cone fitting	136.94	114.89	31.37	0.020305	−0.019289	0.99961	0.44

To evaluate the goodness of the primitive fitting, we compute the distance between every point and both the cylinder and the cone surfaces, respectively. The results are shown where the mean distance is ±2.3 cm from the fitting cylinder surface (Figure 11a), while it is ±1.4 cm from the fitting cone surface (Figure 11b).

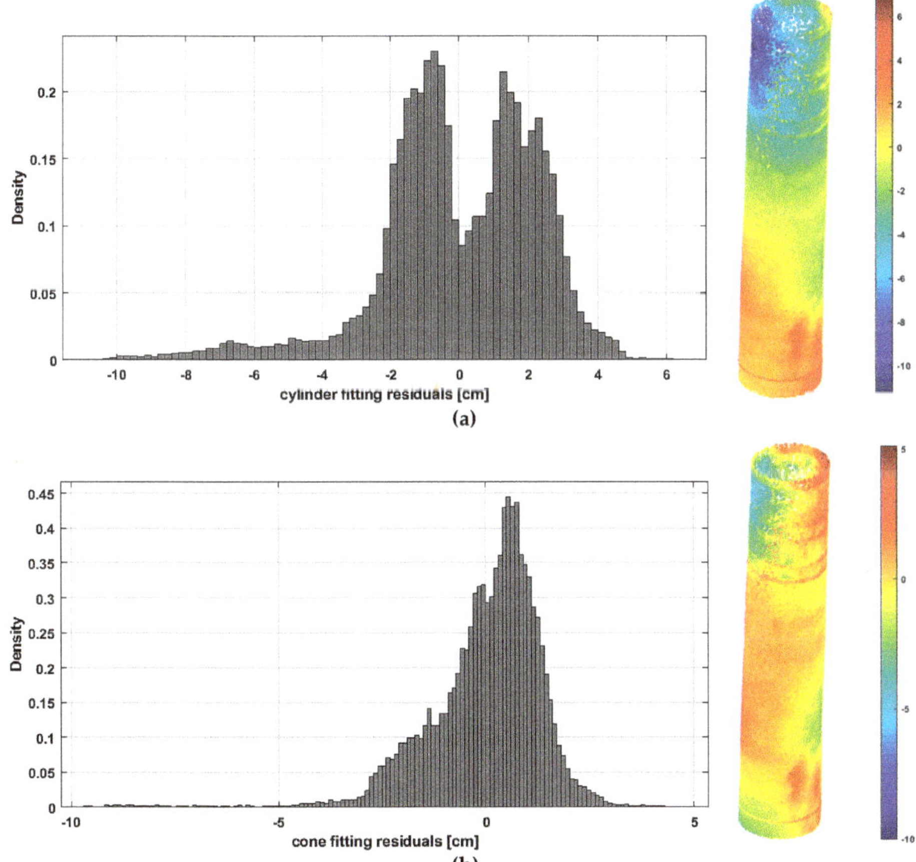

Figure 11. Cylindrical and conical geometric primitive fitting results. (**a**) The distance between the point cloud and the best fit cylinder surface (σ = ±2.3 cm). (**b**) The distance between the point cloud and the best fit conic surface (σ = ±1.4 cm).

The third fitting option is to use a cylinder or a cone fitting at multiple elevations. Accordingly, multiple height sections with a width of 1 m are selected from the point cloud and followed by a cylindrical or conical fitting applied for each level.

Table 3 illustrates the seven parameters of the fitting cylinders in the multi elevation sections shown in Figure 12a. The last column in Table 2 shows the computed out of plumb inclination between every section and the minaret base using the planimetric coordinates of the involved cylinder centers as in Equation (23):

$$\text{out of plumb} = \sqrt{\left(X_{top} - X_{bottom}\right)^2 + \left(Y_{top} - Y_{bottom}\right)^2} \tag{23}$$

It is shown that the computed horizontal shift between the minaret base and its top is 45 cm, which represents the out of plumb inclination of the minaret. The result is similar to the field survey measurements implemented by professional surveyors.

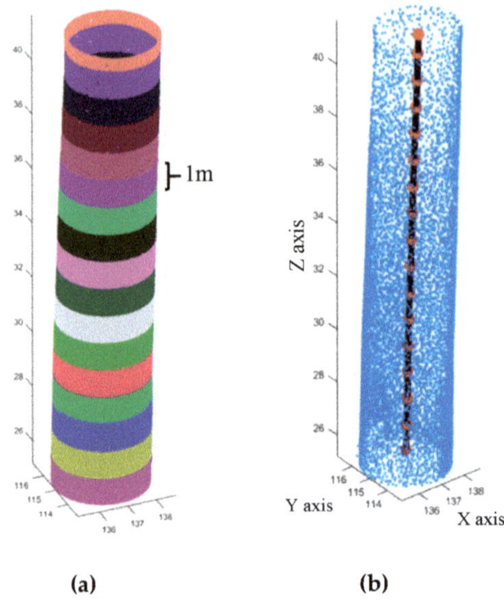

(a) (b)

Figure 12. Multi cylinder primitive fitting results of the shrine minaret. (**a**) The cylindrical minaret body is divided into multiple 1 m sections. (**b**) The barycentric coordinates at each height level is computed through a cylinder fitting.

Table 3. The best fitting cylinder parameters of the multi sections along the minaret structure.

Section	xo [m]	yo [m]	zo [m]	a	b	c	R [m]	Out of Plumb [m]
1st	136.84	114.99	25.72	0.0177	−0.0207	0.9996	1.608	-
2nd	136.86	114.97	26.69	0.0148	−0.0265	0.9995	1.592	0.03
3rd	136.87	114.96	27.69	0.0095	−0.0049	0.9999	1.594	0.04
4th	136.88	114.95	28.67	−0.0007	−0.0132	0.9999	1.585	0.06
5th	136.89	114.93	29.69	0.0193	−0.0176	0.9997	1.577	0.07
6th	136.91	114.91	30.68	0.0231	−0.0249	0.9994	1.574	0.10
7th	136.93	114.89	31.67	0.0251	−0.0241	0.9994	1.571	0.14
8th	136.96	114.87	32.67	0.0262	−0.0223	0.9994	1.568	0.17
9th	136.98	114.85	33.68	0.0210	−0.0169	0.9996	1.567	0.20
10th	137.00	114.83	34.68	0.0201	−0.0223	0.9995	1.563	0.23
11th	137.02	114.81	35.67	0.0205	−0.0154	0.9997	1.557	0.25
12th	137.05	114.79	36.69	0.0288	−0.0294	0.9992	1.552	0.29
13th	137.08	114.77	37.68	0.0305	−0.0241	0.9992	1.544	0.33
14th	137.11	114.74	38.67	0.0182	−0.0308	0.9994	1.534	0.37
15th	137.13	114.71	39.67	0.0163	−0.0229	0.9996	1.525	0.40
16th	137.15	114.69	40.71	0.0228	−0.0093	0.9997	1.515	0.43
17th	137.17	114.68	41.41	0.0074	−0.0139	0.9999	1.530	0.45

As previously stated, Table 3 shows that the radius of the cylindrical minaret tower decreases from bottom to top, and fitting the whole minaret body with one cylinder or cone would not be so accurate. Similar to Figure 11, the multi cylinder fitting is evaluated by computing the distances between the point cloud and the fitting cylinders. Then the standard deviation of 1σ is computed as ±0.8 cm, indicating the high precision of the multi cylinder fitting to the point cloud, as shown in Figure 13.

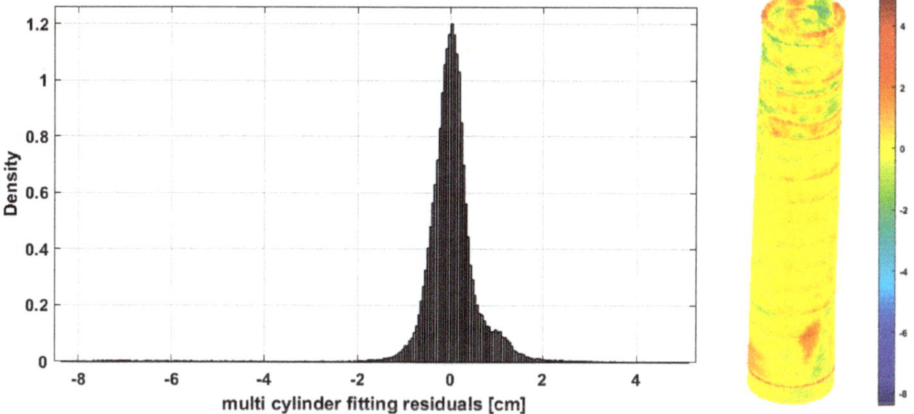

Figure 13. The distances between the multi fitting cylinders surfaces and the related point clouds.

A thorough verification of the computed out of plumb of Table 3 shows that the inclination is not truly linear, but somehow curvilinear. The best fitting polynomial is found to be cubic with a norm of residuals of 2 cm as shown in Figure 14:

$$y = p_1 x^3 + p_2 x^2 + p_3 x + p_4 \tag{24}$$

where

y: the out of plumb in [m].
x: difference of elevations [m].
$p_1 = -7.84 \times 10^{-5}$, $p_2 = 0.003$, $p_3 = 0.006$, and $p_4 = 0.037$

Obviously, the minaret was built from bricks line by line, using the techniques available at the time of construction, which are not accurate according to our current standards; this resulted in the slightly curvilinear plumb axis of the minaret.

Finally, a 3D circle fitting is applied to the selected multiple sections of the minaret point cloud and the computed out of plumb is 45 cm with a fitting residuals standard deviation of ±6 cm. The results of the four fitting scenarios show that the best approach is to apply multiple cylindrical sections fitting.

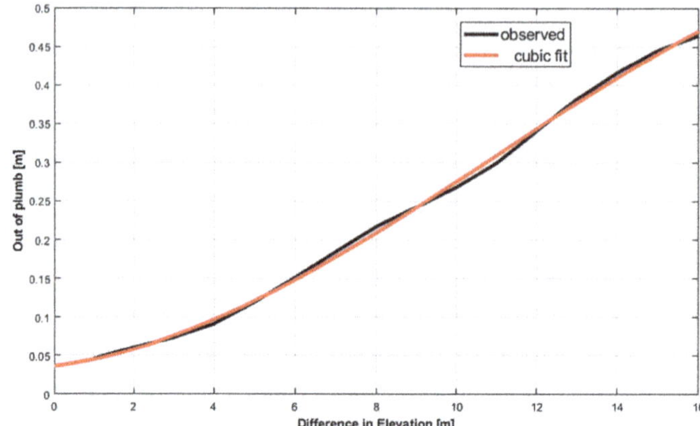

Figure 14. The curvilinear relation between the out of plumb of the minaret body and the successive difference in elevations with respect to the minaret base.

3.2. Second Experiment

The Al-Hadbaa leaning Minaret (Figure 15a) is one of the remarkable monuments in Iraq which was destroyed in 2017 (Figure 15b) during the liberation of the old city of al-Mosul from ISIS [1,27]. The minaret was established in the year 1173 AD (568 Hijri) during the Seljuk rule of Nur ad-Din Zenki.

(a)   (b)

Figure 15. Al-Nuri Grand Mosque in the Old City of Mosul, Iraq. (**a**) before destruction. (**b**) after destruction [27].

The minaret is a part of the construction of the Great Mosque or al-Nuri mosque and is known as Al-Hadbaa which means "the hunchback" in the Arabic language. Since 2003, the image of the minaret has decorated one of the banknotes in the Iraqi currency.

According to [28], the cylindrical body of the minaret, which has an elevation of 47.25 m, starts at a level of 17.31 m with a diameter of 2.62 m. The minaret consisted of three parts: the foundations; a prismoid base with dimensions of 12 m, 2 m below the ground; and the cylindrical-like minaret. The minaret is built out of brick with plaster materials, while the internal stairs inside the minaret are built of stone. The effect of the weather and the weak construction materials caused the leaning of the minaret body over the course of hundreds of years after it was established, with a lean of 2.19 m at a height of 47.25 m, as measured in 1981 [28].

Similar to the lost minaret of Aleppo in Syria [29], and based on the archived multimedia records of the minaret, engineering descriptions in the literature, and published 3D models in SketchUp 3D warehouse [30] and Sketchfab [31], a 3D image-based model was created (Figure 16). The model was scaled to reality based on the published dimensions of the minaret base of 9 × 9 meters [28]. Subsequently, a 3D point cloud was created, and the out of plumb computations were performed similarly to the procedure in the first experiment. It should be noted that the point cloud (≈700,000 points) was of moderate accuracy, since it was mainly based on low-resolution crowdsourced images available to the public. Furthermore, since this minaret object is lost, no field measurements could be applied, and the comparisons are based on the out of plumb reported previously.

Figure 16. The derived image-based 3D model (.stl) of Al-Hadbaa minaret.

From the geometric description of the minaret and the available images, it is obvious that the minaret body was not a completely regular cylinder or cone, since at higher points, the diameter is slightly decreased and leans with a curvilinear axis.

Therefore, we apply two kinds of fitting, similar to in the previous experiment: first, by assuming a conical body for the whole minaret; and second, by assuming multiple cone sections with width of 1 m for fitting.

The out of plumb computations of 1.9 m, as shown in Table 4, result from using Equation (15) for the whole point cloud cone fitting; the least squares results are shown in Figure 17:

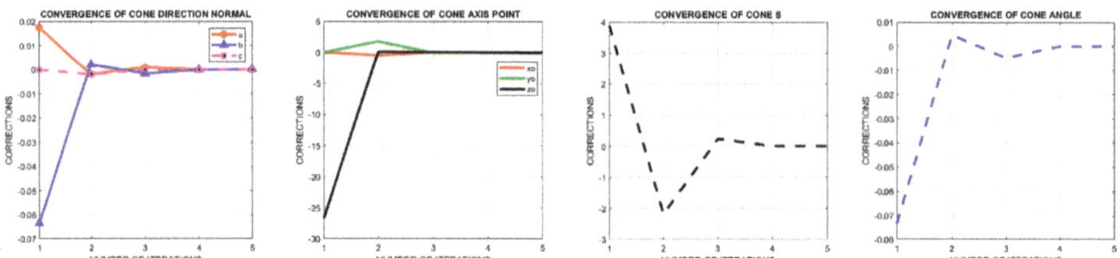

Figure 17. Nonlinear least squares cone fitting parameters converged to optimal.

Table 4. Cone fitting parameters of the whole point cloud of al Hadbaa minaret.

	xo [m]	yo [m]	zo [m]	a	b	c	Out of Plumb [m]
Cone fitting	104.76	104.58	31.477	0.016572	−0.063281	0.99786	1.9

To evaluate the goodness of the primitive fitting, we compute the distance between every point and the cone surface. The results are shown in Figure 18a, where the median of the residuals between the cone surface and the points is −27 cm and the standard deviation is ±9.7 cm of 1σ.

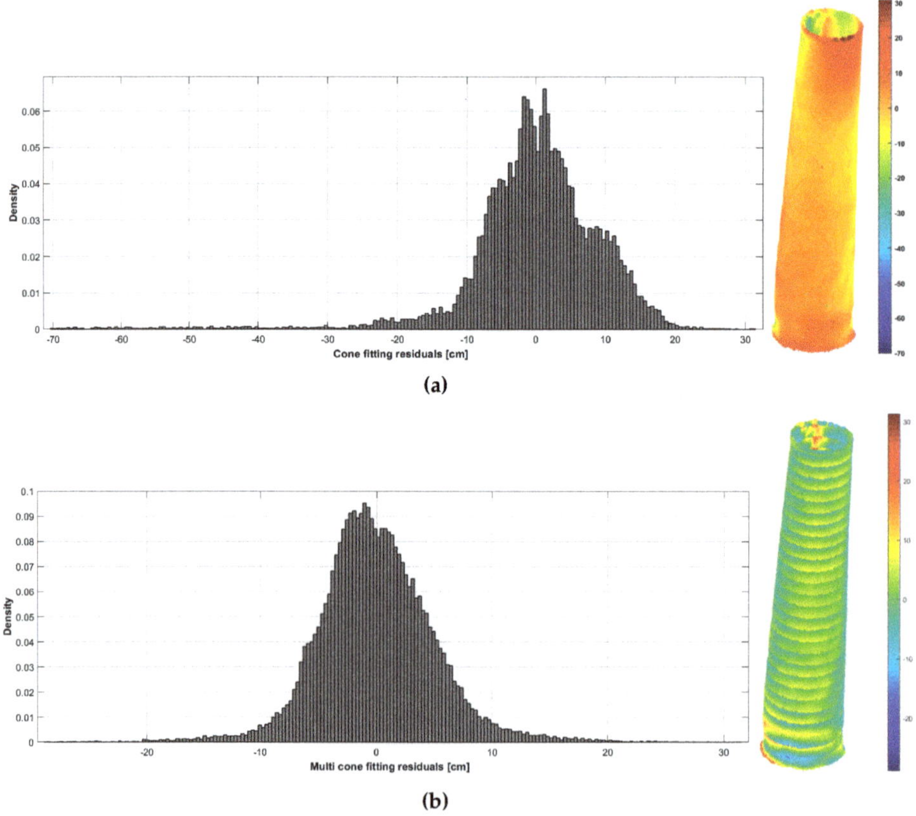

(a)

(b)

Figure 18. The conical geometric primitive fittings applied to the lost heritage minaret. (**a**) One cone fitting for the whole minaret point cloud. (**b**) Multi section cone fitting.

Meanwhile, the out of plumb is computed as 2.3 m using the multi conical section fitting (Figure 18b), with a higher precision for a zero median and ± 5 cm standard deviation of 1σ. This computed out of plumb is in agreement with the measured deviation reported in [28] using field surveying methods in 2007.

Lastly, a 3D circle fitting is applied to multiple selected sections of the minaret point cloud, and the computed out of plumb is 2.27 m with a standard deviation of the fitting residuals of ± 4.8 cm.

4. Conclusions

In this paper, we presented a geometric primitive fitting approach for calculating the out of plumb of cylindrical-like cultural heritage minaret towers. Three types of primitive fittings for the minaret point clouds are suggested, namely: cylinder, cone and 3D circle.

In the first experiment, the approach is applied with respect to the minaret at the holy shrine of Imam AlKadhem by following the photogrammetric engineering techniques of different tasks. The tasks included:

- Preplanning, image capture with a professional camera.
- Automated image orientation using state-of-the-art software tool.
- Using field survey control points for reality scaling.
- Creating a dense point cloud for 3D modeling of the minaret.
- and finally, to compute the out of plumb amount by primitive fitting of the point clouds.

Error analysis showed (Figure 8c) that the oriented images were able to meet centimetric accuracy of the modeled points of the minaret. Furthermore, the RMSE of the control points was less than half a centimeter. In addition to cleaning the noise from the point cloud, which was of minor magnitude, the RANSAC technique was also applied during the fitting calculations to avoid the effect of the noisy points if they were present on the reliable fitting parameters.

The geometric primitive fitting calculations showed the existence of an out of plumb inclination of 45 cm between the top and the bottom of the shrine minaret body, as illustrated in Tables 1 and 2. Statistical analysis showed that multi cylinder fitting to the point cloud of the minaret body was accurate to within 1 cm at 68% probability (1σ), as shown in Figure 13. Furthermore, because the minaret was built from bricks line by line, it was found that the out of plumb inclination of the minaret body is better fitted to a cubic polynomial rather than a straight line (Figure 14). These results have been reported to the shrine administration authorities in order to handle the proper structural treatment and monitoring of the minaret.

The second experiment was applied with respect to a digital model of the lost heritage site of the al Hadbaa minaret (Figure 15). A cone fitting is applied, rather than a cylinder fitting, because of the conical-like shape of the minaret. The computed out of plumb using one cone fitting for the whole point cloud was 1.9 m, with an accuracy of 10 cm (Figure 18a), whereas the computed multiple sections of cone fitting and the 3D circle fitting both resulted in 2.3 m out of plumb (Figure 18b). Using the cone primitive has a higher precision of ± 5 cm, which is closer to the reported inclination results in 2007.

Accordingly, the proposed primitive fitting showed reliable estimation of the out of plumb of the cylindrical-like minarets, which can be automated and integrated with state-of-the-art software tools.

For future work, we will consider the out of plumb computation of other geometrically shaped objects, like squared minarets, using the proposed technique of primitive fitting. Similarly to our second experimental test of the al Hadbaa minaret, the lost heritage of the great mosque minaret of Aleppo in Syria is a recognizable future case study of square-like minarets.

Author Contributions: Field work, Y.H.K. and N.A.A.; Resources, N.A.A., Y.H.K. and B.A.; Software, B.A.; Supervision, B.A.; Visualization, B.A.; Writing, B.A.

Conflicts of Interest: The authors declare no conflict of interest.

ISPRS Int. J. Geo-Inf. **2019**, *8*, 64

References

1. United Nations Education, Scientific and Cultural Organization UNESCO. 2018. Available online: https://en.unesco.org/news/uae-unesco-and-iraq-conclude-historic-50m-partnership-reconstruct-mosul-s-iconic-al-nouri (accessed on 17 November 2018).
2. Beshr, A.A. Structural Data Analysis for Monitoring the Deformation of Oil Storage Tanks Using Geodetic Techniques. *J. Surv. Eng.* **2014**, *140*, 1. [CrossRef]
3. Ayman, H.; Ashraf, O.; Charles, M. Monitoring the structural response of historical Islamic minarets to environmental conditions. In Proceedings of the SMAR—Fourth Conference on Smart Monitoring, Assessment and Rehabilitation of Civil Structures, Zurich, Switzerland, 13–15 September 2017.
4. Controlling Vertical Towers. Available online: https://w3.leica-geosystems.com/media/new/product_solution/Leica_Geosystems_TruStory_Controlling_Vertical_Towers.pdf (accessed on 15 October 2018).
5. Jaafar, H.A. Detection and Localisation of Structural Deformations Using Terrestrial Laser Scanning and Generalised Procrustes Analysis. Ph.D. Thesis, University of Nottingham, Nottingham, UK, 2017.
6. Jatmiko, J.; Psimoulis, P. Deformation Monitoring of a Steel Structure Using 3D Terrestrial Laser Scanner (TLS). In Proceedings of the 24th International Workshop on Intelligent Computing in Engineering, Nottingham, UK, 10–12 July 2017; Available online: https://www.researchgate.net/publication/318658994_Deformation_Monitoring_of_a_Steel_Structure_Using_3D_Terrestrial_Laser_Scanner_TLS (accessed on 15 October 2018).
7. Selbesoglu, M.O.; Bakirman, T.; Gokbayrak, O. Deformation Measurement Using Terrestrial Laser Scanner for Cultural Heritage. In Proceedings of the International Archives of the Photogrammetry, Remote Sensing and Spatial Information Sciences, Istanbul, Turkey, 16–17 October 2016; Volume XLII-2/W1, pp. 89–93.
8. Gordon, S.J.; Lichti, D.D. Modeling terrestrial laser scanner data for precise structural deformation measurement. *J. Surv. Eng.* **2007**, *133*, 72–80. [CrossRef]
9. Park, H.; Lee, H.; Adeli, H.; Lee, I. A new approach for health monitoring of structures: Terrestrial laser scanning. *Comput. Aided Civ. Infrastruct. Eng.* **2007**, *22*, 19–30. [CrossRef]
10. Giuseppina, V.; Fausto, M.; Flavio, S. Terrestrial Laser Scanner for Monitoring the Deformations and the Damages of Buildings. In Proceedings of the XXIII ISPRS Congress, Prague, Czech Republic, 12–19 July 2016; Volume XLI-B5, pp. 453–460. [CrossRef]
11. Hanke, K.; Grussenmeyer, P. Architectural Photogrammetry: Basic theory, Procedures, Tools. September 2002 ISPRS Commission 5 Tutorial. Available online: http://www.isprs.org/commission5/tutorial02/gruss/tut_gruss.pdf (accessed on 25 July 2018).
12. Nocerino, E.; Menna, F.; Remondino, F. Accuracy of typical photogrammetric networks in cultural heritage 3D modeling projects. In *The International Archives of the Photogrammetry, Remote Sensing and Spatial Information Sciences, Proceedings of the ISPRS Technical Commission V Symposium, Riva del Garda, Italy, 23–25 June 2014*; ISPRS: Vienna, Austria, 2014; Volume XL-5, pp. 465–472. [CrossRef]
13. Agisoft LLC. Metashape. Available online: http://www.agisoft.com/ (accessed on 1 October 2018).
14. Photomodeler. Photomodeler Tecnologies. Available online: https://www.photomodeler.com/ (accessed on 11 October 2018).
15. Pix4D. Available online: https://www.pix4d.com/ (accessed on 11 October 2018).
16. Forbes, B.A. *Least-Squares Best-Fit Geometric Elements*; Report Number: NPL Report DITC 140/89; National Physical Laboratory: London, UK, 1989. [CrossRef]
17. Craig, M.S. Least-Squares Fitting Algorithms of the NIST Algorithm Testing System. *J. Res. Natl. Inst. Stand. Technol.* **1998**, *103*, 634–641.
18. Panyam, M.; Kurfessa, T.; Tucker, T. Least Squares Fitting of Analytic Primitives on a GPU. *J. Manuf. Syst.* **2008**, *27*, 130–135. [CrossRef]
19. Ghilani, C.D.; Wolf, P.R. *Adjustment Computations Spatial Data Analysis*, 4th ed.; John Wiley & Sons, Inc.: Hoboken, NJ, USA, 2006; p. 398.
20. Point Cloud Library PCL Documentation, How to Use Random Sample Consensus Model. Available online: http://pointclouds.org/documentation/tutorials/random_sample_consensus.php (accessed on 5 March 2018).
21. Fischler, M.A.; Bolles, R.C. Random sample consensus: A paradigm for model fitting with applications to image analysis and automated cartography. *Commun. ACM* **1981**, *24*, 381–395. [CrossRef]

22. Miki. Fitting a Circle to Cluster of 3D Points. 2016. Available online: https://meshlogic.github.io/posts/jupyter/curve-fitting/fitting-a-circle-to-cluster-of-3d-points/ (accessed on 21 March 2017).

23. Musa_al-Kadhim Shrine. Available online: https://en.wikipedia.org/wiki/Musa_al-Kadhim (accessed on 15 February 2018).

24. Abed, F.; Ibrahim, O.; Jasim, L.; Khalaf, Y.; Hameed, H.; Hussain, Z. Terrestrial Laser Scanning to Preserve Cultural Heritage in Iraq Using Monitoring Techniques. In Proceedings of the 2nd International Conference of Buildings, Construction and Environmental Engineering (BCEE2-2015), Baghdad, Iraq, 17–18 October 2015.

25. Wenzel, K.; Rothermel, M.; Fritsch, D.; Haala, N. Image Acquisition and Model Selection for Multi-View Stereo. In *International Archives of the Photogrammetry, Remote Sensing and Spatial Information Sciences, Proceedings of the 3D-ARCH 2013—3D Virtual Reconstruction and Visualization of Complex Architectures, Trento, Italy, 25–26 February 2013*; ISPRS: Vienna, Austria, 2013; Volume XL-5/W1, pp. 251–258. [CrossRef]

26. Alsadik, B.; Gerke, M.; Vosselman, G.; Daham, A.; Jasim, L. Minimal Camera Networks for 3D Image Based Modeling of Cultural Heritage Objects. *Sensors* **2014**, *14*, 5785–5804. [CrossRef] [PubMed]

27. Al-Hadba' Minaret—World Monument Fund. Available online: https://www.wmf.org/project/al-hadba%E2%80%99-minaret (accessed on 06 November 2018).

28. Al-Gburi, M.; Salih, M.; Taeb, S. Study of Maintenance Al Hadba Minaret Monument in Mosul. 2010. Available online: https://www.researchgate.net/publication/268508608_Study_of_Maintenance_Al_Hadba_Minaret_Monument_in_Mosul (accessed on 15 October 2018).

29. Grussenmeyer, P.; Al Khalil, O. From Metric Image Archives to Point Cloud Reconstruction: Case Study of The Great Mosque of Aleppo in Syria. In *International Archives of the Photogrammetry, Remote Sensing and Spatial Information Sciences, Proceedings of the 26th International CIPA Symposium, Ottawa, ON, Canada, 28 August–1 September 2017*; ISPRS: Vienna, Austria, 2017; Volume XLII-2/W5. [CrossRef]

30. 3D Warehouse. Available online: https://3dwarehouse.sketchup.com/?hl=en (accessed on 15 October 2018).

31. Sketchfab. Available online: https://sketchfab.com/models/adefb309080844c6b78370e7c129d4c4 (accessed on 15 October 2018).

International Journal of *Geo-Information*

Article

Three-Dimensional Digital Documentation of Cultural Heritage Site Based on the Convergence of Terrestrial Laser Scanning and Unmanned Aerial Vehicle Photogrammetry

Young Hoon Jo * and Seonghyuk Hong

Department of Cultural Heritage Conservation Sciences, Kongju National University, Gongju 32588, Korea; h123kr@kongju.ac.kr
* Correspondence: joyh@kongju.ac.kr; Tel.: +82-041-850-8539

Received: 27 November 2018; Accepted: 21 January 2019; Published: 24 January 2019

Abstract: Three-dimensional digital technology is important in the maintenance and monitoring of cultural heritage sites. This study focuses on using a combination of terrestrial laser scanning and unmanned aerial vehicle (UAV) photogrammetry to establish a three-dimensional model and the associated digital documentation of the Magoksa Temple, Republic of Korea. Herein, terrestrial laser scanning and UAV photogrammetry was used to acquire the perpendicular geometry of the buildings and sites, where UAV photogrammetry yielded higher planar data acquisition rate in upper zones, such as the roof of a building, than terrestrial laser scanning. On comparing the two technologies' accuracy based on their ground control points, laser scanning was observed to provide higher positional accuracy than photogrammetry. The overall discrepancy between the two technologies was found to be sufficient for the generation of convergent data. Thus, the terrestrial laser scanning and UAV photogrammetry data were aligned and merged post conversion into compatible extensions. A three-dimensional (3D) model, with planar and perpendicular geometries, based on the hybrid data-point cloud was developed. This study demonstrates the potential for using the integration of terrestrial laser scanning and UAV photogrammetry in 3D digital documentation and spatial analysis of cultural heritage sites.

Keywords: terrestrial laser scanning; unmanned aerial vehicle photogrammetry; integrated three-dimensional modeling; digital documentation; cultural heritage site

---

1. Introduction

There are many cultural heritage sites in the Republic of Korea. However, the sites incurred constant deformation owing to deteriorations and disasters. The acquisition of geospatial information based on numeric data is very important for systematic management to respond to the deformation. The use of three-dimensional (3D) coordinate data for multi-faceted analysis has recently increased in the conservation and management of cultural heritage sites [1–6]. In particular, the precise investigation and monitoring of cultural heritage sites in terms of preventive conservation has gained attention, and digital documentation processes have been recognized as an essential element in conservation, rather than the active repair of cultural heritage sites [7–9].

In addition, the efficiency of survey methods should be considered for the acquisition of 3D numeric data because most heritage sites are fairly large. To achieve this goal, 3D terrestrial laser scanning and unmanned aerial vehicle (UAV) photogrammetry are considered to be representative of documentation technology because they create a digital model from recorded data that is nearly identical to the physical geometry [10–13]. These technologies have relatively cheap operational expenses [14,15] and can rapidly and accurately reacquire high-resolution images [16,17]. Recently, multiple studies pertaining to the lightening of equipment and increasing of precision have been performed [18]. In particular, these technologies have been applied in conservation assessment [19,20], building information modeling (BIM) [21–23], and archaeological documentation [24–26].

Terrestrial laser scanning measures the 3D spatial information of an object within a certain distance of the ground using a laser [27,28]. This method can quickly acquire the geometry of a large cultural heritage site, owing to its high operation speed, mobility, and accessibility. The data acquisition rate for terrestrial laser scanning is optimum when in a perpendicular direction.

UAV photogrammetry provides some alternative advantages when compared with laser scanning [29]. The orthoimage created via UAV photogrammetry allows for distances, angles, plane coordinates, and areas to be directly measured as the relationships between different locations are identical to those on the topographic map [30]. UAV photogrammetry has a higher planar data acquisition rate in upper zones, such as the roof of a building, than terrestrial laser scanning.

In general, temple sites include multiple buildings and are relatively wide, open, and complex. Therefore, documenting the shape of the entire site exclusively through terrestrial laser scanning is difficult. In particular, acquiring data at positons where the scanner is not accessible is difficult, for example the roof of a building. In addition, the acquired data has a low point density owing to the restricted field of view even when the scan is performed for a long duration and from a high position. The superior mobility and accessibility of UAV photogrammetry must be actively utilized to overcome these disadvantages. If terrestrial laser scanning and UAV photogrammetry can be appropriately integrated, multidirectional numerical information together with the arrangements of architectural heritage sites could be acquired [31–34]. Accordingly, the fusion of laser scanning and photogrammetry has been widely used for the 3D modeling of buildings [35,36] and cultural heritages [37,38].

This study applied both terrestrial laser scanning and UAV photogrammetry to the 3D digital documentation of Magoksa Temple in Gongju, a representative temple of the Republic of Korea [39]. Point clouds of the temple site obtained from the two technologies were integrated after their accuracy had been verified to produce an orthoimage and a 3D model using specialized software. In particular, overall workflow of 3D fusion modeling and the possibility of application of the completed terrain model were discussed. The results of this study are expected to contribute to the development of accurate 3D documentation and spatial analysis of heritage sites.

2. Study Area and Method

2.1. Study Area

The Magoksa Temple, located in Gonju, Republic of Korea, was built in the 7th century. The temple was included in the UNESCO World Heritage list in 2018 as one of the "Traditional Buddhist Mountain Temples of Korea." This temple is a good example of how a Korean-style temple layout exhibits harmony between the inner and outer space with the surrounding scenery, which in this case is a mountainous area. This feature led to the temple's recognition as a World Heritage site. The Magoksa Temple can comprehensively exhibit thoughts and the consciousness unique to Korean Buddhism together with the life and culture associated with a mountain temple.

The Magoksa Temple is divided into northern and southern territories based around a central stream, as shown in Figure 1a. The two territories have both horizontal and vertical arrangements. The northern territory primarily contains the Daeungbojeon Hall, the Daegwanbojeon Hall, and a Five-Story Stone Pagoda. The main heritage site of the southern territory is the Yeongsanjeon Hall. The five-story stone pagoda in this temple has a Lamaistic upper part, which is unique in the Republic of Korea; this hybrid style evidences active cultural exchanges with the Yuan Dynasty in the 13th century (Figure 1b).

Figure 1. Field existence of the Magoksa Temple, Gongju: (**a**) location and temple layout; (**b**) five-story stone pagoda.

2.2. Method

The terrestrial laser scanner (Leica, ScanStation C10) used to create the 3D model of the Magoksa Temple is based on the time of flight of the laser pulse (Figure 2a). The scanner used for this study has a maximum scan speed of 50,000 points per second and is accurate to 6 and 4 mm in terms of position and distance between 1 m and 50 m, respectively. A specialized external camera with a fisheye lens (SIGMA, 8-mm F3.5 EX DG Circular FISHEYE) was used for texture mapping. The scanned point cloud data were post-processed using Cyclone 9.3 software produced by Leica Geosystems.

UAV photogrammetry was conducted using a rotorcraft drone (Leica, Aibot X6). The UAV used is an autonomously flying and high-performance hexacopter with stable flight performance, specially designed for demanding tasks in surveying, industrial inspection, agriculture, and forestry. Moreover, the UAV had various on-board sensors, including GPS, a gyroscope, an ultrasonic transducer, and a smart camera system. In addition, to create a 3D model, a 24 MP mirrorless digital camera (SONY, Alpha 6000) with a 20-mm lens was mounted on the drone for photogrammetry. Generally, the camera calibration is performed using the bundle adjustment option of specialized software [40] or developed methods [41]. However, a special camera calibration was not performed because this study used a new camera and lens for photogrammetry.

Based on the previous statement, a gimbal became a mandatory part of the UAV equipment because it smooths the angular movements of a camera and provides advantages for acquiring better images [42]. This UAV is able to provide high-quality data, and it has an exceptional camera mount that has automatic pitch-and-roll compensation of the camera gimbal for smoother pictures and video output (Figure 2b).

PhotoScan 1.3.4 Professional Edition software produced by Agisoft was used for the photogrammetric process. The software allows for professional 3D modeling of the captured aerial photographs. The software automatically processes the photographs to produce a 3D reconstruction of the images. A ground control point (GCP) survey of the Magoksa Temple was conducted across five points, (CP1–CP5) using the virtual reference station (VRS) GPS (Trimble, R6 Model 3) and levelling (Figure 2c). The VRS GPS system used for this survey is accurate to 8 mm + 0.5 ppm and 15 mm + 0.5 ppm in terms of horizontality and verticality, respectively.

Specifically, the survey was performed by considering positions that were not obstructed by nearby obstacles and where the visual field could be secured. The results of the survey points were obtained by fixing four unified control stations (UCS) and conducting a GPS baseline analysis. In addition, the levelling was utilized to improve the vertical accuracy of GPS survey. These GCPs were used to relocate the terrestrial laser scanning and UAV photogrammetry models in the true north direction and obtain further positional information regarding the land surface.

Figure 2. Equipment and methods used for 3D digital documentation: (**a**) terrestrial laser scanning using a specialized external camera; (**b**) hexacopter unmanned aerial vehicle (UAV); (**c**) five ground control points.

3. Result and Discussion

3.1. Terrestrial Laser Scanning

The most important consideration in a field scan is to select a suitable scan position for the object of interest. In this study, the field scan was performed from 82 positions, considering various factors such as the measurable range and overlap as the Magoksa Temple covers a fairly wide area. In addition, the stone pagoda, a key part of the heritage site, was scanned from 24 positions by dividing the pagoda into lower and upper sections based on its height. The 3D shape of the temple was generated by registering, merging, and filtering the scanning data and inputting the GCP data.

Figure 3. Orthoimage and detailed view of the Magoksa Temple created via terrestrial laser scanning: (**a**) reflectance intensity image; (**b**) RGB texture mapping image.

Shape registering was conducted by selecting at least three corresponding points from each point cloud. As a result, all registered point clouds were merged and filtered to generate a complete object model. Finally, the laser scanning models were converted to an absolute coordinate system based on the five GCPs to increase the accuracy and reliability of the topographical survey. Figure 3 shows the point clouds and texture mapping images of the Magoksa Temple obtained by merging and filtering the laser scanning data. Notably, the overall 3D shape reflected the perpendicular geometry of the buildings; however, high-level planar point clouds (e.g., the geometry of the roof) were not acquired sufficiently for geometrical reconstruction.

3.2. UAV Photogrammetry

UAV photogrammetry was used to determine the overall topography of the Magoksa Temple and acquire the 3D numerical data of the upper part, where using the terrestrial laser scanner was not possible. This was achieved by conducting a planning survey suitable for the Magoksa Temple and then performing overall modeling using specialized software. First, two UAV flight plans were determined for the southern and northern territories of the Magoksa Temple. Next, the flight ranges (approximately 250×400 m) and paths were established on the satellite map using specialized software, and the following conditions were input into the software: vertical camera angle; altitude = 80 m; overlap = 80%; sidelap = 60%; flight speed = 3 m/s; and flight time = 20 min. Then, the UAV flight was performed under autonomous control according to the preset path.

Image processing involved aligning the images, inputting GCP, reconstructing the images, building point clouds, and mapping the textures. First, the 213 still photographs obtained by the UAV camera, the alignment of which was based on the shooting sequences and image feature points. Then, geometric reconstruction using the five GCPs was performed to scale and orient the 3D model. Finally, the orthoimage was completed by mapping the texture. The data acquisition rate in the planar direction was higher for UAV photogrammetry as compared to the terrestrial laser scanning (Figure 4).

Figure 4. Orthoimage and detailed view of the Magoksa Temple created via UAV photogrammetry: (a) density image; (b) RGB texture mapping image.

3.3. Accuracy Assessment

The 3D digital documentation of individual architectural heritages and artifacts has been performed in the past. Recent developments in laser scanning technology and UAV photogrammetry have extended the scope of digital documentation to large-scale architectural heritages and sites. In particular, UAV photogrammetry has been used to acquire 3D numerical data across a wide area, which, when combined with terrestrial laser scanning data, provides a large number of point clouds; this reduces the time required for data acquisition. In this respect, studies on the digital documentation of cultural heritage sites using different technologies have been conducted in conjunction with the deployment of UAV [43–47]. In this paper, we focused on evaluating the reliability of the point data obtained via terrestrial laser scanning and UAV photogrammetry to enhance the convergent usability of both datasets by comparing the position accuracy of the point cloud data.

First, we compared the coordinate errors with the laser scanning and photogrammetry based on the five GCPs to evaluate the overall quality and absolute positional accuracy (Table 1, Figure 5). As a result, the X, Y, and Z coordinates of the laser scanning result deviated in the range −0.009–0.005 m (RMS 0.006 m), −0.015–0.001 m (RMS 0.008 m), and −0.011–0.099 m (RMS 0.045 m), respectively. Additionally, the X, Y, and Z coordinates of the photogrammetry result exhibited discrepancies that range from −0.005–0.044 m (RMS 0.024 m), −0.028–0.027 m (RMS 0.021 m), and −0.009–0.079 m (RMS 0.044 m), respectively. Overall, the X and Y coordinates of the terrestrial laser scanning and UAV photogrammetry methods exhibit a good degree of agreement with the GCP values at the millimeter-level accuracy. However, the Z coordinate shows high discrepancies at the centimeter-level accuracy. The comparison results according the survey methods show that the terrestrial laser scanning result is more accurate than UAV photogrammetry.

Table 1. Comparison of coordinate errors of the point data obtained via terrestrial laser scanning and UAV photogrammetry based on the five ground control points.

GCPs	Terrestrial Laser Scanning			UAV Photogrammetry		
	X	Y	Z	X	Y	Z
CP1	0.000 m	−0.004 m	0.000 m	0.030 m	0.005 m	−0.009 m
CP2	0.001 m	0.001 m	0.000 m	−0.003 m	0.001 m	0.024 m
CP3	−0.007 m	0.000 m	−0.001 m	−0.001 m	0.027 m	0.005 m
CP4	0.005 m	−0.008 m	−0.011 m	0.044 m	−0.028 m	0.052 m
CP5	−0.009 m	−0.015 m	0.099 m	−0.005 m	−0.025 m	0.079 m
Mean	−0.002 m	−0.005 m	0.017 m	0.013 m	−0.004 m	0.030 m
RMS	0.006 m	0.008 m	0.045 m	0.024 m	0.021 m	0.044 m

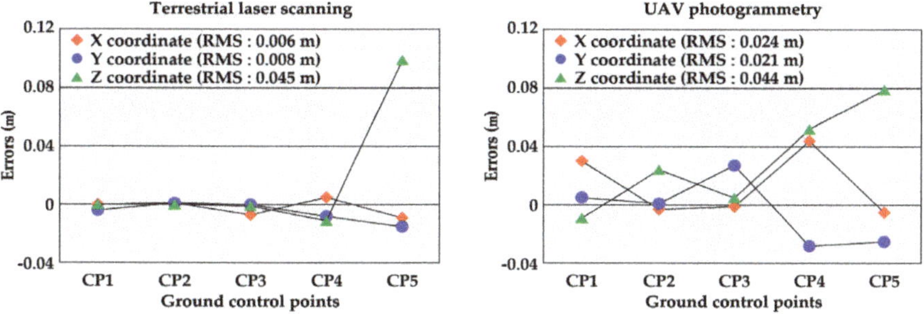

Figure 5. Graphs showing coordinate errors of the terrestrial laser scanning and UAV photogrammetry based on the five ground control points.

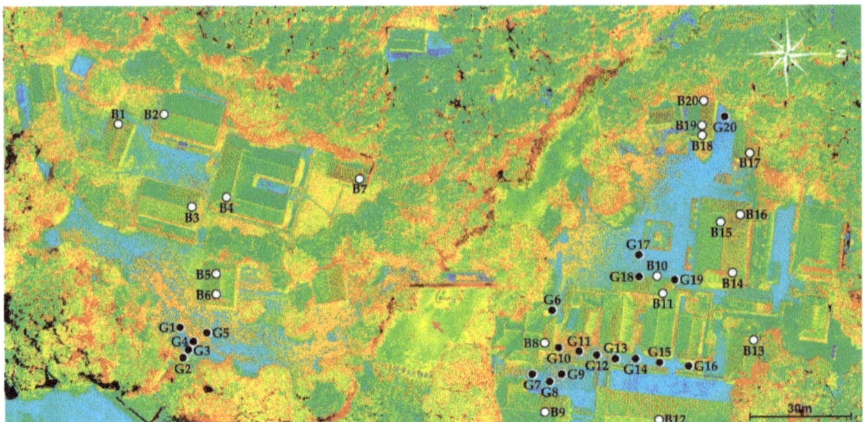

Figure 6. A total of 40 randomly selected points on the ground and buildings for analysis of the relative positional accuracy between terrestrial laser scanning and UAV photogrammetry.

The relative positional accuracy between terrestrial laser scanning and UAV photogrammetry were analyzed using randomly selected points (Figure 6). Coordinate discrepancies were calculated at a total of 40 points on the ground and on buildings (Table 2, Figure 7). The maximum negative and positive difference values on the ground were -0.020 m and 0.030 m in X coordinate (RMS 0.015 m), −0.028 m and 0.025 m in Y coordinate (RMS 0.015 m), and −0.012 m and 0.030 m in Z coordinate (RMS 0.010 m), respectively. Moreover, the X, Y, and Z coordinates on the buildings deviated in the range of −0.037–0.074 m (RMS 0.043 m), −0.049–0.049 m (RMS 0.028 m), and −0.071–0.057 m (RMS 0.037 m), respectively. Overall, the differences of the coordinates were less than ±0.030 m on the ground and ±0.080 m on the buildings. In particular, there were significantly higher deviations on some building points. The RMS of the buildings was approximately two to three times bigger than RMS on the ground. The point cloud comparison was an effective approach to investigate the correspondence of the terrestrial laser scanning and UAV photogrammetry methods. In the future, detailed and macroscopic analyses of overall datasets should be performed using graphical tools.

Table 2. Relative positional coordinate discrepancies between terrestrial laser scanning and UAV photogrammetry.

Point No.	Ground (G)			Buildings (B)		
	X	Y	Z	X	Y	Z
1	0.012 m	0.012 m	−0.002 m	0.061 m	−0.033 m	−0.050 m
2	0.013 m	−0.008 m	−0.010 m	−0.037 m	−0.016 m	0.023 m
3	0.026 m	0.008 m	−0.009 m	0.073 m	−0.001 m	−0.044 m
4	0.016 m	−0.007 m	−0.005 m	0.021 m	0.009 m	−0.016 m
5	−0.001 m	0.004 m	−0.002 m	0.053 m	−0.005 m	−0.028 m
6	−0.004 m	0.002 m	0.019 m	0.047 m	0.040 m	−0.028 m
7	−0.013 m	−0.003 m	0.004 m	0.000 m	0.003 m	0.057 m
8	−0.012 m	0.007 m	0.001 m	0.061 m	−0.035 m	−0.015 m
9	0.009 m	−0.006 m	−0.003 m	−0.012 m	0.012 m	−0.065 m
10	0.008 m	−0.028 m	0.030 m	0.010 m	−0.026 m	0.018 m
11	0.021 m	−0.019 m	0.002 m	0.043 m	−0.047 m	0.037 m
12	0.019 m	−0.022 m	0.000 m	0.062 m	0.041 m	−0.031 m
13	0.004 m	0.006 m	−0.008 m	0.074 m	−0.013 m	−0.049 m
14	−0.004 m	0.013 m	0.002 m	0.037 m	0.031 m	−0.011 m

Table 2. *Cont.*

Point No.	Ground (G)			Buildings (B)		
	X	Y	Z	X	Y	Z
15	0.030 m	−0.014 m	−0.003 m	0.050 m	−0.049 m	−0.021 m
16	0.011 m	−0.021 m	−0.011 m	0.022 m	0.049 m	0.018 m
17	−0.011 m	−0.006 m	−0.010 m	0.003 m	0.012 m	0.015 m
18	−0.010 m	0.011 m	−0.012 m	0.030 m	−0.002 m	0.039 m
19	−0.020 m	0.025 m	0.009 m	−0.007 m	−0.014 m	−0.071 m
20	0.007 m	−0.027 m	0.000 m	−0.032 m	0.018 m	−0.004 m
Mean	0.005 m	−0.004 m	0.000 m	0.028 m	−0.001 m	−0.011 m
RMS	0.015 m	0.015 m	0.010 m	0.043 m	0.028 m	0.037 m

To determine the shape accuracy of individual buildings in addition to the overall deviation analysis of cultural heritage sites, the roof section of the Daegwangbojeon Hall, a representative building in the Magoksa Temple, was examined. Figure 8 shows that the photogrammetric point cloud data is complementary to areas of low point density in the scanning point cloud data. Thus, the level of accuracy of both laser scanning and photogrammetry is high enough to be widely used in survey drawings for buildings as well as in BIM. In this respect, a 3D model that combines terrestrial laser scanning and UAV photogrammetry is deemed a useful technology for the digital documentation of large and complex cultural heritage sites such as the Magoksa Temple, because such models jointly overcome the limitations of data acquisition through a single technology.

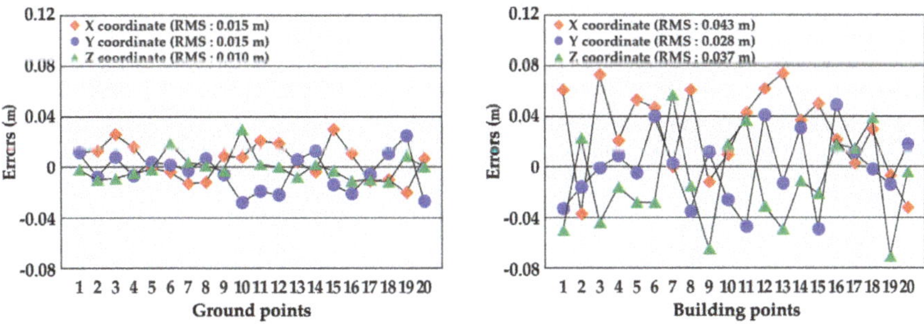

Figure 7. Graphs showing relative positional coordinate discrepancies between terrestrial laser scanning and UAV photogrammetry.

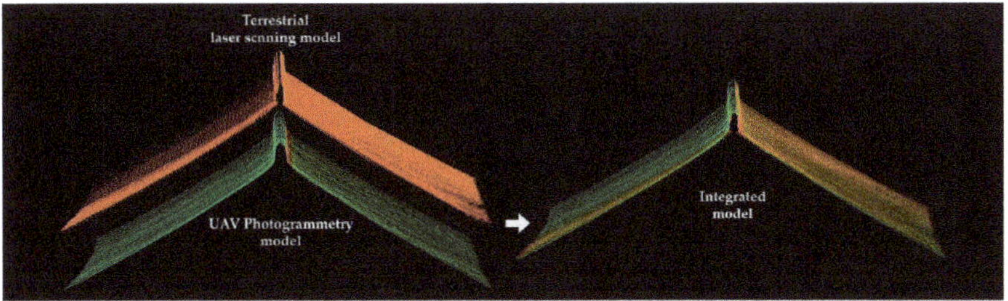

Figure 8. Point clouds showing the roof section of the Daegwangbojeon Hall.

3.4. Integrated Three-Dimensional Modeling

Considering the properties of terrestrial laser scanning and UAV photogrammetry, it is clear that the two technologies provide complementary information. However, the synergic characteristics of both systems can be fully utilized only after the successful registration of the laser scanning and photogrammetry data relative to a common reference frame [48]. In general, a popular mathematical computation is the "Iterative Closet Point" (ICP) algorithm [49,50].

This study created a 3D model of the Magoksa Temple, including the terrain and buildings, by combining laser scanning data that focused on perpendicular point clouds with UAV photogrammetry data that focused on the planar point clouds. To achieve this goal, the two given datasets of the scanning and photogrammetry applied model-based fusion using the five GCPs. Then, the relative position of the laser scanning and photogrammetry results was determined via bundle adjustment and ICP algorithm such that the mean squared error (RMS) given by the sum of the squared distances between the corresponding points was minimized. The process was continued until either the RMS fell down below a given threshold or no further improvement could be achieved. As a result of the overlap error statistics between the laser scanning and photogrammetry results, the RMS is gradually decreased depending on the alignment stages, and the final RMS value is 0.005 m (Table 3).

Table 3. Overlap error statistics between the laser scanning and photogrammetry according to alignment stages.

Alignment Processing	Max Search Distance Parameters	Iterations	RMS	Mean
First stage	0.100 m	12	0.038 m	0.030 m
Second stage	0.050 m	10	0.020 m	0.016 m
Third stage	0.025 m	18	0.012 m	0.010 m
Fourth stage	0.012 m	6	0.006 m	0.005 m

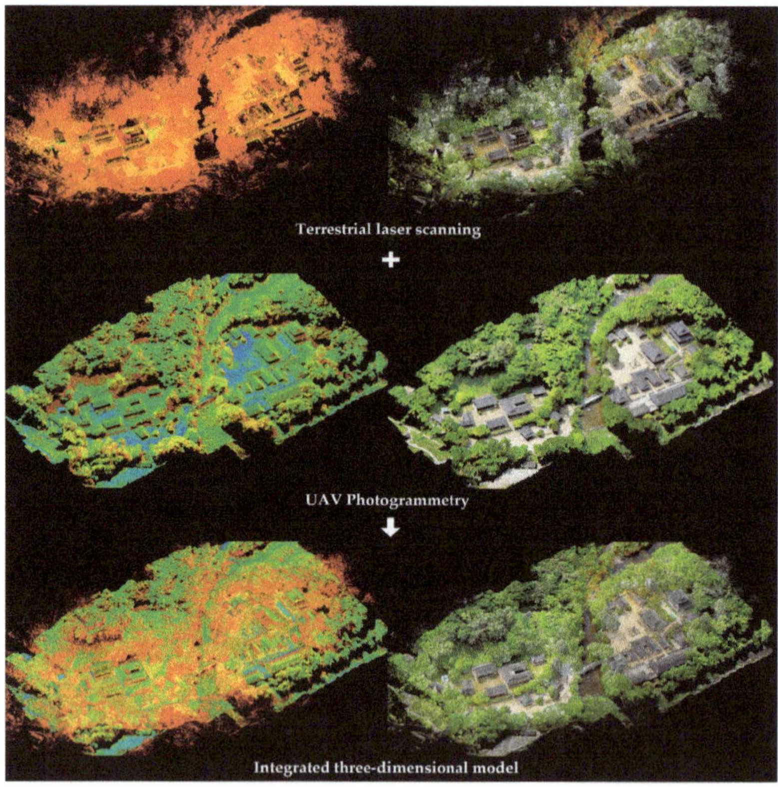

Figure 9. Process and result of integrated 3D modeling.

The completed 3D shape exhibited perfect planar and perpendicular geometries, including wooden buildings and the surrounding environment (Figure 9). As shown in Figure 10, the Magoksa Temple has east–west and south–north lengths of approximately 100 and 230 m, respectively, and the temple is located between 100 and 130 m above sea level. Although the Magoksa Temple seems like a mountain temple in its layout the survey results indicate that the Magoksa Temple is located on a site that is virtually flat. This 3D terrain model of the site is expected to be used as basic data for conservation and management of the Magoksa Temple.

The workflows for the processes involved in the acquisition of the point cloud data on the Magoksa Temple using terrestrial laser scanning and UAV photogrammetry together with the 3D modeling process that combines the data from these two technologies can be summarized as follows (Figure 11). First, laser scanning collects a primary set of 3D shape data through field scans, data registering, merging, filtering, and the input of GCPs. Through feature point matching, GCP inputting, point cloud creating, and texture mapping, UAV photogrammetry produces a 3D model created from the aerial photographs obtained after the setup of a flight plan. If data convergence is not required for both models, laser scanning produces a 3D model based on the perpendicular geometry, and photogrammetry produces an orthoimage model based on the planar geometry. In contrast, if the convergence of the point cloud data is required between these two approaches, generating a 3D model based on the hybrid point cloud data is possible, which includes topography and building shape through shape registering and merging, after changing the data to have compatible extensions.

Figure 10. True orthoimage and cross-section of Magoksa Temple's terrain.

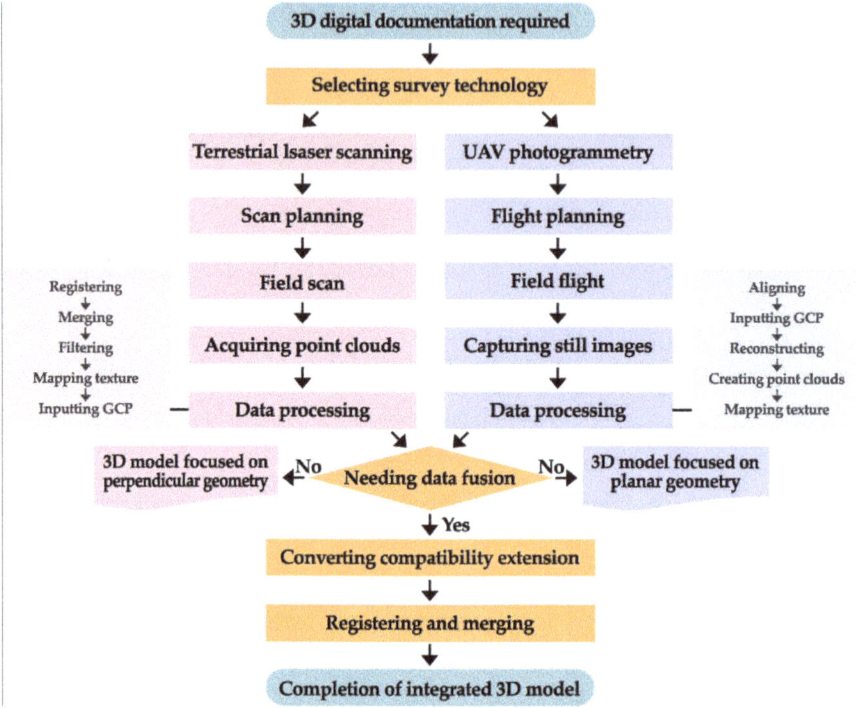

Figure 11. Workflow of integrated 3D modeling using terrestrial laser scanning and UAV photogrammetry.

4. Conclusions

This study established a 3D model and accurate digital documentation of the Magoksa Temple using terrestrial laser scanning and UAV photogrammetry. The two technologies were used to acquire multidirectional numerical information of the temple site. Laser scanning showed a high data acquisition rate in the perpendicular direction. In contrast, photogrammetry generated high-level planar point clouds. Laser scanning or photogrammetry are useful individual techniques for digital documentation of cultural heritage sites to determine the layout conditions and topographical features based on an orthoimage image; however, these techniques are of limited application if precise survey drawings are required. Thus, constructing a 3D model that includes topography as well as building shapes through a hybrid convergence technology, using both terrestrial laser scanning and UAV photogrammetry, is crucial.

The accuracy of the two technologies based on GCPs prior to their convergence was analyzed: laser scanning has higher positional accuracy than photogrammetry, and the overall discrepancy of the two technologies was sufficient to generate convergent data. The photogrammetric point cloud data was aligned and merged based on the laser scanning results. Consequently, photogrammetry could increase the value of a 3D model by complementing the point cloud data for the upper parts of buildings that are difficult to acquire through laser scanning. Furthermore, the accuracy of the overall topography as well as the shape of an individual building is increased, which enhances the composition of survey drawings and the BIM.

Integrated 3D models using both terrestrial laser scanning and UAV photogrammetry are expected to contribute to the improved applicability of hybrid point clouds and digital documentation methodologies for cultural heritage sites and individual monuments. However, future work must focus on reducing the positional discrepancies between the two survey technologies and determine the tendency of positional discrepancies to vary according to the different scales and geomorphic environments of the cultural heritage site. Moreover, cross-validation with various other measurement techniques such as total station is required to increase the reliability of positional accuracy. Although some further studies for data acquisition, fusion workflow, accuracy assessment, and applicability improvement have been conducted, the 3D integrated model of terrestrial laser scanning and UAV photogrammetry seems to be a powerful tool for the digital documentation and conservation management of the cultural heritage sites.

Author Contributions: Conceptualization, Y.H.J.; methodology, Y.H.J.; software, S.H.; validation, Y.H.J. and S.H.; formal analysis, Y.H.J.; investigation, Y.H.J. and S.H.; resources, Y.H.J.; data curation, S.H.; writing—original draft preparation, Y.H.J.; writing—review and editing, Y.H.J.; visualization, S.H.; supervision, Y.H.J.; project administration, Y.H.J.; funding acquisition, Y.H.J.

Funding: This research was funded by the Ministry of Science, ICT and Future Planning (NRF-2016R1C1B2010883).

Acknowledgments: This research was supported by Basic Science Research Program through the National Research Foundation of Korea (NRF).

Conflicts of Interest: The authors declare no conflict of interest.

References

1. Aragóna, E.; Munar, S.; Rodríguez, J.; Yamafunec, K. Underwater photogrammetric monitoring techniques for mid-depth shipwrecks. *J. Cult. Herit.* **2018**, *34*, 255–260. [CrossRef]
2. Discamps, E.; Muth, X.; Gravina, B.; Lacrampe-Cuyaubère, F.; Chadelle, J.; Faivre, J.; Maureille, B. Photogrammetry as a tool for integrating archival data in archaeological fieldwork: Examples from the Middle Palaeolithic sites of Combe-Grenal, Le Moustier, and Regourdou. *J. Archaeol. Sci. Rep.* **2016**, *8*, 268–276. [CrossRef]
3. Erenoglu, R.C.; Akcay, O.; Erenoglu, O. An UAS-assisted multi-sensor approach for 3D modeling and reconstruction of cultural heritage site. *J. Cult. Herit.* **2017**, *26*, 79–90. [CrossRef]
4. O'Driscoll, J. Landscape applications of photogrammetry using unmanned aerial vehicles. *J. Archaeol. Sci. Rep.* **2018**, *22*, 32–44. [CrossRef]
5. Themistocleous, K. Model reconstruction for 3-D vizualization of cultural heritage sites using open data from social media: The case study of Soli, Cyprus. *J. Archaeol. Sci. Rep.* **2017**, *14*, 774–781. [CrossRef]
6. Wilson, L.; Rawlinson, A.; Frost, A.; Hepher, J. 3D digital documentation for disaster management in historic buildings: Applications following fire damage at the Mackintosh building, The Glasgow School of Art. *J. Cult. Herit.* **2018**, *31*, 24–32. [CrossRef]
7. Messaoudi, T.; Véron, P.; Halin, G.; Luca, L.D. An ontological model for the reality-based 3D annotation of heritage building conservation state. *J. Cult. Herit.* **2018**, *29*, 100–112. [CrossRef]
8. Xiao, W.; Mills, J.; Guidi, G.; Rodríguez-Gonzálvez, P.; Barsanti, S.G.; González-Aguilera, D. Geoinformatics for the conservation and promotion of cultural heritage in support of the UN Sustainable Development Goals. *ISPRS J. Photogramm. Remote. Sens.* **2018**, *142*, 389–406. [CrossRef]
9. Zimmer, B.; Liutkus-Pierce, C.; Marshall, S.T.; Hatala, K.G.; Metallo, A.; Rossi, V. Using differential structure-from-motion photogrammetry to quantify erosion at the Engare Sero footprint site, Tanzania. *Quat. Sci. Rev.* **2018**, *198*, 226–241. [CrossRef]
10. Angelo, L.D.; Stefano, P.D.; Fratocchi, L.; Marzola, A. An AHP-based method for choosing the best 3D scanner for cultural heritage applications. *J. Cult. Herit.* **2018**, *34*, 109–115. [CrossRef]
11. Galantucci, L.M.; Pesce, M.; Lavecchia, F. A stereo photogrammetry scanning methodology, for precise and accurate 3D digitization of small parts with sub-millimeter sized features. *CIRP Ann-Manuf. Techn.* **2015**, *64*, 507–510. [CrossRef]
12. Herráez, J.; Martínez, J.C.; Coll, E.; Martín, M.T.; Rodríguez, J. 3D modeling by means of videogrammetry and laser scanners for reverse engineering. *Measurement* **2016**, *87*, 216–227. [CrossRef]
13. Vacaa, G.; Dessì, A.; Sacco, A. The use of nadir and oblique UAV images for building knowledge. *ISPRS Int. J. Geo-Inf.* **2017**, *6*, 393. [CrossRef]
14. Galantucci, L.M.; Piperi, E.; Lavecchia, F.; Zhavo, A. Semi-automatic low cost 3D laser scanning system for reverse engineering. *Procedia CIRP* **2015**, *28*, 94–99. [CrossRef]

15. Reznicek, J.; Pavella, K. New low cost 3D Scanning Techniques for Cultural Heritage Documentation. In Proceedings of the International Archives of Photogrammetry and Remote Sensing, XXXVI-B5, Beijing, China, 3–11 July 2008; pp. 237–240.

16. Tannant, D.D. Review of photogrammetry-based techniques for characterization and hazard assessment of rock faces. *Int. J. Georesour. Environ.* **2015**, *1*, 76–87. [CrossRef]

17. Zlot, R.; Bosse, M. Three-dimensional mobile mapping of caves. *J. Cave Karst Stud.* **2014**, *76*, 191–206. [CrossRef]

18. Reichert, J.; Schellenberg, J.; Schubert, P.; Wilke, T. 3D scanning as a highly precise, reproducible, and minimally invasive method for surface area volume measurements of scleractinian corals. *Limnol Oceanogr. Meth.* **2016**, *14*, 518–526. [CrossRef]

19. Palomar-Vazquez, J.; Baselga, S.; Viñals-Blasco, M.; García-Sales, C.; Sancho-Espinós, I. Application of a combination of digital image processing and 3D visualization of graffiti in heritage conservation. *J. Archaeol. Sci. Rep.* **2017**, *12*, 32–42. [CrossRef]

20. Quagliarini, E.; Clini, P.; Ripanti, M. Fast, low cost and safe methodology for the assessment of the state of conservation of historical buildings from 3D laser scanning: The case study of Santa Maria in Portonovo (Italy). *J. Cult. Herit.* **2017**, *24*, 175–183. [CrossRef]

21. Biagini, C.; Capone, P.; Donato, V.; Facchini, N. Towards the BIM implementation for historical building restoration sites. *Automat. Constr.* **2016**, *71*, 74–86. [CrossRef]

22. Fryskowska, A.; Stachelek, J. A no-reference method of geometric content quality analysis of 3D models generated from laser scanning point clouds for hBIM. *J. Cult. Herit.* **2018**, *34*, 95–108. [CrossRef]

23. Simeone, D.; Cursi, S.; Acierno, M. BIM semantic-enrichment for built heritage representation. *Automat. Constr.* **2019**, *97*, 122–137. [CrossRef]

24. Galeazzi, F. Towards the definition of best 3D practices in archaeology: Assessing 3D documentation techniques for intra-site data recording. *J. Cult. Herit.* **2016**, *17*, 159–169. [CrossRef]

25. Scafuri, M.P.; Rennison, B. Scanning the H.L. Hunley: Employing a structured-light scanning system in the archaeological documentation of a unique maritime artifact. *J. Archaeol. Sci. Rep.* **2016**, *6*, 302–309. [CrossRef]

26. Monna, F.; Esin, Y.; Magail, J.; Granjon, L.; Navarro, N.; Wilczek, J.; Saligny, L.; Couette, S.; Dumontet, A.; Chateau, C. Documenting carved stones by 3D modelling—Example of Mongolian deer stones. *J. Cult. Herit.* **2018**, *34*, 116–128. [CrossRef]

27. Rüther, H.; Chazan, M.; Schroeder, R.; Neeser, R.; Held, C.; Walker, S.J.; Matmon, A.; Horwitz, L.K. Laser scanning for conservation and research of African cultural heritage sites: The case study of Wonderwerk Cave, South Africa. *J. Archaeol. Sci.* **2009**, *36*, 1847–1856. [CrossRef]

28. Fabbri, S.; Sauro, F.; Santagata, T.; Rossi, G.; Waele, J.D. High-resolution 3-D mapping using terrestrial laser scanning as a tool for geomorphological and speleogenetical studies in caves: An example from the Lessini mountains (North Italy). *Geomorphology* **2017**, *280*, 16–29. [CrossRef]

29. Fernández-Lozano, J.; Gutiérrez-Alonso, G. Improving archaeological prospection using localized UAVs assisted photogrammetry: An example from the Roman Gold District of the Eria River Valley (NW Spain). *J. Archaeol. Sci. Rep.* **2016**, *5*, 509–520. [CrossRef]

30. Watanabe, Y.; Kawahara, Y. UAV photogrammetry for monitoring changes in river topography and vegetation. *Procedia Eng.* **2016**, *154*, 317–325. [CrossRef]

31. Assali, P.; Grussenmeyer, P.; Villemin, T.; Pollet, N.; Viguier, F. Surveying and modeling of rock discontinuities by terrestrial laser scanning and photogrammetry: Semi-automatic approaches for linear outcrop inspection. *J. Struct. Geol.* **2014**, *66*, 102–114. [CrossRef]

32. Yang, B.; Zang, Y.; Dong, Z.; Huang, R. An automated method to register airborne and terrestrial laser scanning. *ISPRS J. Photogramm.* **2015**, *109*, 62–76. [CrossRef]

33. Aicardi, I.; Dabove, P.; Lingua, A.M.; Piras, M. Integration between TLS and UAV photogrammetry techniques for forestry applications. *Ifor. Bioengsic. For.* **2016**, *10*, 41–47. [CrossRef]

34. Liang, H.; Li, W.; Lai, S.; Zhu, L.; Jiang, W.; Zhang, Q. The integration of terrestrial laser scanning and terrestrial and unmanned aerial vehicle digital photogrammetry for the documentation of Chinese classical gardens—A case study of Huanxiu Shanzhuang, Suzhou, China. *J. Cult. Herit.* **2018**, *33*, 222–230. [CrossRef]

35. Forkuo, E.; King, B. Automatic Fusion of Photogrammetric Imagery and Laser Scanner Point Clouds. In Proceedings of the International Archives of Photogrammetry and Remote Sensing, XXXV–B4, Istanbul, Turkey, 12–23 July 2004; pp. 921–926.
36. Vosselman, R. Fusion of Laser Scanning Data, Maps, and Aerial Photographs for Building. In Proceedings of the IEEE International Geoscience and Remote Sensing Symposium, Toronto, ON, Canada, 24–28 June 2002; pp. 85–88.
37. Gašparović, M.; Malarić, I. Increase of Readability and Accuracy of 3D Models Using Fusion of Close Range Photogrammetry and Laser Scanning. In Proceedings of the International Archives of Photogrammetry and Remote Sensing, XXXIX–B5, Melbourne, Australia, 25 August–01 September 2012; pp. 93–98.
38. Beraldin, J.-A. Integration of Laser Scanning and Close-Range Photogrammetry—The last decade and beyond. In Proceedings of the International Archives of Photogrammetry and Remote Sensing, XXXV–B4, Istanbul, Turkey, 12–23 July 2004; pp. 12–23.
39. Jo, Y.H.; Kim, J. Three-Dimensional Digital Documentation of Heritage Sites Using Terrestrial Laser Scanning and Unmanned Aerial Vehicle Photogrammetry. In Proceedings of the XXVI International CIPA Symposium, Ottawa, ON, Canada, 28 August–1 September 2017; pp. 395–398.
40. Pérez, M.; Agüera, F.; Carvajal, F. Low Cost Surveying Using an Unmanned Aerial Vehicle. In Proceedings of the International Archives of Photogrammetry and Remote Sensing, XL-1/W2, Rostock, Germany, 4–6 September 2013; pp. 311–315.
41. Gašparović, M.; Gajski, D. Two-Step Camera Calibration Method Developed for Micro UAV's. In Proceedings of the International Archives of Photogrammetry and Remote Sensing, XLI-B1, Prague, Czech Republic, 12–19 July 2016; pp. 829–833.
42. Gašparović, M.; Jurjević, L. Gimbal influence on the stability of exterior orientation parameters of UAV acquired images. *Sensors* **2017**, *17*, 401. [CrossRef] [PubMed]
43. Nikolakopoulos, K.G.; Soura, K.; Koukouvelas, I.K.; Argyropoulos, N.G. UAV vs. classical aerial photogrammetry for archaeological studies. *J. Archaeol. Sci. Rep.* **2017**, *14*, 758–773. [CrossRef]
44. Liu, K.; Ding, H.; Tang, G.; Na, J.; Huang, X.; Xue, Z.; Yang, X.; Li, F. Detection of catchment-scale gully-affected areas using unmanned aerial vehicle (UAV) on the Chinese Loess Plateau. *ISPRS Int. J. Geo-Inf.* **2018**, *5*, 238. [CrossRef]
45. Liu, Y.; Zheng, X.; Ai, G.; Zhang, Y.; Zuo, Y. Generating a high-precision true digital orthophoto map based on UAV images. *ISPRS Int. J. Geo-Inf.* **2018**, *7*, 333. [CrossRef]
46. Salach, A.; Bakuła, K.; Pilarska, M.; Ostrowski, W.; Górski, K.; Kurczyński, Z. Accuracy assessment of point clouds from LiDAR and dense image matching acquired using the UAV platform for DTM creation. *ISPRS Int. J. Geo-Inf.* **2018**, *7*, 342. [CrossRef]
47. Moon, D.; Chung, S.; Kwon, S.; Seo, J.; Shin, J. Comparison and utilization of point cloud generated from photogrammetry and laser scanning: 3D world model for smart heavy equipment planning. *Automat Constr.* **2019**, in press. [CrossRef]
48. Habib, A.; Ghanma, M.; Mitishita, E. Co-registration of photogrammetric and LIDAR data: Methodology and case study. *Rev. Brasileira Cartogr.* **2004**, *56*, 1–13.
49. Besl, P.; McKay, N. A method for registration of 3-D shapes. *IEEE Trans. Pattern Anal. Mach. Intell.* **1992**, *14*, 239–256. [CrossRef]
50. Chen, Y.; Medioni, G. Object modelling by registration of multiple range images. *Int. J. Comput. Vis. Image Underst.* **1992**, *10*, 145–155. [CrossRef]

 International Journal of
Geo-Information

Article

Multi-Sensor UAV Application for Thermal Analysis on a Dry-Stone Terraced Vineyard in Rural Tuscany Landscape

Grazia Tucci [1,*], **Erica Isabella Parisi** [1], **Giulio Castelli** [2,3,*], **Alessandro Errico** [2,3], **Manuela Corongiu** [1], **Giovanna Sona** [4], **Enea Viviani** [5], **Elena Bresci** [2,3] and **Federico Preti** [2,3]

[1] SCHEMA—Joint Laboratory between DICEA—Department of Civil and Environmental Engineering, Italian Military Geographic Institute, University of Florence and IMGI, Via di S. Marta, 3-50139 Florence, Italy; ericaisabella.parisi@unifi.it (E.I.P.); manuela.corongiu@unifi.it (M.C.)

[2] DAGRI—Department of Science and Technology of Agriculture, Food, Environment and Forestry—University of Florence, Via di S. Bonaventura, Quaracchi, 13-50145 Florence, Italy; alessandro.errico@unifi.it (A.E.); elena.bresci@unifi.it (E.B.); federico.preti@unifi.it (F.P.)

[3] WaVe (Water and Vegetation) Research Unit, University of Florence, Via di S. Bonaventura, Quaracchi, 13-50145 Florence, Italy

[4] DICA—Department of Civil and Environmental Engineering, Polytechnic University of Milan, P.zza Leonardo da Vinci, 32-20133 Milan, Italy; giovanna.sona@polimi.it

[5] IMGI—Italian Military Geographic Institute, Via C. Battisti, 10/12-50122 Florence, Italy; eneaviviani@hotmail.com

[*] Correspondence: grazia.tucci@unifi.it (G.T.); giulio.castelli@unifi.it (G.C.); Tel.: +39-055-275-8874 (G.T.); +39-055-275-5647 (G.C.)

Received: 31 January 2019; Accepted: 13 February 2019; Published: 15 February 2019

Abstract: Italian dry-stone wall terracing represents one of the most iconic features of agricultural landscapes across Europe, with sites listed among UNESCO World Heritage Sites and FAO Globally Important Agricultural Heritage Systems (GIAHS). The analysis of microclimate modifications induced by alterations of hillslope and by dry-stone walls is of particular interest for the valuation of benefits and drawbacks of terraces cultivation, a global land management technique. The aim of this paper is to perform a thermal characterization of a dry-stone wall terraced vineyard in the Chianti area (Tuscany, Italy), to detect possible microclimate dynamics induced by dry-stone terracing. The aerial surveys were carried out by using two sensors, in the Visible (VIS) and Thermal InfraRed (TIR) spectral range, mounted on Unmanned Aerial Vehicles (UAVs), with two different flights. Our results reveal that, in the morning, vineyard rows close to dry-stone walls have statistically lower temperatures with respect to the external ones. In the afternoon, due to solar insulation, temperatures raised to the same value for each row. The results of this early study, jointly with the latest developments in UAV and sensor technologies, justify and encourage further analyses on local climatic modifications in terraced landscapes.

Keywords: drones; photogrammetry; thermal imaging; thermal dynamics; dry-stone walls; terraces; heritage; landscape; geomatics; UNESCO

---

1. Introduction

Dry-stone wall terracing represents one of the most iconic features of agricultural landscapes across Europe, as well as in other rural regions of the world. Italian and other dry-stone walls were added in November 2018 to UNESCO's Representative List of the Intangible Cultural Heritage of Humanity (Decision 13.COM 10.b.10). Dry stone walling was recognized to be still in use in several countries, playing a fundamental role in maintaining the environment and landscape.

In addition, many terraced landscapes of the world are already inserted among UNESCO World Heritage Sites list, as well as in the newly established FAO Globally Important Agricultural Heritage Systems (GIAHS), including cases of Italian dry-stone wall terracing, such as "Portovenere, Cinque Terre, and the Islands terraced vineyards", (UNESCO Heritage Site), or the "Olive Groves of the Slopes between Assisi and Spoleto" (FAO GIAHS).

In Italy, and in particular in the Tuscany region, rural landscape is characterized by the presence of stone wall bench terraces [1]. It is evident how terracing can alter the physical landscape, influencing factors such as soil moisture, erosion control and soil conservation [2]. The different ecosystem services provided by agricultural terraces have been object of a large number of studies in recent years [2–4]. These structures can induce significant modifications to local microclimate, such as a shading, as well as effects on local winds, inducing a potential thermal effect.

Due to the growing availability of short- (cameras) to long-range (satellite) sensors, remote sensing-based methodologies have been applied to the analysis of agricultural terraced fields, focusing on geo-morphological aspects, including analysis of preferential drainage patterns [5], and large scale [6] to medium scale terraces sites mapping [7]. Furthermore, remote sensing methodologies can be used to detect relevant bio-geo physical phenomena in these unique agricultural landscapes, such as microclimatic effects induced by the geo-morphological shape of bench terraces, as well as the presence of anthropic artifacts, such as dry-stone walls.

Nowadays, Unmanned Aerial Vehicle (UAV) systems can compete with traditional remote-sensing platforms (i.e., aircrafts and satellites) for agricultural mapping, thanks to their high flexibility of use, low operational cost and very high spatial resolution (cm-level). In Matese et al. [8] a comparison between vegetation indices obtained by satellite, aircraft and UAV platforms for intra-vineyard variability monitoring, highlights pros and cons of the different technologies: (i) satellite survey allows to simultaneously cover a wider area but is affected by some limitations, such as low spatial resolution for precision agriculture purposes, meteorological- and orbit-dependent acquisition, high economic costs; (ii) aircraft surveys are characterized by higher planning flexibility although noticeable organizational efforts and costs needs to be considered; (iii) UAVs are suitable systems for small field applications, but are still limited on wider areas because of their short flight endurance and limited payload for sensor carriage.

UAV photogrammetry has been extensively used for mapping and 3D modeling in the geomatics field, as evidenced by the huge increase in this research field in the last few years [9–12]. These platforms, equipped with digital RGB cameras, represent now a common alternative to traditional aerial manned photogrammetry. The application fields of image-based photogrammetric products from UAVs, such as 3D models or orthoimages, relate to forestry and agriculture [13], archaeology [14], architecture [15], cultural heritage [16–18], environmental surveying [19], traffic monitoring [20], and many others. Considering the above-mentioned features, and the possibility of carrying different type of payloads, ranging from RGB to Thermal InfraRed (TIR) sensors, UAVs appears to represent a key technology allowing a deeper understanding of small- to medium-scale bio-geophysical processes in agricultural landscapes.

The acquisition of the emitted radiation from natural surfaces in the range of TIR, thanks to airborne mounted sensors, allows us to obtain useful indication about soil and crops status by the analysis of temperature distribution across the field. For the specific case of vineyards, temperature patterns play a crucial role in regulating plant physiology, influencing phenology [21] and photosynthesis [22]. Furthermore, temperature trends were found to significantly impact on the berry developmental processes, such as the accumulation of dry matter and sugar [23], or the accumulation of anthocyanins [24], thus potentially determining the quality of the final product, wine. Costa et al. [25] reported a research on the evaluation of canopy temperature to generate maps for irrigation scheduling; other thermal-related indices were proposed by Idso et al. [26] and Jones [27,28] for the evaluation of crop water stress and stomatal conductance, respectively. Applications of thermal imaging on vineyard has been tested by García-Tejero et al. [29], Grant et al. [30], Jones et al. [28,31]

and Moller et al. [32], who researched the correlation between thermal indices and different water regimes of vineyards. Further tests with thermal camera mounted on UAV systems are reported in Baluja et al. [33], for the evaluation of water status variability; Berni et al. [34] estimated biophysical parameters and water stress from thermal imagery; Turner et al. [35] presented a UAV system equipped with a thermal camera for soil moisture assessment to improve irrigation efficiency; Bellvert et al. [36] reported the evaluation of crop water stress index (CWSI) for precision irrigation in a Spanish vineyard.

To our knowledge, few, early experiences of thermal monitoring of dry-stone agricultural terraces have been carried out. Warren et al. [37] tested the use of TIR imagery for the detection of dry-stone walls features. They showed how thermal imagery can be used to detect wall features such as the depth of backfill, areas of high moisture and the presence of particular structural elements not visible from conventional surveys. Barbera et al. [38] analyzed the microclimatic modifications induced by dry stone walls in particular structures called "jardinu" (circular dry-stone walls protecting single or groups of trees), located in the Pantelleria Island terraced fields. They found out that the thermal inertia of the stones may greatly enhance dewfall occurrence, both on stones and on leaves, suggesting a thermal gradient between the stones and the nearby environment. Moreover, they observed that the dissipation of the heat cumulated during the day by the stones reduced temperature excursion at night, positively affecting citrus adaptation. The importance of microclimatic monitoring through remotely sensed methods for a more reliable and advanced management of eco-, as well as agricultural systems, is advancing in parallel to technological innovations of sensors and UAV platforms, as pointed out by Zellweger et al. [39]. For instance, the recent study of Romboli et al. [40] showed how microclimate spatial variability interacts with vine vigor in determining grape and wine characteristics. This latter one, however, was carried out in one non-terraced vineyard, with no anthropic modification to the surrounding landscape apart from cultivations. From the perspective of better understanding all the implications connected to dry-stone walls traditional structures over microclimate variations, further investigations are necessary.

Taking advantage by recent progresses in UAVs technologies, this paper presents a preliminary test carried out using a TIR sensor mounted on a UAV, coupled with a common RGB survey, to detect the thermal dynamics of a vineyard grown on a dry-stone wall terraced land, with the specific purposes of:

- Investigate strengths and weaknesses of the use of TIR sensor mounted on UAV for thermal analysis of terraced crops;
- conduct a preliminary test on the possible thermal effect that dry-stone walls can have on the vineyard microclimate, testing the hypothesis that stones have an influence on the temperature patterns of the field which can influence grape ripening and quality.

2. Study Area

The terraced area of Lamole, within the municipality of Greve in Chianti (Florence, Italy; 43°32′34.73″ N, 11°21′29.14″ E), represents one of the most relevant examples of agricultural terracing sites of the country. Despite not yet being present in FAO GIAHS and UNESCO Heritage lists, the site was included in the Rural Historic Landscape Catalogue [41] of the Italian Ministry of Agricultural, Food, Forestry Policies in February 2018. As reported in the related Ministerial Decree (n. 6415 of 20 February 2018), the Lamole area was included in the Catalogue because it represents "an historical landscape where soil protection measures, represented by dry-stone wall terraces, together with policoltures represented by olive groves and vineyards, the woodland and the historical settlements preserve the Chianti cultural identity. Moreover, Lamole represents a remarkable example of dynamic recover and conservation of historical agricultural practices, which play nowadays a peerless added value for the development of the entire territory."

The study has been carried out in the Grospoli terraced vineyard, a site that has been object of a significant number of studies related to hydrological processes and failures on terraces (among others,

Preti et al. [42,43]). The area lies in the Chianti Classico production region. Sangiovese, Petit Verdot, Cabernet Sauvignon cultivars are grown in the vineyard (Figure 1).

Figure 1. Vigna Grospoli vineyard in Lamole. Photo by G. Castelli.

The vineyard has an area of 1.76 ha, with slopes ranging from from 12 to 22% [43] and an average altitude of 545 m a.s.l. The average terrace width is around 11 m, with wall ranging from 1 to 2.7 m height. For the remote sensing analysis, rows of vines were divided in two groups, the North and the South part, that are shown in Figure 2.

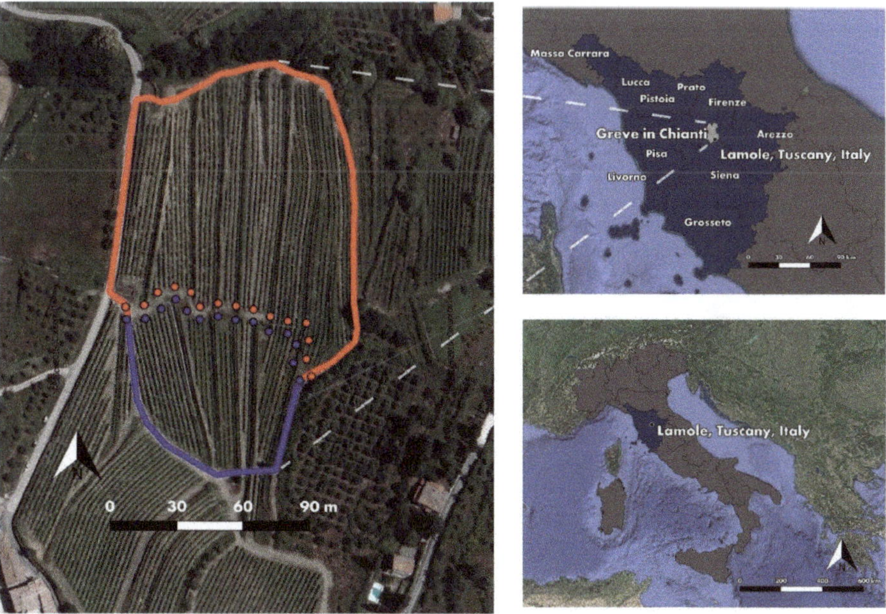

Figure 2. Location of the study area. Red border—North area; blue border—South area of the Grospoli vineyard in Tuscany region (Italy).

3. Materials and Methods

An aerial survey campaign was performed on 8 September 2017 over the Grospoli terraced vineyard in Lamole, a few weeks before grape harvesting, by employing two different UAVs, equipped with different sensors. The operations were carried out by the research group of the Survey, Cultural Heritage, Engineering, Monitoring, Analysis (SCHEMA) Laboratory, a joint collaboration between the University of Florence and the Italian Military Geographic Institute (IGMI). The first photogrammetric test was performed with a classic aerial approach in the visible (VIS) range, to produce a Digital Elevation Model (DEM) of the area and the respective projected orthophoto. The second test was performed in the TIR range in order to study the influence of stone walls upon the thermal properties of the neighboring rows. The TIR campaign was originally designed by planning three flights, in the morning, half day and evening, to monitor the thermal range over the entire day. Two flights were finally performed, due to the adverse climate conditions, in the morning (08:50 CET) and at 15:00 CET in the afternoon. Each flight was performed in terrain-following mode, which ensures the correct value of overlap and Ground Sample Distance (GSD) on uneven ground is maintained. The surveys with UAVs respected the regulation of the Italian Civil Aviation Authority (Ente Nazionale Aviazione Civile) in terms of non-critical operations, thus in Visual Line of Sight (VLoS).

3.1. VIS Data Capture and Processing

A single flight was performed with a DJI Phantom 4 Pro over the entire area of the Grospoli vineyard, for the photogrammetric data capture in the VIS range. The used UAV platform is a multirotor quadcopter, which can be remotely controlled and programmed for automatic navigation, provided by Global Positioning System (GPS) waypoints. The aircraft is characterized by limited weight of about 1400 g (battery, propellers and sensor included) and size (350 mm diagonal size), with max flight time of approximately 30 minutes and max speed of 72 km/h. The system supports two operating frequencies, at 2.4 GHz and 5.8 GHz and the maximum transmission range is of 3.5 km (according to CE standard). The platform is also equipped with a sensor system, with dual rear front optical sensors and side infrared sensors, which allow a total of 5 directions obstacle sensing and 4 direction obstacle avoidance. Dual Inertial Measurement Units (IMU) and compasses design provides redundancy in case of errors. Other technical characteristics are reported in Table 1.

Table 1. Technical specifications of DJI Phantom 4 Pro.

DJI Phantom 4 Pro Specifications	
Typology	Quadricopter
Weight	1388 g
Diagonal size	350 mm
Max flight time	Approx. 30 min
Max speed	72 kph (S-mode)
Power source	LiPo 4S
Satellite positioning system	GPS-GLONASS
Gimbal stabilization	3-axis (pitch, roll, yaw)
Range pitch	120° ($-90°$ to $+30°$)
Operating frequencies	2.4 GHz and 5.7 GHz
Max Transmission distance	3.5 km (CE)
Operating temperature	0–40 °C
Obstacle Sensory Range	0.2–7 m

The DJI Phantom 4 Pro is equipped with DJI FC6310 integrated digital RGB camera, stabilized with a 3-axis (pitch, roll, yaw) gimbal. The onboard camera uses a 1-inch CMOS 20 MP sensor, which can take images of 5272 × 3648 pixels as maximum resolution. The fixed focal length is 8.8 mm (35 mm format equivalent of 24 mm) with 84° FOV and f/2.8-f/11. The mechanical shutter, with 1/2000 s max speed, as well as the lens arrangement allows to obtain good image quality in terms of both sharpness and vividness. Table 2 lists the relevant technical properties of the RGB sensor (left column).

Table 2. Technical specifications of the two sensors: DJI FC6310 (RGB) and OPTRIS PI450 (TIR).

Camera	DJI FC6310	OPTRIS PI450
Spectral range	RGB [2]	TIR [4] (7.5–13 μm)
Sensor	1″ CMOS	FPA [5], uncooled
Sensor size	13.1 × 8.7 mm	25 × 25 μm
Resolution	20 MP (5472 × 3648 px)	382 × 288 px
Focal length	8.8 mm (f/2.8–f/11)	8 mm
FOV [1]	84°	62° × 49°
Output	JPEG [3] image	.RAVI [6] video
Weight	300 g	320 g
Temperature resolution	-	± 2 °C
Operating temp. range	0–40 °C	−20–250 °C

Acronyms explication: [1] Field Of View (FOV), [2] Red, Green, Blue, [3] Joint Photographic Experts Group (JPEG), [4] Thermal InfraRed (TIR), [5] Focal Plane Array (FPA), [6] RAdiometric VIdeo sequence (RAVI).

The flight plan was realized with UgCS software (SPH Engineering UgCS), by setting the suitable parameters in order to have a GSD of approximately 2 cm. The flight was designed according to a classic photogrammetric criterion, with a flight pattern consisting of 9 swipes, with forward overlapping of 80% and sidelap of 70%. The images were acquired with nadiral camera at a constant speed of 5 m/s and altitude of 70 m Above Ground Level (AGL). More details of the flight plan are reported in Table 3 (left column).

The georeferencing of the photogrammetric survey was made by using 10 Ground Control Points (GCP), which guarantee also the metric accuracy of the survey. The GCPs were materialized as targets functional for both visible and thermal survey: 60 × 60 cm² panels covered with aluminium foil and marked with black sprayed triangles. These targets were homogeneously distributed around the surveyed vineyard, to cover all the interested surface. The positioning and measuring activities for each target, were carried out by the Geodetic Direction of the IGMI, with the Global Navigation Satellite System (GNSS) TOPCON GR3. Two multi-frequency receivers were used for the coordinates acquisition, one as Master (base receiver) and the other as Rover receiver, working in Real Time Kinematic (RTK) radio mode. The coordinate system used in all data processing is ETRS89/UTM32N (EPSG:25832). The GCPs accuracy is about 1 cm.

The 206 images acquired for the VIS flight were processed with the photogrammetric software Agisoft Photoscan (version 1.2.6 build 2834). After image alignment, thanks to common Structure from Motion (SfM) procedures, GCPs were collimated and associated to their GNSS coordinates, for georeferencing and scale the photogrammetric model. A set of 7 of the 10 total collimated targets were used as GCPs, while 3 of them were used as Check Points (CP). Multi-View Stereo (MVS) algorithms were then used to obtain the dense cloud, from where the DEM of the surveyed area was generated. Finally, an orthophoto was projected as texture on the 3D model.

Table 3. The three flight plans in the visible (VIS) and Thermal InfraRed (TIR) ranges. The pictures represent camera locations (black dots) of each swipe and image overlap (colors).

Flight Plans	VIS Range	TIR Range
Time of acquisition	13:30	08:50 [1] /15:00 [2]
Flight altitude AGL	70 m	40 m
Forward overlap	80%	80%
Side overlap	70%	60%
GSD	2 cm/pix	12 [1,2] cm/pix
Number of pictures	206	578 [1] /603 [2]
Speed	5 m/s	3 m/s

TIR values in the [1] morning and [2] afternoon.

3.2. TIR Data Capture and Processing

The flight campaign for the thermal data capture was carried out on the same day with two distinct flights, at 08.50 CET and at 15:00 CET, by using an hexacopter customized by MicroGeo s.r.l. as UAV.

The flight plan was realized with the open source Mission Planner software, by setting the suitable parameters in order to have a GSD of 11 cm. The flights were designed according to a classic photogrammetric criterion, with a flight pattern consisting of 12 swipes, with forward overlapping of 80% and sidelap of 63%. The radiometric video sequences were acquired with nadiral orientation at a constant speed of 3 m/s and altitude of 40 m AGL. More details of the flight plan are reported in Table 3 (right column).

The UAV platform mounted the thermal camera OPTRIS PI450, with maximum optical resolution of 382 × 288 pixels at 80 Hz frame rate. The fixed focal length is 8 mm, with an angular FOV of 62° × 49°. The focal plane array (FPA) image sensor is made of uncooled microbolometers (25 × 25 μm each), with a spectral TIR response in the range 7.5–13 μm and accuracy of ±2 °C. Table 2 (right column) shows a summary of thermal camera specifications. The camera can be connected to a computer via a USB 2.0 protocol to download the acquired data in radiometric video sequences format (.RAVI). Single frames had consequently to be extracted from the video to be suitable for a classical photogrammetric project.

The same targets used for RGB photogrammetry were used for georeferencing the two thermal orthomosaics: since aluminium foil has a very low emissivity (0.09 ε) it can be easily seen in the TIR range. Some other panels were painted in black and white, onto the aluminium foil, and distributed all around the test field, as delimited test areas for pre-flight calibration. In fact, before the TIR flights, a calibration with the thermal sensor was made by detecting the thermal response over the GCP targets and the b/w panels and by comparing them with temperature measurements taken with contact thermometers.

The final values of ambient temperature and emissivity were kept as default, T_{amb} 23 °C and $\varepsilon = 1$ (vegetation usually has high values of emissivity, around 0.98 [34]) during the data processing step. For this reason, all the resulting thermal values must be considered only as qualitative information, because for absolute values a previous radiometric calibration would have been made.

The output of the thermal survey were, as mentioned, .RAVI radiometric videos, managed with the camera related software Optris PI Connect. The procedure for the frame extraction, in order to obtain pictures combined with the thermal information to use in the photogrammetric project, consisted of the following steps:

- automatic extraction of single frames from the thermal video, by setting the proper time of acquisition to maintain a suitable overlapping (Auto Key Presser was used as software). The result is a series of .csv datasheet files for each screenshot;
- conversion of the .csv files into 16-bit TIF images, through a point-to-point conversion (software ThermoVision_JoeC v. 1.0.6.0), to obtain three kinds of files: i) temperature 16-bit raster images (thermograms) with black and white values scaled according to the min/max temperature values of the overall set of images, ii) pictures with color palette set according to their own min/max temperature values, iii) an overview .txt file which reports the overall temperature range;
- evaluation of thermal outliers for each frame: a temperature range too large results in an insufficient contrast for the identification of homologous points by the photogrammetric software. In this case, the frames containing people were removed, because body temperature (~36 °C) increased the temperature maximum overall value. On the contrary, the aluminium targets, which reflect sky temperature, gave outliers of about −30 °C. In fact, the solar rays enter the thermal camera as reflected instead of emitted radiation, thus compromising the measures. For this reason, a default threshold of 0 °C was assigned, with an automatic script made with MathWorks MATLAB, to all values < 0 °C;
- after this normalization, the extracted thermograms were finally processed with the photogrammetric software Agisoft Photoscan (version 1.2.6 build 2834). The photogrammetric workflow, to obtain the DEM and the final thermal orthomosaics, is the same as reported in Section 3.1 for VIS processing, with the difference of having grey-scaled thermograms instead of RGB pictures;
- finally, a linear transformation was applied to the thermal orthomosaics, with a GIS software (ESRI ArcGIS) to re-calibrate the 16-bit raster values as a function of the min/max values of temperatures, by considering the overall data set.

This procedure of data processing has been developed and tested at the Polytechnic University of Milan, by Sona, as reported in Reference [44].

3.3. GIS Processing

The obtained thermal orthomosaics, were then processed in GIS software (open source QGIS). Every single vineyard row included in the test area has been acquired as vector feature, by manual photointerpretation of the RGB orthophoto, in order to classify only the vegetation component. Each row has been identified considering the position on the right and on the left in respect to the dry-stone wall of that terraced bench. Then a statistical zonal analysis was performed on each row, resulting from the overlay with the two thermal orthomosaic raster, to classify each row by their own thermal characteristics.

3.4. Statistical Analysis

Vineyard row thermal behaviour was finally analysed starting from the differences in average temperatures between the ones in external and the ones in internal position. For each bench terrace, we considered as external row the one at the edge of the terrace, and as internal the ones closest to the dry-stone wall supporting the overlying bench, as shown in Figure 3. All the rows placed in between the external and internal rows of each bench are identified as intermediate. For both the morning and the afternoon flights, the mean and the standard deviation of the temperatures of the internal rows and the one of external rows were computed, and the difference of the means was then calculated and analysed. In both cases, the level of significance of the differences in mean temperatures of the two samples (internal rows and external rows) was tested with Student's t-test for the comparison of two means.

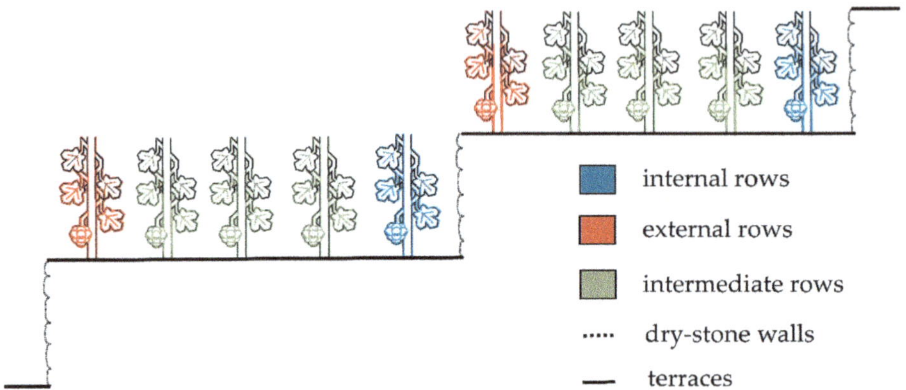

Figure 3. Each vineyard row has been labelled as: internal (blue) for the closest rows to the dry-stone wall (dotted line), external rows (red) for the ones at the edge of the terrace (black lines). All the rows comprised between internal and external have been identified as intermediate (green).

4. Results

4.1. Results of VIS Image Analysis

The processing of the 206 pictures from the VIS flight resulted in a 2 mm GSD orthophoto with a pixel size of 2.41×2.41 μm^2 (see Figure 4). The first features-matching procedure implemented in the SfM algorithm, produced a point cloud consisting of 1,308,933 features in a local coordinate system over an area of 114,000 m^2. By inserting the GCPs' coordinates, a bundle adjustment was performed to register the model in the UTM reference system, as reported in Section 3.1. The statistical values of the external orientation are listed in Table 4, reported as total RMSE values of GCPs and CPs. After the matching process which generated a sparse cloud, a denser cloud of 110,536,245 points was generated. Successively, a DEM surface model was created with a resolution of 4 cm/pix and point density of 772 points/m^2.

Figure 4. The orthophoto of the study area produced by the photogrammetric processing of the images acquired from the RGB flight over the Grospoli vineyard.

4.2. Results of TIR Image Analysis

The different spectral range and the lower spatial resolution of the TIR sensor compared to the VIS one resulted in orthomosaics with ground resolution of 12 cm/pix both. In fact, the lack of information in the grey-scaled thermograms for the feature-matching procedure of the software gave a lower number of tie points as a result, compared to the VIS orthophoto, respectively of 76,287 for the morning orthomosaic and of 105,949 for the afternoon one.

The same GCPs used to scale and georeferencing the VIS 3D model, gave different values of RMSE for the TIR projects, as reported in Table 4. The higher error values associated to the TIR CPs could be attributed to the difficulties in exactly picking the centre of the targets, mainly because of the low image quality, thus resulting in wrong identifications. In fact, the calculated Instantaneous Field Of View (IFOV) is of 91 mm, by considering the lens characteristics (FOV) of the thermal camera and the distance from the ground, thus giving a spatial resolution of about 1 cm, i.e., the minimum detail that can be seen on the thermal map.

The two generated dense clouds had similar sizes, of 964,066 points for the morning and 880,995 points for the afternoon. The DEM surface model was finally created with a resolution of 23 cm/pix and 18 points/m^2 for both.

Table 4. Statistics of Ground Control Points (GCPs) and Check Points (CPs) for RGB and TIR photogrammetric projects provided by Agisoft PhotoScan.

	Control Points		Check Points	
	RMSE [cm]	RMSE [pix]	RMSE [cm]	RMSE [pix]
RGB	1.59	0.13	2.52	0.15
TIR morning	0.34	0.05	5.43	0.05
TIR afternoon	0.45	0.06	11.4	0.06

Figure 5 reports the two orthomosaics scaled differently, in function of their maximum and minimum temperature, to enhance visualization. In particular the morning orthomosaic shows a temperature range of 14.8–18.3 °C while in the afternoon the T range is between 21 °C and 25 °C.

From a visible comparison, it appears that the external rows are warmer compared to the internal ones in the morning, while the same have similar temperature in the afternoon. The subtraction of the morning orthomosaic from the afternoon thermal map is shown in Figure 6, where the different temperature range of internal rows in the morning compared to the afternoon is clearly visible.

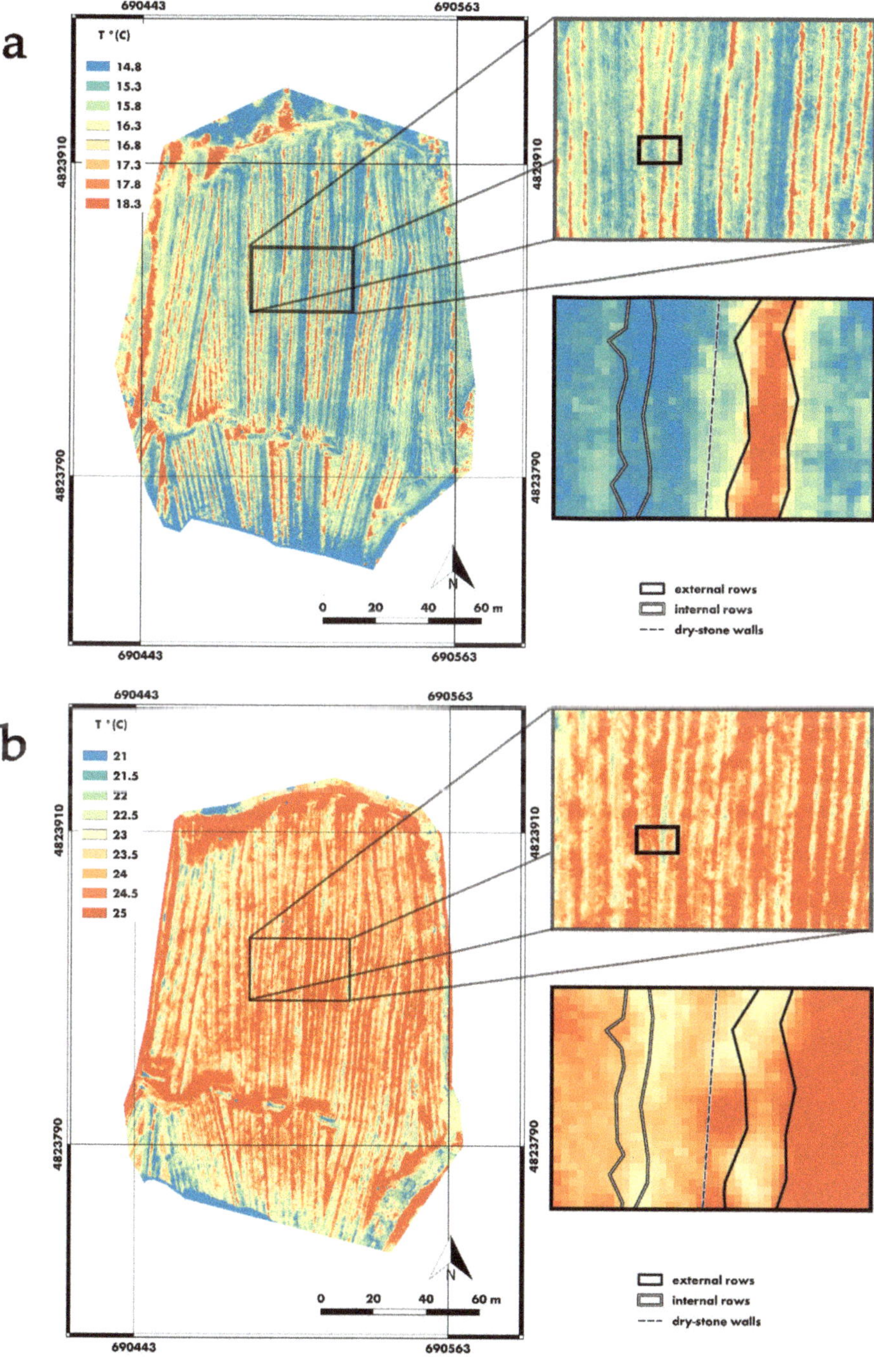

Figure 5. Thermal maps of the Grospoli vineyard in (**a**) morning (08.50 CET) and (**b**) afternoon (15:00 CET). The enlarged windows highlight internal and external rows, showing their respective thermal behaviours.

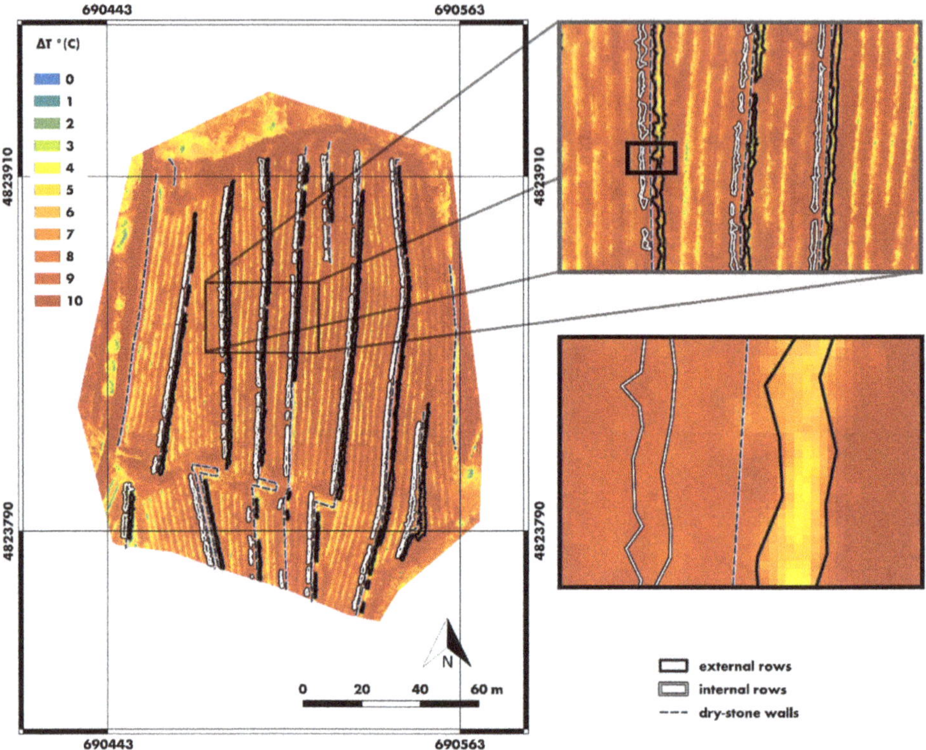

Figure 6. Difference between the afternoon thermal map and the morning orthomosaics highlighting, in the enlarged windows, the behaviour of the internal and external rows.

4.3. Results of Thermal Behaviour Analysis

The analysis of morning and afternoon temperatures showed how, on average, internal rows were colder then external ones of more than 2 °C, with a statistical significance of >99%, while temperatures recorded at 15:00 CET were almost equal for the two groups, around 23 °C (Table 5). Furthermore, the scatter plot of the morning and afternoon temperatures (Figure 7) showed how, in the morning, the whole population of internal rows exhibited temperatures lower than the external ones. On the other hand, afternoon temperatures were at the same level. The two populations appeared then as clearly separated.

To further investigate spatial and temporal temperatures patterns in the terraced hillslope, a full analysis of all rows of Grospoli vineyard was developed, considering also the values of temperatures of intermediate rows (Figure 8). The analysis showed how, for both the North and the South areas (see Figure 2), temperatures progressively decreased from the external edge to the internal position in the morning, while they grew up to the same value after 5 hours of insulation.

Table 5. Statistical analysis of averages of morning and afternoon temperatures, for external and internal rows (°C).

	Morning	Afternoon
External	17.6 (0.9)	23 (0.4)
Internal	15.4 (0.4)	23.1 (0.5)
P-value	7E-09	0.919
Statistical significance	>99%	NO

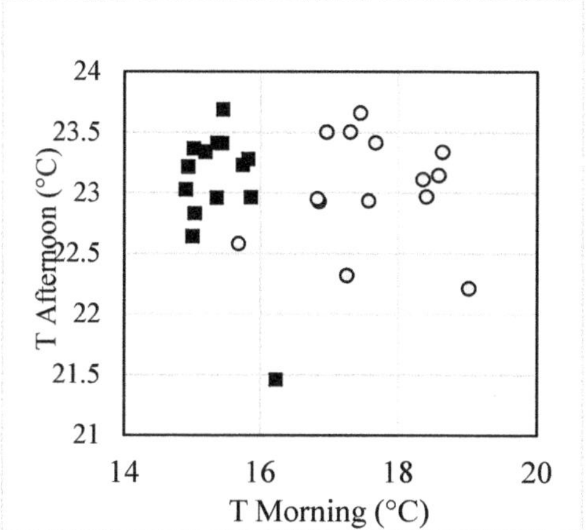

Figure 7. Scatter plot of morning and afternoon temperatures for external (circles) and internal (black squares) rows.

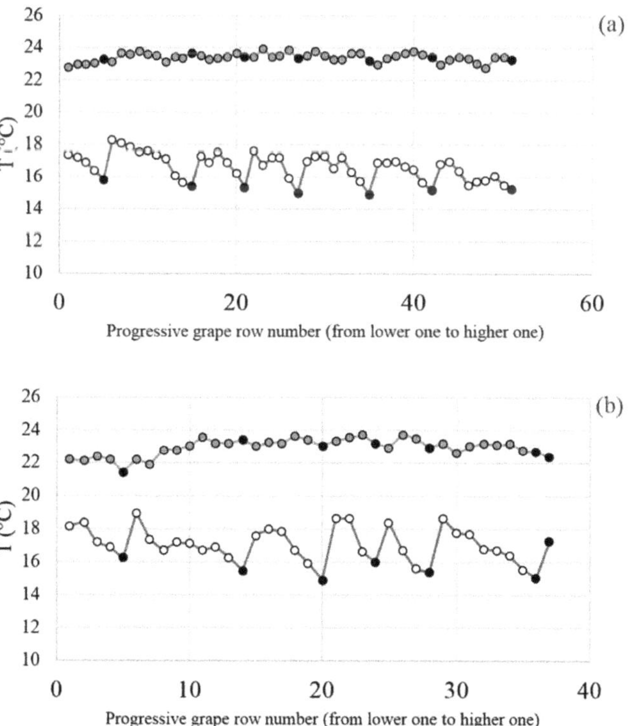

Figure 8. Temperature average values along the hillside terraced Grospoli vineyard. (**a**) North area; (**b**) South area. White and grey timeseries represent morning and afternoon temperatures respectively. Black dots represent internal rows temperature mean values.

5. Discussion

Agricultural terraces are a fundamental feature of Mediterranean landscapes, which are threatened (among other causes) by the risk of abandonment and consequent degradation due to the progressive depopulation of marginal areas and the industrialization of agriculture [2]. In some cases, these ancient structures allowed the settlement and the survival of entire peoples, which through the practice of terracing have allowed cultivation of the steepest slopes and, thanks to the enhanced slope stability, the building of villages at the valley bottoms, containing the risk of landslides. In these situations, the abandonment of such anthropic structures can even imply the increase of hazards related to geo-hydrological processes that could be triggered by extreme rainfall events. In this context, a dramatic example is represented by the tremendous event occurred in the Cinque Terre area in 2011 [45,46]. However, the aspect of civil protection is limited to the few specific cases where settlements lie under terraced slopes. Beyond this, agricultural terraces represent a cultural heritage which should be maintained as part of a culture strongly connected to every territory where it is, or it was, present [47]. In this perspective, an economical sustainability of agriculture on terraced land should be encouraged and helped to keep on going despite the higher costs that it involves. For this reason, understanding and quantifying all the benefits produced by a terraced landscape becomes of paramount importance to address effective policies for its conservation.

In this context, the availability of a cost-affordable and efficient methodology for monitoring of terraces evolutions, in terms of both spatial morphology (RGB) or thermal behaviour induced by the presence of dry-stone walls (TIR), can be a useful tool for risk prevention and managing purposes.

Our preliminary tests carried out by TIR surveys showed some influence on the temperature distribution across the vineyard due to the presence of terraces.

The thermal behaviour of the rows can be well explained by considering the sun position in respect to the vineyard. Since the slope of the terraced vineyard is aligned to the E-W position, at the sunset the internal rows are shaded by the presence of overlying dry-stone walls. From 08:50 CET to 15:00 CET, the sun moves along its trajectory, irradiating progressively the intermediate rows, in the middle of each bench, and then the internal ones. At 15:00 CET, the sun's position allowed the irradiation of all the rows, as reported in Figure 9.

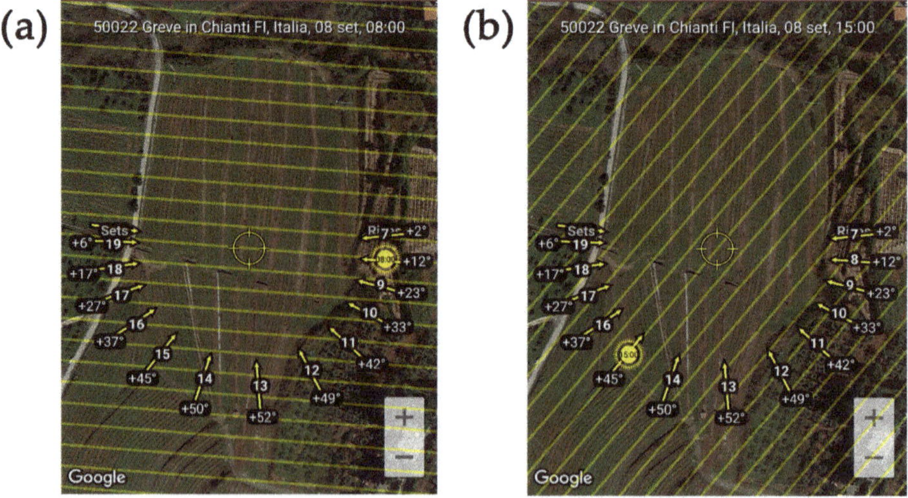

Figure 9. Sun position over the Grospoli vineyard on 8 September 2017, at (**a**) 08:00 CET (solar rays perpendicular to the rows) and (**b**) 15:00 CET (solar rays irradiating all the rows).

Therefore, thermal differences appeared to be mainly influenced by the shading effect of the terrace risers, but we do not exclude the possibility that the walls themselves could have other effects related to the thermal behavior of stones.

A small fluctuation between the average temperature of internal and external rows can still be observed in the afternoon thermal data (see Figure 8). This variation might be explained by considering the high thermal inertia associated to stone wall. In fact, the medium value of specific heat for stone, is around 840 J/kg·°C (at 25 °C and atmospheric pressure) [48], thus making it a very common material in historical buildings of the Mediterranean area, typically characterized by thick and massive walls [49]. In that context the main function was related to prevent summer overheating by keeping low and comfortable indoor temperature. The poor heat conductivity of rocks, which exact value depends on stone porosity, mineral composition and specific heat conductivity, usually entails low heat transfer from the irradiated surface to the inner part of the rock [48]. These properties can be applied also on the dry-stone walls of the vineyard, which will warm up slowly during the day and they will passively release heat during the night, thus influencing also the facing internal rows temperature and microclimate.

Barbera et al. [38] analysed a wide range of environmental parameters to investigate the role of particular dry-stone wall structures in modifying the microclimate for crops. Among other interesting findings, they showed how the different thermal inertia of stones result in temperature differences (stones are colder in the morning and warmer in the evening, also showing very high temperatures when heated by the sun) which might influence the local microclimate. Working in a totally different context, Argyle and Stevens [50] investigated the distribution and growth patterns of a spontaneous shrub, the netleaf hackberry, growing around boulders in the cold desert biome of Wasatch Mountains (Utah). Unexpectedly, they found that hackberries associated with boulders were more likely to grow near the south side, suggesting that big stones may influence patterns of growth providing thermal radiation that melts snow thus providing a higher water availability for seedlings. Despite the fact that their research is far from the case of terraced vineyards, we hypothesize that the thermal inertia of dry-stone walls can have an influence, at least at the microclimatic scale, on temperature patterns of the crop and thus on its growth, phenology, ripening, and consequently on quality. However, neither works available in literature nor our findings have not yet shown it clearly. We believe that repeated measurements taken in different days at equal or comparable solar insulation conditions would provide a reliable time series to confirm these patterns of temperatures.

Nevertheless, the UAV-mounted TIR sensor, coupled with a RGB camera, proved to be a suitable monitoring method to reveal thermal variations induced by both natural sources (sun) as by anthropic artifacts (dry-stone walls). The main advantages of this approach consist in: (i) contained operative costs (ii) higher spatial resolution compared to traditional remote sensing platforms for TIR sensors (aircrafts and satellites); (iii) higher spatial covering in respect to ground-based TIR surveys; (iv) possibility of georeferencing spatial (RGB) and radiometric (TIR) information in a GIS software to maximize the data available for managing and prevention purposes; (v) reduced time required for data acquisition, thus allowing to repeat the survey multiple times during the day/month/year for monitoring of daily/seasonal variations; (vi) sufficient radiometric accuracy to discriminate qualitative daily thermal variations.

Some limitations, however, need to be pointed out. As Conte et al. reported in their work [51], having both high spatial and radiometric resolution in aerial thermal imagery is currently not possible. Despite the above-mentioned advantages in using UAVs as platforms and the progressively miniaturizing process of multi-spectral sensors, the two resolutions are not comparable yet, even if an exponential technological growth is happening.

Another aspect to consider, is related to calibration. A radiometric calibration should be carried out in order to obtain absolute values of temperature. In this work, a relative thermal information was enough to obtain some preliminary information and to test the adopted methodology. For this reason, statistical analysis on exact temperature values (reported in Section 4.3) has to be considered as a relative information about the different thermal behaviour between morning and afternoon. Furthermore, the temperature resolution of the TIR camera does not allow us to make quantitative considerations. A proper radiometric calibration with a black body in laboratory, as well as setting adequately other factors as ambient temperature, emissivity, and spot size, might reduce the possible margin of error for more accurate temperature evaluations.

A geometric calibration, instead, concerns the metric accuracy of the orthomosaics, which is fundamental when RGB and TIR data are overlapping in GIS. In fact, TIR orthomosaics geometric resolution is significantly lower than the RGB, where accuracies of 0.1–0.2 pix can be achieved. The main source of error is due to the radial distortion of the TIR sensor, deriving from refraction variations and other factors, as reported by Luhmann [52]. For this reason, a pre-flight calibration for the thermal camera is suggested, with special targets to enhance contrast in thermal images. In fact, the reduced contrast and the blurring effects on objects edges make very difficult a precise collimation of a point in thermal frames. The authors suggest using algorithms for radiometric image enhancement and increasing the number of overlapping images (i.e., redundancy) to ease the photogrammetric workflow (SfM algorithm). Furthermore, proper accuracy evaluations should be made by considering fixed calibration certificates, obtained in laboratory, and the analysis of computed ones, in addition the sole evaluation of check point residuals.

Further analysis should be considered for evaluating the potential thermal inertia effect of dry-stone walls on the surrounding vegetation, as shown by Barbera et al. [38]. Depending on its strength, this additional effect may have the potential of influencing local microclimate, as well as the characteristics of cultivations grown on bench terraces, such as grapevine. Considering the results of the present experiment, the use of TIR remote sensing based on UAV appears to be a reliable technological solution to assess this additional effect, for instance by mapping temperatures during or after the sunset. Further tests should be also performed to improve the photogrammetric process, as reported by Arandjelović et al. [53] for the registration of time-separated aerial images, to improve the overall geometric accuracy of the produced outputs (orthomosaics). Moreover, applications of TIR remote sensing that can be useful for analysis and monitoring of terraced landscape are not limited to thermal analysis. In fact, as shown by Glaser et al. [54], TIR analysis can highlight surface and subsurface water saturation dynamics. This application can provide additional insights to the studies carried out on the generation of preferential drainage patterns in terraced hillslopes, fundamental to understanding terraces degradation dynamics, that so far were based only on geo-morphological analyses and modelling [42,43,55].

Another future extension of this contribution might be the integration of different sensors on the same UAV platform, operating in different ranges of the electromagnetic spectrum, such as multi- or iper-spectral sensors [33,56,57]. In this way, additional information coming from selected or wider band of the spectral range would complete and extend the data collected, with biophysical vegetation indexes useful also for precision agriculture purposes.

6. Conclusions

The present work describes an early application of the use of combined multi-sensor surveys, in the VIS and TIR range mounted of UAV platforms, to detect thermal dynamics of dry-stone terraced areas, which represent one of the most iconic features of agricultural landscapes in Italy and Europe. The use of UAVs platforms for remote sensing presents advantages in terms of operative cost, high planning flexibility and higher spatial resolution if compared to common remote-sensing platforms (such as aircrafts and satellites) [8]. The methodology tested in this work, with an application in the historic Chianti area of Tuscany, proved to be an efficient tool for monitoring and managing these traditional rural landscapes. In particular, the UAV-mounted TIR sensor, coupled with an RGB camera approach allowed us to:

- obtain sufficient radiometric accuracy to reveal daily thermal variations induced by both natural sources (sun) as by anthropic artifacts (dry-stone walls)

- contain the operative costs and have higher spatial resolution compared to traditional remote sensing platforms for TIR sensors (aircrafts and satellites)

- cover a wider area in respect to ground-based TIR surveys

- georeferencing spatial (RGB) and radiometric (TIR) information in a GIS software

- repeat the survey multiple times during the day, thanks to reduced time needed for data acquisition

The preliminary results proposed in this contribution highlight a different thermal behavior of rows in the morning, when the internal rows show lower temperature than the external rows, due to solar position and shading effect. After some hours of direct irradiation, in the afternoon, temperature of both internal and external rows increased, reaching almost the same values.

These outcomes, as well as the rapid evolution in sensors and UAV technologies, encourage further tests with the above-mentioned methodology to detect microclimatic effects of dry-stone walls on the bench-growing vegetation. For instance, repeating thermal flights more frequently within the entire 24 hours could help to better understand temperature patterns across the entire terraced field. Other future activities might concern the integration with other sensors, operating in different bands of the electromagnetic spectrum to obtain biophysical indexes of vegetation, as well as monitoring at different time of the day/month/year.

Author Contributions: Conceptualization, F.P., G.C., G.T., and E.B.; methodology, G.S. and G.T.; software, M.C., G.S. and E.V.; validation, E.I.P., G.C. and A.E.; formal analysis, G.C., E.V., M.C.; investigation, E.I.P., G.C. and A.E.; resources, G.T. and F.P.; data curation, E.I.P., G.C. and E.V.; writing—original draft preparation, E.I.P.; writing—review and editing, E.I.P., G.C. and A.E.; visualization, E.I.P. and G.C.; supervision, G.T., G.S. and. F.P.; project administration, G.T., E.B. and. F.P.; funding acquisition, G.T. and F.P.

Funding: The work has been partially supported as part of the Italian Research Project of National Interest "GAMHer—Geomatics Data Acquisition and Management for Landscape and Built Heritage in a European Perspective" (PRIN2015 n.2015HJLS7E), of which DICEA is partner. A.E. and F.P. acknowledge the financial support provided by Programma Interreg Italia-Francia marittimo 2014-2020 ("Adapt-assistere l'adattamento ai cambiamenti climatici dei sistemi urbani dello spazio trasfrontaliero", CUP B19J16002890007). Publication costs were covered by the University of Florence DICEA and DAGRI Departments.

Acknowledgments: For providing the access to the study area of Grospoli vineyard and for his friendliness and hospitality, we would like to thank Paolo Socci, owner of the Fattoria di Lamole farm. The authors wish to thank the MicroGeo s.r.l. company for having kindly provided the tested UAV platform coupled with the TIR sensor and their flight operator Arch. Mattia Ventimiglia. Special thanks go also to Filippo Fiaschi, DICEA flight operator for the RGB flight. We also want to thank Dr. Valentina Bonora, for planning data collection in both VIS and TIR range and to Arch. Lidia Fiorini for field operations. Our gratitude also goes to Geodetic Direction of the IGMI for the GNSS operations support and to Brig. Gen. Enzo Santoro, Chief of the Geospatial Information Department, for coordinating the activities. Finally, we express our gratitude to Prof. Michael Hensel (Vienna University of Technology) for having encouraged this study through interdisciplinary collaborations.

References

1. Preti, F.; Tarolli, P.; Dani, A.; Calligaro, S.; Prosdocimi, M. LiDAR derived high resolution topography: The next challenge for the analysis of terraces stability and vineyard soil erosion. *J. Agric. Eng.* **2013**, *44*. [CrossRef]

2. Wei, W.; Chen, D.; Wang, L.; Daryanto, S.; Chen, L.; Yu, Y.; Lu, Y.; Sun, G.; Feng, T. Global synthesis of the classifications, distributions, benefits and issues of terracing. *Earth-Sci. Rev.* **2016**, *159*, 388–403. [CrossRef]

3. Arnáez, J.; Lana-Renault, N.; Lasanta, T.; Ruiz-Flaño, P.; Castroviejo, J. Effects of farming terraces on hydrological and geomorphological processes. A review. *CATENA* **2015**, *128*, 122–134. [CrossRef]

4. Socci, P.; Errico, A.; Castelli, G.; Penna, D.; Preti, F. Terracing: From Agriculture to Multiple Ecosystem Services. In *Oxford Research Encyclopedia on Agriculture*; in press.

5. Tarolli, P.; Sofia, G.; Calligaro, S.; Prosdocimi, M.; Preti, F.; Dalla Fontana, G. Vineyards in Terraced Landscapes: New Opportunities from Lidar Data: Vineyards in terraced landscapes. *Land Degrad. Dev.* **2015**, *26*, 92–102. [CrossRef]

6.   Eckert, S.; Tesfay Ghebremicael, S.; Hurni, H.; Kohler, T. Identification and classification of structural soil conservation measures based on very high resolution stereo satellite data. *J. Environ. Manag.* **2017**, *193*, 592–606. [CrossRef] [PubMed]

7.   Diaz-Varela, R.A.; Zarco-Tejada, P.J.; Angileri, V.; Loudjani, P. Automatic identification of agricultural terraces through object-oriented analysis of very high resolution DSMs and multispectral imagery obtained from an unmanned aerial vehicle. *J. Environ. Manag.* **2014**, *134*, 117–126. [CrossRef]

8.   Matese, A.; Toscano, P.; Di Gennaro, S.; Genesio, L.; Vaccari, F.; Primicerio, J.; Belli, C.; Zaldei, A.; Bianconi, R.; Gioli, B. Intercomparison of UAV, Aircraft and Satellite Remote Sensing Platforms for Precision Viticulture. *Remote Sens.* **2015**, *7*, 2971–2990. [CrossRef]

9.   Eisenbeiss, H.; Sauerbier, M. Investigation of uav systems and flight modes for photogrammetric applications: Investigation of UAV systems and flight modes. *Photogramm. Rec.* **2011**, *26*, 400–421. [CrossRef]

10.   Neitzel, F.; Klonowski, J. Mobile 3D mapping with a low-cost UAV system. *ISPRS Int. Arch. Photogramm. Remote Sens. Spat. Inf. Sci.* **2012**, *XXXVIII-1/C22*, 39–44. [CrossRef]

11.   Chiabrando, F.; Lingua, A.; Piras, M. Direct photogrammetry using UAV: Tests and first results. *ISPRS Int. Arch. Photogramm. Remote Sens. Spat. Inf. Sci.* **2013**, *XL-1/W2*, 81–86. [CrossRef]

12.   Dominici, D.; Alicandro, M.; Massimi, V. UAV photogrammetry in the post-earthquake scenario: Case studies in L'Aquila. *Geomat. Nat. Hazards Risk* **2017**, *8*, 87–103. [CrossRef]

13.   Torresan, C.; Berton, A.; Carotenuto, F.; Di Gennaro, S.F.; Gioli, B.; Matese, A.; Miglietta, F.; Vagnoli, C.; Zaldei, A.; Wallace, L. Forestry applications of UAVs in Europe: A review. *Int. J. Remote Sens.* **2017**, *38*, 2427–2447. [CrossRef]

14.   Rinaudo, F.; Chiabrando, F.; Lingua, A.; Spanò, A. Archaeological site monitoring: UAV photogrammetry can be an answer. *ISPRS Int. Arch. Photogramm. Remote Sens. Spat. Inf. Sci.* **2012**, *XXXIX-B5*, 583–588. [CrossRef]

15.   Murtiyoso, A.; Grussenmeyer, P. Documentation of heritage buildings using close-range UAV images: Dense matching issues, comparison and case studies. *Photogramm. Rec.* **2017**, *32*, 206–229. [CrossRef]

16.   Lo Brutto, M.; Garraffa, A.; Meli, P. UAV platforms for cultural heritage survey: First results. *ISPRS Ann. Photogramm. Remote Sens. Spat. Inf. Sci.* **2014**, *II–5*, 227–234. [CrossRef]

17.   Achille, C.; Adami, A.; Chiarini, S.; Cremonesi, S.; Fassi, F.; Fregonese, L.; Taffurelli, L. UAV-Based Photogrammetry and Integrated Technologies for Architectural Applications—Methodological Strategies for the After-Quake Survey of Vertical Structures in Mantua (Italy). *Sensors* **2015**, *15*, 15520–15539. [CrossRef] [PubMed]

18.   Tucci, G.; Bonora, V. Geomatics and management of at-risk cultural heritage. *Rendiconti Lincei* **2015**, *26*, 105–114. [CrossRef]

19.   Gonçalves, J.A.; Henriques, R. UAV photogrammetry for topographic monitoring of coastal areas. *ISPRS J. Photogramm. Remote Sens.* **2015**, *104*, 101–111. [CrossRef]

20.   Kanistras, K.; Martins, G.; Rutherford, M.J.; Valavanis, K.P. A survey of unmanned aerial vehicles (UAVs) for traffic monitoring. In Proceedings of the 2013 International Conference on Unmanned Aircraft Systems (ICUAS), Atlanta, GA, USA, 28–31 May 2013; IEEE: Atlanta, GA, USA, 2013; pp. 221–234.

21.   Martínez-Lüscher, J.; Kizildeniz, T.; Vučetić, V.; Dai, Z.; Luedeling, E.; van Leeuwen, C.; Gomès, E.; Pascual, I.; Irigoyen, J.J.; Morales, F.; et al. Sensitivity of Grapevine Phenology to Water Availability, Temperature and CO_2 Concentration. *Front. Environ. Sci.* **2016**, *4*. [CrossRef]

22.   Ferrini, F.; Mattii, G.B.; Nicese, F.P. Effect of Temperature on Key Physiological Responses of Grapevine Leaf. *Am. J. Enol. Viticult.* **1995**, *46*, 5.

23.   Greer, D.; Weedon, M. Temperature-dependent responses of the berry developmental processes of three grapevine (*Vitis vinifera*) cultivars. *N. Z. J. Crop Hortic. Sci.* **2014**, *42*, 233–246. [CrossRef]

24.   Gaiotti, F.; Pastore, C.; Filippetti, I.; Lovat, L.; Belfiore, N.; Tomasi, D. Low night temperature at veraison enhances the accumulation of anthocyanins in Corvina grapes (*Vitis vinifera* L.). *Sci. Rep.* **2018**, *8*. [CrossRef] [PubMed]

25.   Costa, J.M.; Grant, O.M.; Chaves, M.M. Use of Thermal Imaging in Viticulture: Current Application and Future Prospects. In *Methodologies and Results in Grapevine Research*; Delrot, S., Medrano, H., Or, E., Bavaresco, L., Grando, S., Eds.; Springer: Dordrecht, The Netherlands, 2010; pp. 135–150.

26.   Idso, S.B.; Jackson, R.D.; Pinter, P.J.; Reginato, R.J.; Hatfield, J.L. Normalizing the stress-degree-day parameter for environmental variability. *Agric. Meteorol.* **1981**, *24*, 45–55. [CrossRef]

27. Jones, H.G. Use of infrared thermometry for estimation of stomatal conductance as a possible aid to irrigation scheduling. *Agric. For. Meteorol.* **1999**, *95*, 139–149. [CrossRef]

28. Jones, H.G. Use of infrared thermography for monitoring stomatal closure in the field: Application to grapevine. *J. Exp. Bot.* **2002**, *53*, 2249–2260. [CrossRef]

29. García-Tejero, I.F.; Costa, J.M.; Egipto, R.; Durán-Zuazo, V.H.; Lima, R.S.N.; Lopes, C.M.; Chaves, M.M. Thermal data to monitor crop-water status in irrigated Mediterranean viticulture. *Agric. Water Manag.* **2016**, *176*, 80–90. [CrossRef]

30. Grant, O.M.; Tronina, L.; Jones, H.G.; Chaves, M.M. Exploring thermal imaging variables for the detection of stress responses in grapevine under different irrigation regimes. *J. Exp. Bot.* **2006**, *58*, 815–825. [CrossRef]

31. Jones, H.G.; Serraj, R.; Loveys, B.R.; Xiong, L.; Wheaton, A.; Price, A.H. Thermal infrared imaging of crop canopies for the remote diagnosis and quantification of plant responses to water stress in the field. *Funct. Plant Biol.* **2009**, *36*, 978. [CrossRef]

32. Moller, M.; Alchanatis, V.; Cohen, Y.; Meron, M.; Tsipris, J.; Naor, A.; Ostrovsky, V.; Sprintsin, M.; Cohen, S. Use of thermal and visible imagery for estimating crop water status of irrigated grapevine. *J. Exp. Bot.* **2006**, *58*, 827–838. [CrossRef]

33. Baluja, J.; Diago, M.P.; Balda, P.; Zorer, R.; Meggio, F.; Morales, F.; Tardaguila, J. Assessment of vineyard water status variability by thermal and multispectral imagery using an unmanned aerial vehicle (UAV). *Irrig. Sci.* **2012**, *30*, 511–522. [CrossRef]

34. Berni, J.; Zarco-Tejada, P.J.; Suarez, L.; Fereres, E. Thermal and Narrowband Multispectral Remote Sensing for Vegetation Monitoring from an Unmanned Aerial Vehicle. *IEEE Trans. Geosci. Remote Sens.* **2009**, *47*, 722–738. [CrossRef]

35. Turner, D.; Lucieer, A.; Watson, C. Development of an Unmanned Aerial Vehicle (UAV) for hyper resolution vineyard mapping based on visible, multispectral, and thermal imagery. In Proceedings of the 34th International Symposium on Remote Sensing of Environment, Sydney, Australia, 10–15 April 2011; p. 4.

36. Bellvert, J.; Zarco-Tejada, P.J.; Girona, J.; Fereres, E. Mapping crop water stress index in a 'Pinot-noir' vineyard: Comparing ground measurements with thermal remote sensing imagery from an unmanned aerial vehicle. *Precis. Agric.* **2014**, *15*, 361–376. [CrossRef]

37. Warren, L.A.; Briggs, K.M.; McCombie, P.F. Advances in the assessment of drystone retaining walls—Some case studies. In Proceedings of the XVI European Conference on Soil Mechanics and Geotechnical Engineering, Edinburgh, UK, 13–17 September 2015; p. 7.

38. Barbera, G.; Chieco, C.; Georgiadis, T.; Motisi, A.; Rossi, F. The "jardinu" of Pantelleria as a paradigm of resource-efficient horticulture in the built-up environment. *Acta Hortic.* **2018**, 351–356. [CrossRef]

39. Zellweger, F.; De Frenne, P.; Lenoir, J.; Rocchini, D.; Coomes, D. Advances in Microclimate Ecology Arising from Remote Sensing. *Trends Ecol. Evol.* **2019**. [CrossRef] [PubMed]

40. Romboli, Y.; Di Gennaro, S.F.; Mangani, S.; Buscioni, G.; Matese, A.; Genesio, L.; Vincenzini, M. Vine vigour modulates bunch microclimate and affects the composition of grape and wine flavonoids: An unmanned aerial vehicle approach in a Sangiovese vineyard in Tuscany: Vine vigour affects grape and wine flavonoids. *Aust. J. Grape Wine Res.* **2017**, *23*, 368–377. [CrossRef]

41. Agnoletti, M. *Paesaggi Rurali Storici. Per un catalogo nazionale*; Laterza: Bari, Italy, 2010.

42. Preti, F.; Errico, A.; Caruso, M.; Dani, A.; Guastini, E. Dry-stone wall terrace monitoring and modelling. *Land Degrad. Dev.* **2018**, *29*, 1806–1818. [CrossRef]

43. Preti, F.; Guastini, E.; Penna, D.; Dani, A.; Cassiani, G.; Boaga, J.; Deiana, R.; Romano, N.; Nasta, P.; Palladino, M.; et al. Conceptualization of Water Flow Pathways in Agricultural Terraced Landscapes: Water Flow Pathways in Agricultural Terraced Landscapes. *Land Degrad. Dev.* **2018**, *29*, 651–662. [CrossRef]

44. Mezzanzanica, M. Use of Multispectral and Thermal Images for the Characterization of Agricultural Lands. Master's Thesis, Polytechnic University of Milan, Milan, Italy, 2016.

45. Brandolini, P.; Cevasco, A.; Capolongo, D.; Pepe, G.; Lovergine, F.; Del Monte, M. Response of Terraced Slopes to a Very Intense Rainfall Event and Relationships with Land Abandonment: A Case Study from Cinque Terre (Italy): Agricultural Terraces and Slope Instability at Cinque Terre (NW Italy). *Land Degrad. Dev.* **2018**, *29*, 630–642. [CrossRef]

46. Agnoletti, M.; Errico, A.; Santoro, A.; Dani, A.; Preti, F. Terraced Landscapes and Hydrogeological Risk. Effects of Land Abandonment in Cinque Terre (Italy) during Severe Rainfall Events. *Sustainability* **2019**, *11*, 235. [CrossRef]

47. Yehong, S.; Qingwen, M.; Junchao, S.; Yabing, J. Terraced Landscapes as a Cultural and Natural Heritage Resource. *Tour. Geogr.* **2011**, *13*, 328–331. [CrossRef]
48. Siegesmund, S.; Snethlage, R. (Eds.) *Stone in Architecture Properties, Durability*, 5th ed.; Springer: Berlin/Heidelberg, Germany, 2014.
49. Evola, G.; Marletta, L.; Natarajan, S.; Maria Patanè, E. Thermal inertia of heavyweight traditional buildings: Experimental measurements and simulated scenarios. *Energy Procedia* **2017**, *133*, 42–52. [CrossRef]
50. Argyle, A.; Stevens, M.T. Influence of Boulders on Netleaf Hackberry (*Celtis reticulata*) Growth and Distribution in the Wasatch Foothills. *West. North Am. Nat.* **2013**, *73*, 525–529. [CrossRef]
51. Conte, P.; Girelli, V.A.; Mandanici, E. Structure from Motion for aerial thermal imagery at city scale: Pre-processing, camera calibration, accuracy assessment. *ISPRS J. Photogramm. Remote Sens.* **2018**, *146*, 320–333. [CrossRef]
52. Luhmann, T.; Piechel, J.; Roelfs, T. Geometric Calibration of Thermographic Cameras. In *Thermal Infrared Remote Sensing: Sensors, Methods, Applications*; Kuenzer, C., Dech, S., Eds.; Springer: Dordrecht, The Netherlands, 2013; pp. 27–42.
53. Arandjelović, O.; Pham, D.-S.; Venkatesh, S. Efficient and accurate set-based registration of time-separated aerial images. *Pattern Recognit.* **2015**, *48*, 3466–3476. [CrossRef]
54. Glaser, B.; Antonelli, M.; Chini, M.; Pfister, L.; Klaus, J. Technical note: Mapping surface saturation dynamics with thermal infrared imagery. *Hydrol. Earth Syst. Sci. Discuss.* **2018**, *22*, 5987–6003. [CrossRef]
55. Tarolli, P.; Preti, F.; Romano, N. Terraced landscapes: From an old best practice to a potential hazard for soil degradation due to land abandonment. *Anthropocene* **2014**, *6*, 10–25. [CrossRef]
56. Sona, G.; Passoni, D.; Pinto, L.; Pagliari, D.; Masseroni, D.; Ortuani, B.; Facchi, A. UAV multispectral survey to map soil and crop for precision farming applications. *ISPRS Int. Arch. Photogramm. Remote Sens. Spat. Inf. Sci.* **2016**, *XLI-B1*, 1023–1029. [CrossRef]
57. Stroppiana, D.; Villa, P.; Sona, G.; Ronchetti, G.; Candiani, G.; Pepe, M.; Busetto, L.; Migliazzi, M.; Boschetti, M. Early season weed mapping in rice crops using multi-spectral UAV data. *Int. J. Remote Sens.* **2018**, *39*, 5432–5452. [CrossRef]

 International Journal of
Geo-Information

Article

Diachronic Reconstruction and Visualization of Lost Cultural Heritage Sites

Pablo Rodríguez-Gonzálvez [1,*], **Ángel Guerra Campo** [2], **Ángel L. Muñoz-Nieto** [3], **Luis J. Sánchez-Aparicio** [2] **and Diego González-Aguilera** [3]

[1] Department of Mining Technology, Topography and Structures, Universidad de León, Avda. Astorga, s/n, 24401 Ponferrada (León), Spain
[2] TIDOP Research Group, University of Salamanca, Avda. de los Hornos Caleros, 50, 05003 Ávila, Spain; agc@usal.es (Á.G.C.); luisj@usal.es (L.J.S.-A.)
[3] Department of Cartographic and Land Engineering, University of Salamanca, Avda. de los Hornos Caleros, 50, 05003 Ávila, Spain; almuni@usal.es (Á.L.M.-N.); daguilera@usal.es (D.G.-A.)
* Correspondence: p.rodriguez@unileon.es; Tel.: +34-987-442-055

Received: 10 December 2018; Accepted: 27 January 2019; Published: 29 January 2019

Abstract: Cultural heritage (CH) documentation is essential for the study and promotion of CH assets/sites, and provides a way of transmitting knowledge about heritage to future generations. The integration of the fourth dimension into geospatial datasets enables generating a diachronic model of CH elements, namely, a set of three-dimensional (3D) models to represent their evolution in various historical phases. The enhanced four-dimensional (4D) modeling (3D plus time) pursues a better understanding of the CH scenario, enriching historical hypotheses as well as contributing to the conservation and decision-making process. Although new geomatic techniques have reduced the amount of fieldwork, when put together, the geometric and temporal dimensions imply the interpretation of heterogeneous historical information sources and their integration. However, this situation could reach a critical point when the study elements are no longer present. The main challenge is to harmonize the different historical and archaeological data sources that are available with the current remains in order to graphically rebuild and model the lost CH assets with a high degree of reliability. Moreover, 4D web visualization is a great way to disclose the CH information and cultural identity. Additionally, it will serve as a basis to perform simulations of possible future risks or changes that can happen during planned or hypothetical restoration processes. This paper aims to examine the study case of a diachronic reconstruction by means of a mobile laser system (MLS) and reverse modeling techniques for a lost urban CH element: the citadel or *Alcázar* gate of Ávila. Within this aim, the final model is evaluated in terms of the consistency of the historical sources to assess its suitability considering the constructive interpretations that are required to integrate heterogenous data sources. Moreover, geometric modeling is evaluated regarding the current remains and its surroundings. Finally, a web 4D viewer is presented for its dissemination and publicity. This paper is an extended and improved version of our paper that was published in the 2018 ISPRS Technical Commission II Symposium, Riva del Garda, Italy, 3–7 June 2018.

Keywords: 4D modeling; cultural heritage; data fusion; monitoring; visualization

---

1. Introduction

Geomatics techniques, which are also referred as geotechnologies, can be defined as a set of sensors and computer algorithms that allow the acquisition, modeling, and analysis of spatial features (individually or together) that are focused on knowledge generation for any level of detail and discipline that has to manage changes in space and/or time [1]. The increasing emergence of different geomatics techniques applied to cultural heritage (CH) modeling save countless fieldwork hours.

However, models based on the current state of buildings and remains are commonly insufficient to explore and explain their historical background and evolution. This issue proves that the management and valorization of CH is a complex task in which scientists, developers, and final users aim to encourage cooperation and practice sharing.

Three-dimensional (3D) modeling and visualization techniques are being employed to document CH sites and assets that are threatened by natural influences (e.g., flooding, earthquakes, etc.) and/or human factors (e.g., mass tourism, pollution, etc.) [2,3]. Digital reconstruction implies the representation of the artefact or monument in its original state [4]. Within this aim, different geomatics techniques are being used to study the current state and the geometry of CH elements as a reference for modeling processes from airborne to ground level, such as laser scanning, photogrammetry, or global navigation satellite systems [5].

Due to the inherent complexity of CH elements, the use of complementary technologies [5–9] is usually required. Their fusion and hybridization are being studied, and in spite of the several approaches that are available [10], this issue still represents a hot topic for the scientific community [11]. Data fusion in CH is biased to the actual acquisition technologies [10], and is purpose-oriented to generate new data from the raw source. Moreover, the approaches are highly based on the data dimensionality [12,13]. The registered point clouds can be reconstructed to provide a parametric modeling of the historical building [14], or by means of non-parametric shapes [15]. However, the main problem remains when these CH elements only exist partially due to different reasons. In this case, 3D modeling or anastylosis approaches [16] are the best way to reconstruct the whole CH element [17,18], but require human expertise and knowledge about available historical documentation, as well as the verification of the suggested reconstruction through additional information, such as historical documents, images, records, etc. The expert and informal knowledge underlying the interpretation of historical data and its circumstances is vital for a correct reconstruction of the CH element [19].

Sources of historical information are really heterogeneous, encompassing texts, paintings, and engravings, to old photographs, maps, etc. [20]. Moreover, the 3D modeling process has to deal with the lack of objectivity and mutual coherence of historical data, due to errors in ancient surveys and/or conservations distortions [21]. Complementarily, archaeological data sources, which include excavated material, standing monuments, and inscribed records, among other sources, can be used as an objective reference to shape historical data sources. Last but not least, the alignment of components in a Cartesian space presents various challenges. It is not only difficult from a practical point of view, but it is complicated by the original shape of the object to be reconstructed only being known within a certain probability. To date, the virtual reconstruction of destroyed buildings has required a manual process that is based on full historical knowledge [22]. The reconstruction demands the consideration of expert knowledge and evaluation. The three-dimensional (3D) digitalization of the current remains serves a dual purpose: i) it could establish a transition to the missing parts of the diachronic reconstruction, and ii) it allows assessing the likelihood of the constructive hypotheses derived from the interpretation of the historical sources.

Once the 3D modeling has been carried out, a dynamic evolution of CH elements across different spatial scales is accessible, allowing the possibility of better understanding the present status of the CH asset/site according to its history [23]. This paper presents an approach to four-dimensional (4D) modeling and visualization through the diachronic reconstruction of lost cultural heritage sites. In the present case, the site is the *Alcázar* gate and its surroundings, which belongs to the Medieval Wall of Ávila (Spain), and was declared a Good of Cultural Interest by UNESCO (United Nations Educational, Scientific and Cultural Organization) in 1982.

The paper is organized as follows. After the introduction, Section 2 is focused on the materials and methods employed; Section 3 describes the study case, the experimental results obtained, and the diachronic viewer implementation. Finally, Section 4 summarizes the conclusions together with some future insights.

2. Materials and Methods

The present section describes both data sources for diachronic reconstruction—historical and geometrical, the methodology followed to obtain the 4D modeling; and finally, the considerations related to the 4D viewer implementation.

2.1. Historical Sources

In this subsection, a description of the different historical sources that were used is presented. Through these sources, the CH transformations through time, including the evolution of the uses of the space and urban changes, are identified. To put this in context, Figure 1 shows a simplified timeline of the historical evolution of the study case (see Section 3.1 for further details), and the sources that were employed.

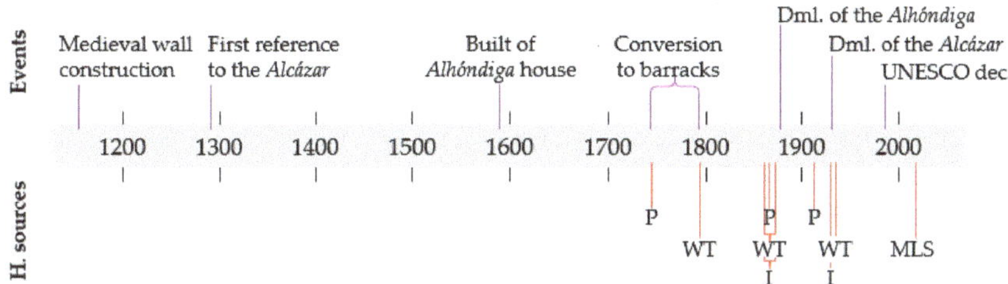

Figure 1. Chronology of *Alcázar* gate evolution and historical sources employed. Abbreviations: P: plan; WT: written testimony; I: old image; MLS: mobile laser system scan; Dml: demolition.

The Medieval Wall of Ávila has had different uses through its history. Most of them were related to defensive purposes. From c. 1291, a citadel (or *Alcázar* in Spanish) was built inside and attached to the southeast corner of the wall enclosure. Through its life, the citadel and near buildings suffered from several changes of use, as shown in Figure 1. It is likely that the demolishment of the citadel in 1930 was the largest transformation.

The most reliable historical sources of the citadel or *Alcázar* gate of Ávila were referred to in the plans made by Juan Gómez Parral in 1749 (Figure 2a,b), and in the plan of Pedro Moreau of 1750 (Figure 2c) related to the works of the conversion of the citadel into barracks. According to the information provided by these plans, a new plan design was built inside for the barracks with masonry arches. Also, a courtyard and a guard portal were created, as well as bedrooms and a gallery. We must highlight the opening of windows and the construction of an officers' gallery. This latter room is of great importance since, as a consequence, two towers belonging to the *Alcázar* were demolished on the south canvas, and the height was reduced in six rows of the *Polvorín* tower. The *Alcázar* was again in a ruinous state according to a report dated in 1780, and the reforms that were carried out in 1806 and 1813 did not solve the serious problem of deterioration. From this moment, the 19th and 20th centuries lead to the total destruction of the *Alcázar* (1930).

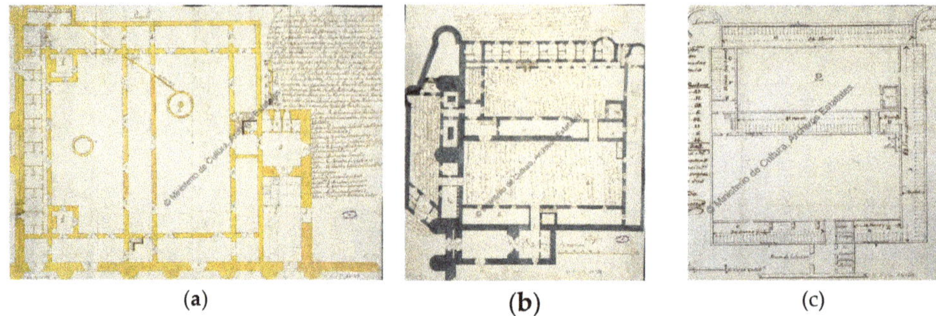

(a)   (b)   (c)

Figure 2. Different historical sources employed: (**a**) Juan Gómez Parral, 1749 (AGS. MPD. 14-027); (**b**) Juan Gómez Parral, 1749 (AGS. MPD, 14, 028); (**c**) Pedro Moreau, 1750 (AGS. MPD, 14, 029). Source: National Archive of Simancas (Spain). Adapted from [1].

Additionally, for the diachronic reconstruction assessment, two documents were used, despite their simplicity: a longitudinal profile of the street entrance to the *Alcázar* square, coming from the municipal architect Angel Cossín in 1864 (Ref. 134–53/27), and the plan made by the municipal architect Emilio González in 1911 (Ref. AHPAv. Ayto. 00070028).

Among all of the data sources that were used for the diachronic reconstruction, the best are those based on ancient drawings, which can provide a large amount of suitable information. However, their reliability is typically unknown. These documents present great contradictions among them, as well as with reality, especially the ones dated from the 18th century.

Other types of historical documents that were employed and analyzed included:

- Historical photographs.
- Maps.
- Written testimonies.

The knowledge coming from these sources was added and weighted in the analysis by means of constraints regarding the buildings materials, constructive techniques, and geometric shape and boundary of those buildings that do not currently exist.

The role played by the different data sources depends on their antiquity, since intrinsic inconsistencies were shown, especially further back in time. In the case of the graphical material, old photographs are the most reliable, although they only cover a small temporal interval. These historical photographs used to be oblique, and thus enable the single-view reconstruction approach developed by [24,25], being able to extract geometric proportions, and even a partial reconstruction depending on the geometry of the shot.

A total of three individual old images from c. 1860–1870 were considered (Figure 3). They cover a temporal keyframe, where some extramural buildings were modified. They played a key role along with the plans, since they provided objective information, in spite of the lack of scale or another direct information metric. There is no available information about the accuracy of these images.

(a)   (b)   (c)

Figure 3. Old images of extramural buildings. (**a**) Mariano Moreno, from Moreno archive (00004_C); (**b**) Laurent, J. from Ruiz Vernacci archive (VN-06348); (**c**) Laurent, J. from Ruiz Vernacci archive (VN-17214). Source: Spanish Cultural Heritage Institute.

The other graphical source was a city plan from c. 1865, whose use in the project is marginal. Its low resolution, due to the small scale (around 1:10,000), has limited the accurate location of the surroundings of extramural buildings, e.g., the circular fountain. As in the previous source, there is no available information about the accuracy of this source.

Finally, the written testimonies, which cover a temporal framework from c. 1860 to 1870, and an isolated one from c. 1935, describe the geometry of the citadel, the defensive system, or the transformation suffered by the asset. In spite of being non-graphical, they were used as a constraint source when there was not graphical information available, or there were discrepancies between the graphical information, the constructive hypothesis, and/or the current remains.

2.2. Geomatics Techniques

Geomatics techniques and technologies have long played an important role in CH documentation and preservation [26–29]. They range from red–green–blue (RGB) and multispectral cameras to terrestrial laser scanners (TLS), airborne laser scanners (ALS), drones [30], and even mobile laser systems (MLS) [31]. Due to the nature of the study case (see Section 3.1), which is placed in an urban center, and its significant extension, the MLS is employed as a data source for recording the present state of the current remains. The suitability of MLS for the documentation and dissemination of CH is described in [32,33]. The technical specifications of the employed MLS are listed in Table 1. This system is composed of two light detection and ranging (LiDAR) sensors, four RGB cameras, and an Applanix POS LV 520 Inertial Navigation System (INS). The system is configured to take 500,000 points per second with a scan frequency of 200 Hz. The maximum range of the sensors is 200 m, with a precision of eight mm (one sigma) and permission to obtain up to four echoes of the signal and the intensity reflected by the objects at a 1550-nm wavelength. The MLS was boarded in a car, achieving speeds of up to 40 km/h in the data acquisition.

The MLS measurement process is not completely straightforward, and the final results will be constrained by the initial planning phase. MLS calibration uncertainty is directly propagated to the recorded object, which could cause a lack of overlapping between scans of the same scene acquired from different trajectories. The completeness of the final scenario is directly linked with the complexity of the CH site, which is usually high, as well as where the vehicle is allowed to go. Regarding the final precision, the dependence of MLS on the GNSS (Global Navigation Satellite System) signal will cause a precision degradation in scenarios with weak satellite geometry, such as narrow streets in the case of urban CH assets [32]. Besides, the precision of the acquired point cloud is related to the laser angle of incidence [34]. All of these issues imply a constant control of the scanning process and a rigorous planning phase. An alternative geomatic technique, the portable mobile mapping systems [35,36], will dismiss some of the previously mentioned constraints (such as completion and GNSS dependence).

Table 1. Technical specifications of Optech LYNX Mobile Mapper. [1].

Parameter	Value
X, Y position	0.020 m
Z position	0.050 m
Roll and pitch	0.005°
True heading	0.015°
Measuring principle	Time of Flight (ToF)
Maximum range	200 m
Range precision	8 mm, (1σ)
Range accuracy	±10 mm, (1σ)
Laser measurement rate	75–500 kHz
Measurement per laser pulse	Up to 4 simultaneous
Scan frequency	80–200 Hz
Laser wavelength	1550 nm (near infrared)
Scanner field of view	360°
Operating temperature	10–40 °C
Angular resolution	0.001°

2.3. 4D Reconstruction Methodology

According to [10], when planning a 3D documentation of a heritage object, it is necessary to observe the data fusion requirements, the scene/object characteristics, and the availability of the equipment and sensors. The main requirements for data fusion are: accuracy level, radiometry importance, documentation purpose, and final use of the 3D products. The accuracy level enables ensuring that accuracy and details are preserved when mixing several data. A recommended approach is based on considering the geometric resolution differences among datasets, defining several accuracy levels, and adapting the level of information of each artefact to its acquisition methodology. However, in the present case study, the datasets that were employed not only have different spatial resolution, but some of them have unknown specifications. So, it was not possible to apply the proposed data fusion pipeline. As a result, the final application of the 4D model was used as a main constraint to providing a diachronic model with a high degree of reliability.

From the different geometrical and historical data sources stated previously, the reconstruction process was divided in the following steps:

- 3D recording of the current remains of the *Alcázar* gate by means of a MLS.
- Reverse modeling of the current state of the *Alcázar* gate and its intramural and extramural sections.
- Diachronic reconstruction or 4D modeling of the *Alcázar* gate based on the historical documents for two different temporal intervals.
- Diachronic reconstruction or 4D modeling of the extramural and intramural buildings of the *Alcázar* gate prior to their demolishment.

Obtaining a 3D recording of the current state by MLS requires an initial planning phase, where not only the aimed spatial resolution and average distance to the study object (constrained by the transit area) is set, but also the trajectory and GNSS availability, since the precision properties are highly dependent on them. Finally, details such as the scanning day and hour will condition the completeness, since this CH asset is located in a heavily trafficked area.

The reconstruction of the current state will provide the basis for the anastylosis process, and will also be essential to anchoring the plausible reconstruction of the lost building elements. This first phase required the classical steps of 3D modeling: filtering the non-desirable elements of the point cloud (e.g., pedestrians, cars, etc.) and extracting the geometrical data of the study area. The extraction of the basic primitives was done on the basis of cross-sections. This process involves a generalization operation and a loss of accuracy due to the idealization of regular shapes (i.e., planes, cylinders, etc.) [37]. The addition of certain constraints, such as the parallelism of the façades with the plumb line, also contributed to idealization and a loss of reliability. The fidelity of the 3D reconstruction is addressed more deeply in Section 3.4. For an in-depth description of the methodological procedure and a description of the different phases of multi-source data integration, please refer to [38].

2.4. 4D Visualization Methodology

Visualization and analysis use to be the final steps in geomatics. The management of different information sources, such as stratigraphic and mensiochronologic analysis or diachronic reconstructions, among others, require the design of specific solutions to allow a metric and quality approach to the scientific analysis of the CH assets/sites, improving the benefits of an ordinary visualization [39]. The integration of 3D models in a timeline or a 4D environment requires the development of an ad hoc software for visualizing and managing large digital datasets in real-time. With the development of the JavaScript API Web Graphics Library (WebGL), Beaverton, OR, USA, the possibility of visualizing and consulting 3D models through the use of a web browser has opened new possibilities in CH [40].

Under this basis, there are several open-source 3D web viewers that are able to deal with complex 3D environments. Inside this group, it is possible to highlight the viewer 3DHOP (3D Heritage Online Presenter) developed by the Visual Computing Lab of ISTI-CNR (Information Science and Technologies of the Italian National Research Council) [41]. This viewer, which is based on the WebGL engine, is able to deal with 3D models by means of meshes that are loaded following a multiresolution approach. This viewer is highlighted due to its easy combination with other web-page tools, allowing the integration of layers or even the linking of information through the use of hotspots [41]. In this framework, the comparison between two models is done by means of integrated links. Another interesting open-source solution for rendering complex 3D scenarios is the framework that Potree developed by the Institute of Computer Graphics and Algorithms [42]. This tool, based on the WebGL technology, enables rendering point clouds through the use of an octree visualization system [42]. Potree is highlighted due to its capacity for loading large point clouds datasets, making measurements (i.e., distances, areas, and volumes), and including annotations (hotspots). In line with the 3DHOP tool, Potree provides the possibility of comparing several models through the use of integrated links. Within commercial software, it is worth mentioning ArcGis CityEngine [43]. This tool is specially designed for urban purposes, accepting the simultaneous comparison of two 3D models made in different epochs (i.e., the evolution of a city through time) using a scrollbar. However, none of these platforms allow a 4D navigation where the user can navigate through a timeline and dynamically consult the 3D appearance of an architectural scenario over time.

Four-dimensional data visualization can be stated as the logical evolution of 3D dataset visualization in which time-varying visual products can be achieved. This evolution can be explained because time aspects become increasingly important in different application fields, whether analyzing or visualizing temporal changes due to anthropic activities, or simply mapping the deterioration caused by the passage of time. The advantages that 4D visualization offer are [38]: (i) a non-destructive way for archaeologists, art historians, and other scholars to analyze and monitor CH sites, monuments, and objects through time; (ii) the user's interaction with objects and information from the past to the present; and (iii) the possibility of building complex virtual scenarios with digitally reconstructed non-existent artefacts within their contextual background, to recreate either a real area or a non-existent environment based on a conceptual hypothesis, where the user can navigate in an interactive way.

Four-dimensional (4D) viewer implementation requires the initial task of requirements analysis; this has been grouped into functionals and non-functionals analyses, which define the specific system behavior and the criteria to judge the system operation, respectively. The functional requirements that are defined are:

- It must be possible to visualize data in a three-dimensional environment in known file formats.
- The data of each file must be organized through a system of layers, which can be displayed and hidden easily.

- A dedicated tool that helps see the data in different epochs of history, namely the fourth dimension, must be available.
- 3D models may have hotspots to provide additional information about the element selected.
- The viewer must be a web application in order to be consulted by different types of users and experts through the Internet.

In order to fulfill all the functional requirements previously shown, the render engine WebGL was chosen. This API enables rendering a 3D environment through the combination of HTML-5 and JavaScript codes. Among the different advantages that this library can offer, highlights include [40]:

- The capacity of rendering dynamically 3D environments.
- Hardware acceleration to render a complex environment in a quicker way.
- Compatibility with most of the desktop and mobile web browsers.

However, the main limitation of this library is the absence of a format for loading 3D models (such as .obj or .ply), as well as the absence of events (e.g., rendering actions or camera events) that allow the creation of interactive scenes. Taking this into consideration, the WebGL rendering engine was implemented through the JavaScript library Threejs (https://threejs.org/). This open-source library includes events for the camera, illumination patterns, and the ability of rendering models. Another important feature of this library is the possibility of using the Graphical Programming Unit (GPU) due to the weight of the different 4D models. As a final advantage, it works easily with other JavaScript libraries. This render engine was complemented by HTML-5, CSS-3, JSON, and standard JavaScript codes through the use of the Bootstrap environment for the interface (https://getbootstrap.com/), and the JavaScript library React (https://reactjs.org/) for the creation of classes. All of these components were implemented following a single page application approach (SPA). This programming strategy, the SPA, enables working in a unique page, avoiding the need to reload the application every time the user executes a function (e.g., loading a different epoch or taking measurements). To this end, the SPA follows a twofold strategy that firstly downloads the main HTML, CSS, and JavaScript codes into the client (user), and then progressively executes the different actions demanded by the user (such as the rotation of the scene or the loading of different scenes) by means of an AJAX protocol (Asynchronous Javascript and XML). In parallel to this, and in order to make the application scalable for further improvements, all of the libraries and frameworks were compiled inside the open source environment Visual Studio Code (https://code.visualstudio.com/) with the help of the library WebPack (https://webpack.js.org/). As a result, it was possible to create an application with the following architecture (Figure 4).

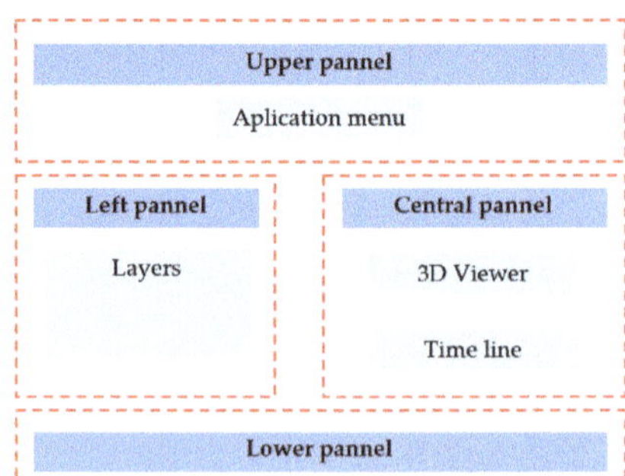

Figure 4. Schematic representation of the architecture used in the developed 4D viewer.

The combination of the previously shown libraries and frameworks enables the visualization of complex 3D scenes by means of a web browser. In order to provide the 4D dimension to the model, the open-source library visjs (http://visjs.org/docs/timeline/) was used. This library enables creating, editing, and deleting items (in our case 3D models) according to a specific date or range of dates. Under these bases, the following sequential process was used in order to visualize the different models along the timeline (Figure 5).

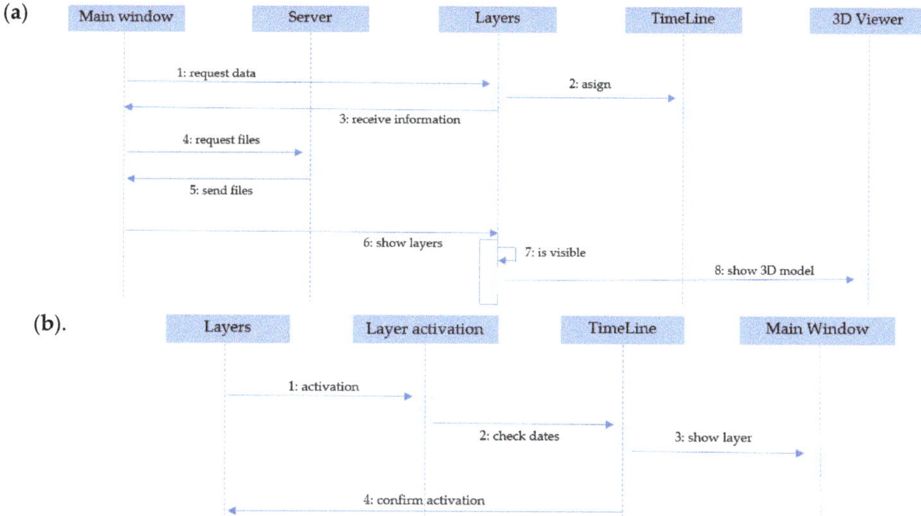

Figure 5. Diagram of the sequences followed by the four-dimensional (4D) viewer: (**a**) sequence to load the data into the three-dimensional (3D) viewer; and (**b**) sequence used for the layer activation.

Apart from the main architecture of the application, two additional tools were included, namely: (i) a distance measuring tool, and (ii) hotspots. On the one hand, the distance measuring tool was implemented with the aim of exploiting the metric capabilities of the diachronic reconstructions. On the other hand, the hotspot viewer was designed with the purpose of showing non-geometrical data such as historical photographs, postcards, and historical/didactical texts. This information appears in the form of an emergent window, which is also called a modal window. This window is loaded in an asynchronous way by means of an AJAX protocol that is executed when the user clicks on the hotspot. In the same way as the 3D models, this information appears in accordance with the date chosen by the user in the timeline.

According to the information exposed above, the application is able to deal with two types of data: (i) geometrical data coming for the 3D models obtained during the diachronic reconstruction, and (ii) non-geometrical data, such as historical photographs or texts, by means of the use of hotspots and modal windows. On the one hand, the geometrical data is uploaded to the viewer in *.obj* format. Then, the data is linked to the layer and timeline layouts by means of a JSON (JavaScript Object Notation) file. This file has the following main fields: (i) name of the layer; (ii) start date; (iii) end date; and (iv) path of the file. Additionally, to these fields, the JSON file includes information about the visibility of the 3D model in the layer and timeline bars. On the other hand, the non-geometrical data is uploaded to the viewer through the use of a HTML-5 modal page, linking this information with the timeline and the 3D environment through the use of a JSON file. This JSON file has the same fields as those defined for the geometrical data.

3. Results

3.1. Case Study

The Medieval Wall of Avila is the universal symbol and the most outstanding monument of the city of Ávila (Spain). Its importance derives from being the best-preserved medieval walled enclosure in Spain. The walls are a very important factor in shaping the urban planning of the city, and historically, they have participated in the distribution of urban space among the various social groups that have inhabited Ávila. In particular, the *Alcázar* gate is the most solemn element of the entire wall. This gate is constituted by two large towers joined by a bridge (which is a unique and a singular element between the European walls), which reinforces the defence of access.

There is no record of an exact date of the beginning of the construction of the *Alcázar* gate (or citadel). However, the oldest graphic document that can be found is a painting belonging to the painter Anton Van der Wyngaerde (1570). The *Alcázar* gate was restored and modified, and already in the 18th century, the *Alcázar* gate definitely lost its status as a fortress and went through different uses. In 1749, with the cession of Fernando VI, the *Alcázar* gate became a fortress, and numerous reforms, such as opening windows and even demolishing two cubes of the wall and reducing the height of another, were realized.

In the outer walls of the *Alcázar* gate, patrimonial elements were found that today no longer exist, such as the house of the *Alhóndiga*, which was built in 1590, and was formerly an establishment where people sold, bought, and even stored grain. This building throughout its history had different uses, including as a cereal market, jail of notables, and police headquarters, before its demolition in 1882.

The 3D point cloud of the current remains by MLS (Table 1) was acquired with a spatial resolution of 60 mm for an average scanning distance of 25 m. The georeferencing of the 3D point cloud into a global coordinate system was done by the integrated INS (Applanix POS LV 520) on the MLS. The total number of points of the current remains was 12.9 million (Figure 6).

(a)   (b)

Figure 6. Color-coded mobile laser system (MLS) data acquisition: (**a**) point cloud; (**b**) schematic view of the path followed during the scanning.

In Figure 7, the reconstruction of the current state of the *Alcázar* gate and its surroundings by means of MLS is shown, as stated in the previous section.

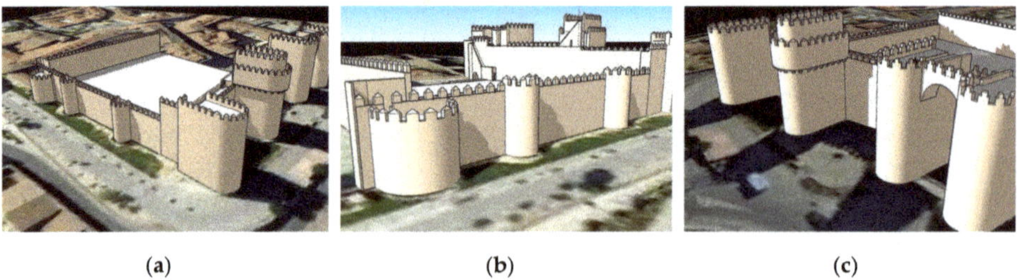

(a)   (b)   (c)

Figure 7. Different view of the reverse modeling process corresponding to the current state: (**a**) southeast view; (**b**) southwest view; (**c**) northeast view. [1].

3.2. Intramural Buildings

The reconstruction of intramural buildings involved two temporal stages according to the amount and type of historical information available. The first diachronic reconstruction corresponded to the most modern, after 1750, when the *Alcázar Gate* (or citadel) was converted to barracks (Figure 8a). In this stage, the ancient drawings were vectorized and fitted according to the present remains.

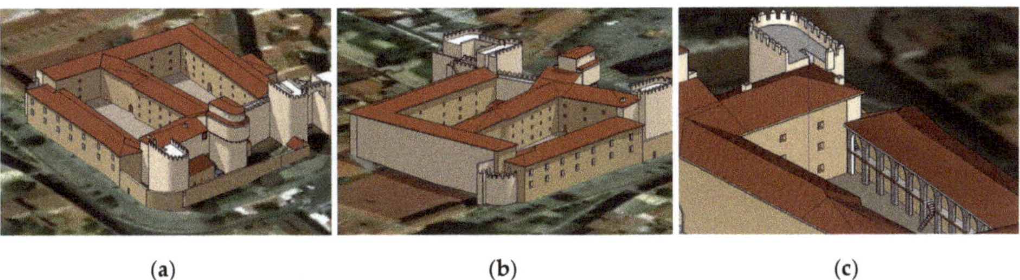

(a)	(b)	(c)

Figure 8. Intramural buildings: different views of the *Alcázar* after 1750: (**a**) southeast view; (**b**) southwest view; (c) detail of the officer's gallery from Northwest viewpoint. [1].

The ancient drawings were already digitalized, but the scalebar was in ancient units, so the vectorization was fitted to the current remains. In this step, the spatial invariants of the old drawings do not verify a simple geometric transformation, since the discrepancies were very large and not homogeneous. The adaptation was done scaling on one dimension, while maintaining the proportions. As a result, they were used in a relative way, namely, reconstructing the relative position of lost elements starting from the current remains. Thanks to old photographs of the demolition process, the width of the interior walls was incorporated. Also, the written testimonies informed about the demolition of the two south towers c. 1792, regarding the amount of anchor elements that were reduced.

One of the main key issues of this stage was the lack of information about the heights of buildings. The written documentation indicated that the building consisted of three floors except for the officers' gallery, which only had two. According to historical data, the height of a single floor was 11 feet. This measurement is not accurate, but it was enough to provide a supported hypothesis. Moreover, in the old photographs a passage concordant with the written testimonies about the officer's gallery that allowed the communication among the buildings avoiding the courtyards was detected (Figure 8c). During the reverse modeling operation, regarding the constraint of the officers' gallery ground floor with the passage existing on the wall, it was more plausible that the 11 feet refered to the distance between the floor and the beams of the ceiling (free span), instead of the distance between floors. Besides, the city of Ávila in general, and particularly the *Alcázar* because of its geographical position, is subject to great inclement weather. During its transformation into military barracks, the passage was built in the gallery so that the officers could traverse from one building to another, and even to the guard posts of the wall without going outside. This hypothesis is reasonable in relation to both military strategy and climatological conditions.

Another example of the aforementioned inconsistency among data sources is the west wall (Figure 8b), which was reconstructed after the demolishment of *Alcázar* (1930). As a result, the outer boundary was kept, but the original width, and subsequently the inner boundary, was modified with regard to the original construction. The final plausible reconstruction is shown in Figure 8.

The second diachronic reconstruction refers to the "real" *Alcázar* gate, when it was used as a citadel. The main problem of this period is the lack of reliable historical information. We have a few written testimonials and some stones that give us clues as to the volume of the *Alcázar*. It was necessary to employ a constructive hypothesis based on similar medieval military constructions, such as the interior defensive walls, or the existence of the northwest turret (supported by the presence of a foundation).

Thanks to a written testimony from 1792, we knew that two towers of the south canvas of the original *Alcázar* were demolished and reconstructed. So, not all of the towers of the current remains (Figure 7b) can be used as geometric anchors. The irregular interval of the towers (Figure 7b) implies that it was reconstructed in an inaccurate position. Due to the internal division of the courtyard into two spaces, the tower was repositioned in a more plausible location (Figure 9a). In addition to these interpretable sources, the remains on which the barracks were built can be analyzed in the latter plan of E. Gonzalez (1911), where two walls of great thickness can be identified.

The likelihood of the 4D model generated (Figure 9) was reduced, as expected, by the combination of surveyed data and philological analysis [23].

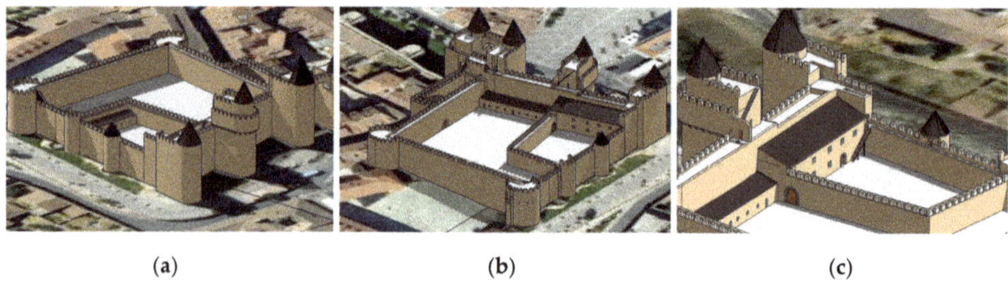

(a) (b) (c)

Figure 9. Intramural buildings: different views of the CAD (Computer-Aided Design) modeling of *Alcázar* previous to the mid-18th century: (**a**) southeast view; (**b**) southwest view; (**c**) interior detail from the northwest viewpoint. [1].

3.3. Extramural Buildings

On the basis of the 4D modeling of the current remains, the reconstruction process was carried out based on several photographic sources, which considerably reduced the uncertainty associated with the final 4D model. The reconstruction process was similar to that described in Section 3.2, with the most remarkable difference caused by the nature of the historical source, which in this case was old photographs.

The first issue was that the extramural buildings disappeared, and there were no current remains. The only anchor elements were the walls, which will be used to estimate the photographs' scale. Secondly, the historical images had vanishing points, and their identification was necessary in order to carry out the reconstruction [24]. The modeling was carried out in two phases; the first was the coarse modeling of building blocks as idealized prisms (e.g., orthogonality between façades, round angles, etc.). They were scaled on the basis of the identifiable elements of the wall. The older the photograph, the higher the uncertainty due to the changes suffered by the wall, especially by the rehabilitation and conservation interventions. For the older images, the distance between some stones was employed, which allowed a reasonable reconstruction of the 4D model.

For the most representative extramural building, the *Alhóndiga*, the old photographs were used to vectorize and reconstruct the façade (e.g., elements doors, windows, columns, etc.). The vanishing points and vanishing lines were used to rectify the images, and therefore enabled the possibility of vectorizing the different façades without distortion. Due to the different points of view, the same historical photographs were not used for all of the buildings. Besides, the historical photographs were used to map the texture, improving the realism of the final result (Figures 10 and 11). In the case of overlapping between different photographs, no blending process or radiometric modification was carried out.

Since the historical images were registered in relation to the 3D coordinate system, it was possible to replicate the original point of view and compare both or add a current image in a similar way based on rephotographing techniques [44].

(a)　　　　　　　　　　　　**(b)**

Figure 10. Extramural buildings: (**a**) old photograph texture integration (Laurent, c. 1865, Ruiz Vernacci Archive, VN-17214); (**b**) with 4D modeling for the extramural section of the *Alcázar* gate. Source: Cultural Heritage Institute of Spain (IPCE). [1].

(a)　　　　　　　　　　　　**(b)**

Figure 11. Extramural buildings; different views of the 4D modeling of the *Alcázar* gate surroundings, mid-19th century: (**a**) overview of the *Alhóndiga* and the wall; (**b**) detailed view of the gate [1].

3.4. Reconstruction Assessment

3.4.1. Diachronic Reconstruction

The five historical sources stated in Section 2.2 were assessed against the final 4D reconstruction to highlight the data source uncertainty. The period between 1750–1930 was evaluated. For this task, four kinds of spatial invariants were selected: distance, wall width, enclosure area, and angle. This way, the uncertainty related to the georeferencing is avoided. The results are listed in Table 2 in terms of the signed discrepancy, and the absolute percentage is in brackets. The absolute percentage was chosen to provide a quick view of the discrepancy magnitude. For example, the first value of –4.0 m implies that the north limit of the north courtyard is four meters shorter in Parral's plan in relation to the final reconstruction, while in the second plan of Parral (1749b), this limit is almost seven meters larger.

The higher percentage error corresponds to the plan of P. Moreau (1750) for three of four invariants (distance, width, and angle). There is not a distinguishable pattern among the discrepancies. Even though area discrepancies are related to the distance and angle discrepancies, the values are listed for a quick assessment of the main enclosures. It is significant that for the four-sided enclosures, two of the angles tend to be coherent among all the sources, but they are not always the same, e.g., the southeast (SE) and northeast (NE) corners of the north courtyard, but the southwest (SW) and northwest (NW) corner of the exterior limits. In the following table (Table 3), the results of the above table (Table 2) are summarized.

Table 2. Geometrical invariant discrepancies for the *Alcázar* 4D modeling for the period that was used as barracks.

			Parral (1749)		Parral (1749b)		Moreau (1750)		Cossín (1864) and González (1911)	
Distance invariant	North courtyard	North	−4.0 m	(8.2%)	6.7 m	(13.8%)	0.4 m	(0.9%)	−3.5 m	(7.2%)
		East	0.4 m	(2.1%)	4.7 m	(26.7%)	7.6 m	(42.8%)	0.0 m	(0.1%)
		South	−2.8 m	(5.8%)	8.0 m	(16.9%)	1.4 m	(3.0%)	0.3 m	(0.6%)
		West	−6.1 m	(25.2%)	−1.7 m	(7.2%)	0.7 m	(3.0%)	0.1 m	(0.4%)
	South courtyard	North	−2.4 m	(5.2%)	2.3 m	(4.9%)	−5.6 m	(12.0%)	0.0 m	(0.1%)
		East	−2.7 m	(13.1%)	−3.8 m	(18.5%)	−2.2 m	(10.5%)	3.0 m	(14.7%)
		South	−6.0 m	(12.0%)	−2.8 m	(5.6%)	−8.4 m	(16.7%)	0.5 m	(1.0%)
		West	2.7 m	(17.6%)	2.7 m	(17.7%)	3.2 m	(21.2%)	2.9 m	(18.7%)
	Guardhouse courtyard	North	-	-	-	-	-	-	-	-
		East	-	-	-	-	-	-	-	-
		South	1.8 m	(9.6%)	−3.5 m	(18.4%)	-	-	0.0 m	(0.0%)
		West	−1.7 m	(14.6%)	−1.6 m	(13.8%)	−2.6 m	(23.1%)	0.0 m	(0.0%)
	Exterior limits	North	5.3 m	(9.4%)	6.2 m	(11.0%)	−0.9 m	(1.7%)	0.7 m	(1.2%)
		East	−3.6 m	(5.1%)	2.4 m	(3.5%)	−13.1 m	(18.9%)	−3.1 m	(4.4%)
		South	−6.0 m	(8.9%)	−1.1 m	(1.6%)	−2.6 m	(3.8%)	3.8 m	(5.6%)
		West	−7.1 m	(10.4%)	−2.3 m	(3.3%)	−2.7 m	(4.0%)	−0.1 m	(0.1%)
	Corridors	North	−1.9 m	(19.8%)	−0.9 m	(8.9%)	−2.6 m	(26.3%)	−1.8 m	(17.9%)
		In-between	−1.5 m	(15.6%)	−0.9 m	(9.4%)	−2.4 m	(25.1%)	0.0 m	(0.4%)
		East	−0.6 m	(5.9%)	−0.2 m	(2.0%)	−2.4 m	(24.0%)	−1.1 m	(11.6%)
		South	−2.0 m	(17.8%)	4.2 m	(37.9%)	−4.0 m	(36.1%)	−2.9 m	(26.3%)
		West	−0.1 m	(1.6%)	−0.2 m	(2.8%)	−1.1 m	(13.9%)	−0.1 m	(1.5%)
	Wall width	Gate	-	-	−0.7 m	(24.0%)	-	-	-	-
		Dividing	-	-	−0.6 m	(22.3%)	−1.7 m	(69.3%)	-	-
Area Courtyards		North	−145.3 m^2	(15.4%)	304.8 m^2	(32.3%)	270.5 m^2	(28.7%)	26.4 m^2	(2.8%)
		South	−159.3 m^2	(18.6%)	−62.0 m^2	(7.2%)	−116.6 m^2	(13.6%)	148.2 m^2	(17.3%)
		Guardhouse	−16.1 m^2	(7.4%)	−61.7 m^2	(28.6%)	-	-	0.9 m^2	(0.4%)
Angular invariants	North courtyard	NE corner	−9.4°	(9.5%)	−9.4°	(9.5%)	−9.4°	(9.5%)	0.6°	(0.6%)
		SE corner	1.2°	(1.4%)	1.2°	(1.4%)	1.2°	(1.4%)	−0.8°	(0.9%)
		SW corner	8.1°	(9.9%)	8.1°	(9.9%)	8.1°	(9.9%)	3.1°	(3.8%)
		NW corner	0.0°	(0.0%)	0.0°	(0.0%)	0.0°	(0.0%)	−3.0°	(3.3%)
	South courtyard	NE corner	−3.0°	(3.2%)	−4.0°	(4.3%)	−4.0°	(4.3%)	0.0°	(0.0%)
		SE corner	10.9°	(13.8%)	11.9°	(15.0%)	11.9°	(15.0%)	0.9°	(1.1%)
		SW corner	2.3°	(2.6%)	2.3°	(2.6%)	2.3°	(2.6%)	2.3°	(2.6%)
		NW corner	−8.3°	(8.4%)	−8.3°	(8.4%)	−7.3°	(7.4%)	−1.3°	(1.3%)
	Guardhouse courtyard	NE corner	-	-	-	-	-	-	-	-
		SE corner	−3.2°	(3.4%)	−3.2°	(3.4%)	-	-	−2.2°	(2.4%)
		SW corner	0.0°	(0.0%)	0.0°	(0.0%)	0.0°	(0.0%)	2.0°	(2.2%)
		NW corner	-	-	-	-	-	-	-	-
	Exterior limits	NE corner	−10.4°	(10.5%)	−9.4°	(9.5%)	−9.4°	(9.5%)	−2.4°	(2.4%)
		SE corner	10.9°	(13.8%)	10.9°	(13.8%)	10.9°	(13.8%)	−0.1°	(0.1%)
		SW corner	0.4°	(0.4%)	−0.6°	(0.7%)	0.4°	(0.4%)	−0.6°	(0.7%)
		NW corner	0.0°	(0.0%)	1.0°	(1.1%)	0.0°	(0.0%)	0.0°	(0.0%)

Table 3. Summary of geometrical invariant discrepancies for the *Alcázar* 4D modeling.

Discrepancy		Parral (1749)	Parral (1749b)	Moreau (1750)	Cossín (1864) and Gonzalez (1911)
Length	Min.	−7.1 m	−3.8 m	−13.1 m	−3.5 m
	Mean	−2.3 m	1.1 m	−1.9 m	0.3 m
	Max.	5.3 m	8.0 m	7.6 m	3.8 m
Width	Min.	−2.0 m	−0.9 m	−4.0 m	−2.9 m
	Mean	−1.2 m	0.1 m	−2.4 m	−1.2 m
	Max.	−0.1 m	4.2 m	−1.1 m	0.0 m
Area	Min.	−159.3 m^2	−62.0 m^2	−116.6 m^2	0.9 m^2
	Mean	−106.9 m^2	60.4 m^2	77.0 m^2	58.5 m^2
	Max.	−16.1 m^2	304.8 m^2	270.5 m^2	148.2 m^2
Angle	Min.	−10.4°	−9.4°	−9.4°	−3.0°
	Mean	0.0°	0.0°	0.4°	−0.1°
	Max.	10.9°	11.9°	11.9°	3.1°

As expected, the more modern the source, the lesser the geometric discrepancy. Besides, it is worth highlighting the high values of uncertainty of this kind of historical source, although the maps should come from survey operations, and they have a scalebar. The figures obtained enable estimating the final error composition from the different sources involved: survey and measurement errors, conservation deformations, idealization processes, etc. The values listed in Tables 2 and 3 should be considered based on the context of the present case study: that is, a CH asset lost approximately 90 years ago, where the current remains are scarce and not necessarily completely reliable. For other CH elements, they can be better.

3.4.2. Current State

The original MLS point cloud was compared against the current stated CAD modeling, focusing on the extramural and intramural sections of the Alcázar gate. The accuracy analysis was carried out with CloudCompare [45] to compute the signed discrepancies against the reconstructed mesh and the point cloud (Figure 12). The non-modeled elements (e.g., rocks, artificial elements attached to the wall, etc.) were manually segmented.

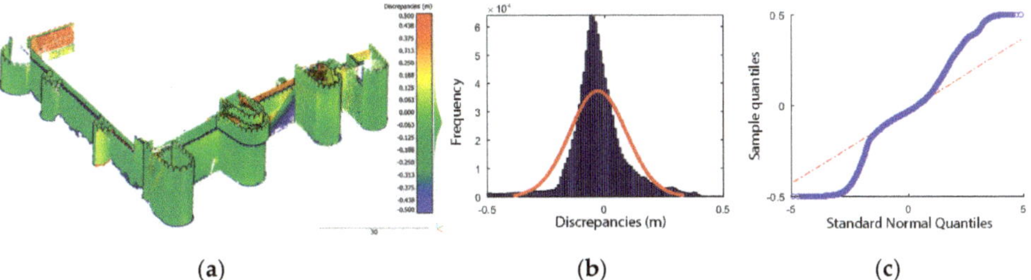

(a) (b) (c)

Figure 12. Reconstruction assessment of current remains: (**a**) spatial view of the signed discrepancies in the *Alcázar gate* and its surroundings; (**b**) histograms of the signed discrepancies with the superimposed curve for the normal distribution; (**c**) Q–Q plot of the distribution of the signed discrepancies. Adapted from [1].

The large amount of points involved avoids the use of numerical normality tests to determine whether the discrepancies (Figure 12b) follow a Gaussian distribution [46]. Instead, a graphical normality test is employed. In Figure 12c, the quantiles of the empirical distribution plotted against the theoretical quantiles of the normal, or Q–Q plot [47], are shown. The deviations from the straight line indicate that the distribution of the errors is not normal.

In Table 4, the robust assessment, as well as the Gaussian ones, are summarized for comparative purposes. The non-parametric estimator that was adopted for the central tendency was the median, while for the dispersion, the normalized median absolute deviation (NMAD) and the square root of the biweight midvariance (BWMV) are employed [35].

$$NMAD = 1.4826 \cdot MAD \tag{1}$$

$$BWMV = \frac{n\sum_{i=1}^{n} a_i(x_i - m)^2 (1 - U_i^2)^4}{\left(\sum_{i=1}^{n} a_i (1 - U_i^2)(1 - 5U_i^2)\right)^2} \tag{2}$$

$$a_i = \begin{cases} 1, if |U_i| < 1 \\ 0, if |U_i| \geq 1 \end{cases} \tag{3}$$

$$U = \frac{x_i - m}{9 \cdot MAD} \tag{4}$$

Table 4. Statistical results of the anastylosis process for the current remains. Adapted from [1]. BWMV: biweight midvariance, MAD: median absolute deviation, NMAD: normalized median absolute deviation.

	Parameter	Value
	Number of points	1,122,289
Gaussian assessment	Mean	−0.026 m
	Standard deviation	±0.121 m
	Kurtosis	2.65
	Skewness	0.11
Robust assessment	Median	−0.036 m
	MAD	±0.053 m
	NMAD	±0.078 m
	Sqrt (BWMV)	±0.096 m
	Quartile 0.25	−0.084 m
	Quartile 0.75	0.024 m
	Percentile 0.025	−0.303 m
	Percentile 0.975	0.266 m
	Interpercentile range 95%	0.570 m

The expect value of the central tendency of the discrepancies is zero. As shown in Table 4, there is a small bias (less than four centimeters) in the reconstruction due to the idealization process to regular shapes, but it is within the expected error for this kind of diachronic reconstruction. The dispersion or deviation between the reconstructed shapes and the ground truth (current remains), is usually assessed by the standard deviation, which refers to the inflection points of the Gaussian distribution. However, the error dispersion is overestimated by the Gaussian assessment (±12 cm versus ±8 cm); thus, the robust assessment is more appropriate to characterize it.

In view of the results and according to the distribution of discrepancies in the histogram, 95% of the modeled points have an error in the range of [−0.30, +0.27] meters. Besides, it can be observed that the hypothesis of a Gaussian distribution for the 3D modeling error at a 95% confidence level is ±0.15 m.

3.5. Visualization through Time

Sharing multi-temporal information via the Internet is another one of the research objectives, since it allows a large dissemination and an easy connection between people and CH. The 4D model described in the previous subsections is presented on a 4D viewer that was developed according to the description stated in Section 2.4. It shows the 3D visualization of a 3D model associated with a specific historical phase of the monument, giving the user the opportunity to navigate the space in 3D and change the historical period with a time-slider (Figure 13). The 4D viewer is available online at http://tidop.usal.es/cht2/.

(a)

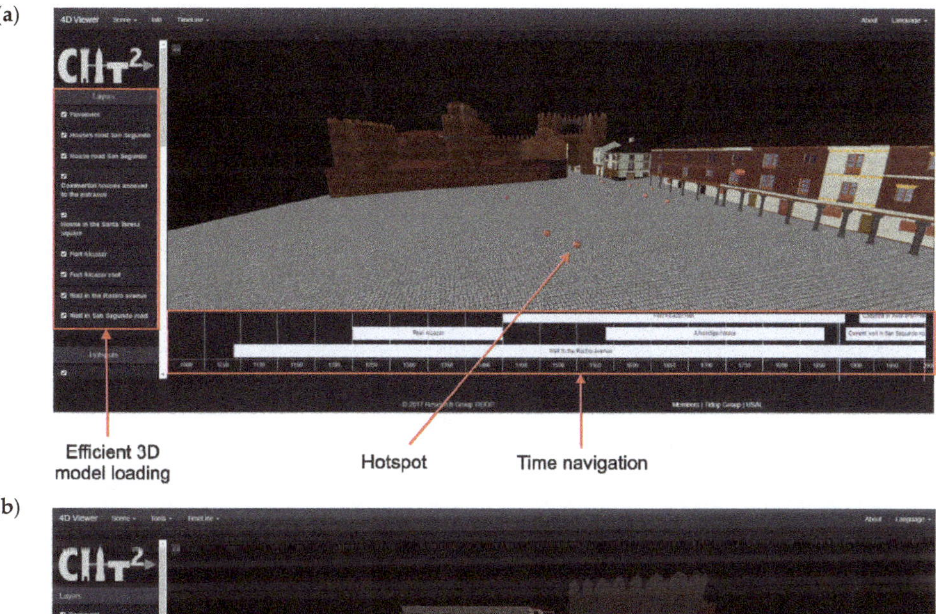

Efficient 3D   Hotspot   Time navigation
model loading

(b)

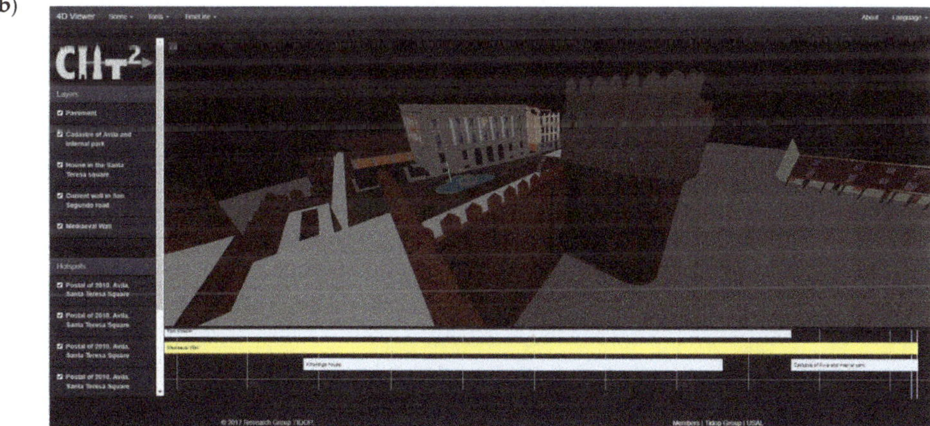

Figure 13. Front-end interface of the viewer developed: (**a**) appearance of the *Alcázar* gate of Ávila around 1900; (**b**) current appearance of the *Alcázar* gate of Ávila (2018).

For better navigation, the 3D models were simplified as lightweight meshes coming from the modeling step. However, for other types of CH elements that cannot be modeled on the basis of parametric shapes (i.e., basic primitives), it can be possible to provide a light representation that keeps the required fidelity and, at the same time, fulfills the required geometric precision [48].

Apart from the consideration exposed above, the 4D viewer includes an asynchronous loading scheme. This system allows progressively loading the different 3D models without blocking the load of the document object model (DOM) of the application. Thanks to this, the user can start consulting the application while in a second plane, all of the 3D models that are required to represent the scene are being loaded (Figure 14).

(a)

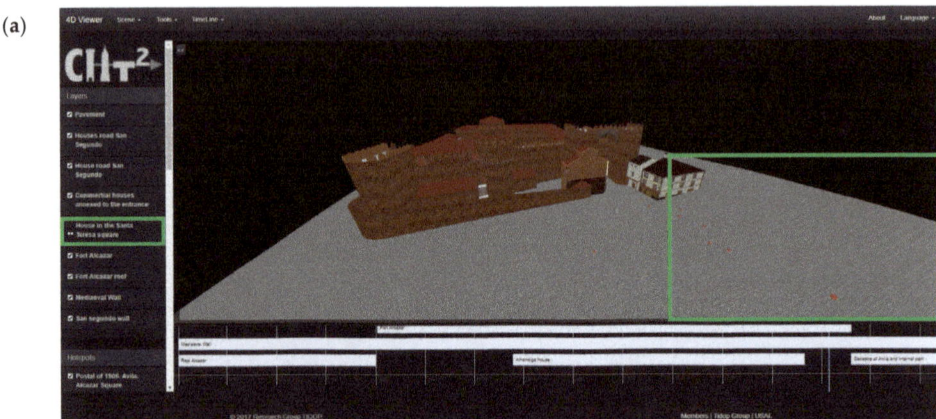

(b)

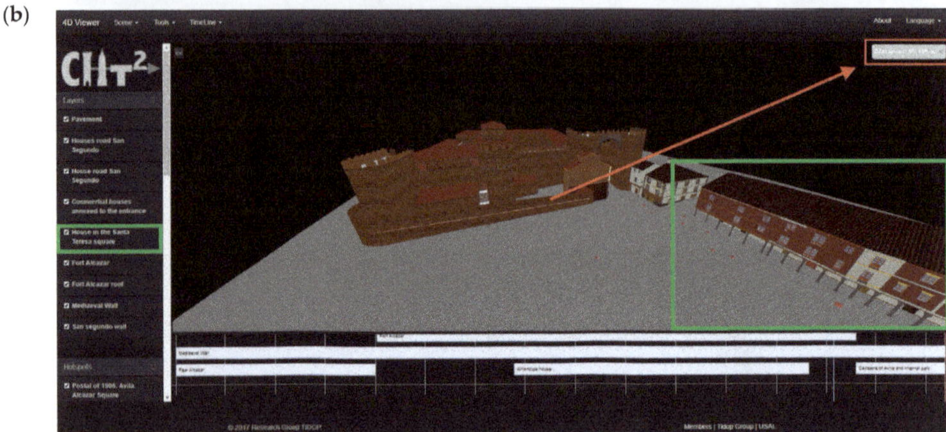

Figure 14. Screenshots of the application on which it is possible to observe the asynchronous loading scheme used: (**a**) previously to the loading of the model called "House in the Santa Teresa Square" and; (**b**) after the loading of all the models. The capacity of taking measurements on the different 3D models is highlighted in red.

Complementary to the ability to consult the evolution of architectural scenarios along the time, the application that was developed was able to take measurements (Figure 14a,b), expanding the applicability of the tool.

3.5.1. Improving the User Experience: Blending Geometrical and Non-Geometrical Data

Additionally, regarding the capacity of visualizing the evolution of the architectural scenario along the time by means of showing and hiding the 3D models, the capabilities of the 4D viewer were improved through the ability to integrate non-geometrical data. To this end, the application uses the concept of the hotspot (see Section 2.4) as a way to link historical non-geometrical data (i.e., historical photographs, postcards, or even historical and didactical texts) into the 3D scene. This source of information appears in form of modal windows subordinated to the application's main window, using HTML-5 and CSS-3 code. The user can access this information by clicking on the different red spheres that appears along the scene. Each one of them contains information related to a specific location during a specific epoch (Figure 15), which can be used to improve the values transmitted by the 3D models (i.e., intangible values such as manners or the evolution of the constructive techniques disappear). This capacity is in line with the suggestions provided by the International Charter of Kraków [49] and the Nara Document on Authenticity [50].

(a)

(b)

Figure 15. Examples of non-geometrical data included inside the modal windows: (**a**) an historical photograph taken during the market day in the *Alcázar* gate; and (**b**) a postal notice about the *Alcázar* gate printed in 1930. It is worth mentioning that the information that appears in both modal windows (hotspots) enhances the capacities of the viewer, allowing the transmission of intangible values (**a**) or even useful information for the study of the evolution of the restoration theories along with the history (**b**).

3.5.2. Performance Assessment

Apart from the previously shown aspects, i.e., the presence of a timeline to consult different 3D architectural scenarios through time or the possibility of using hotspots to link non-geometrical data, a good user experience passes through the use of a viewer that is able to render these scenarios in a short period of time. Taking this into consideration, several loading tests were carried out with the aim of evaluating the influence of several inputs on the time required to load the whole scene. These inputs were: (i) the Internet bandwidth; (ii) the asynchronous loading scheme; (iii) the size of the model; and (iv) the number of hotspots. All of these tests were carried out considering the different plausible scenarios obtained from the combination of the 3D models created during the different diachronic stages (Figure 16) (Table 5): (i) Scenario A was made up of the *Alcázar* previous to the mid-18th century; (ii) Scenario B was composed of the *Alcázar* after the mid-18th century and the *Alhondiga* house; (iii) Scenario C was made up of *Alcázar* after the mid-18th century, the *Alhondiga* house, and some residential buildings annexed to the medieval wall, and (iv) Scenario D was composed of the Medieval Wall, some residential buildings placed on the square, the cadaster building, and the public park built after the demolition of the *Alcázar*. Table 5 shows the sizes of the geometries and the textures used to render the different 3D scenarios. From the results of this table, it was possible to highlight the texture size of the *Alhondiga*, House I, Medieval Wall_1, cadaster, and park models. In particular, the first two models were texturized with historical photos, whereas the medieval wall and the cadaster building were texturized with high-resolution images, which was especially important for the case of the model Medieval Wall_1 due to the need of having a detail stratigraphy of this element for further archaeological studies.

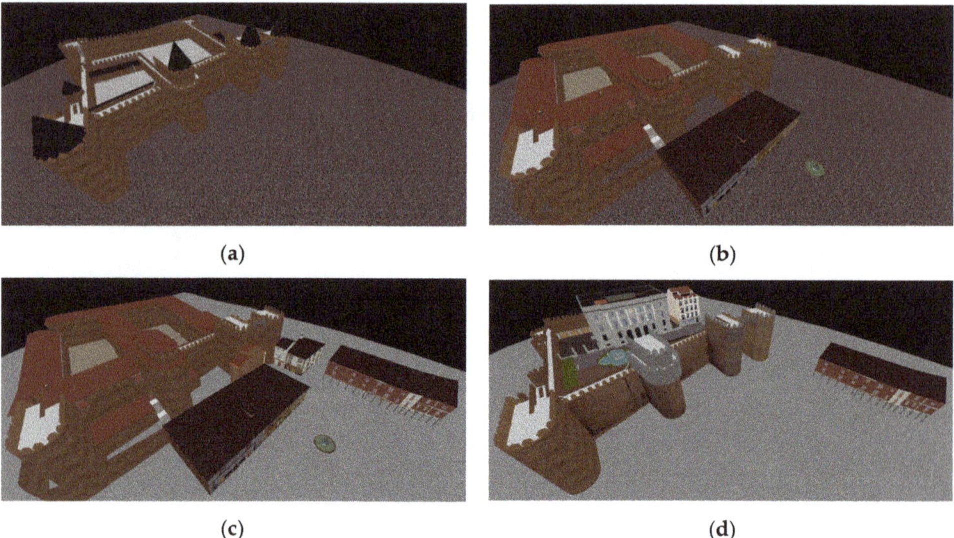

(a) **(b)**

(c) **(d)**

Figure 16. Rendering of the scenarios considered for the different performance tests carried out: (**a**) Scenario A; (**b**) Scenario B, (**c**) Scenario C; and (**d**) Scenario D.

With the aim of avoiding possible Internet interferences during the loading processes, all the tests were carried out on a local network with a controlled bandwidth internet connection as well as the same hardware and software equipment. This equipment was a desktop PC MSI GE70®equipped with an Intel Core i7 4000MQ processor at 2.40 GHz, 16 Gb RAM DDRIII, NVidia GeForce GTX 765M 2GB GDDR5, OS Windows 10, and Google Chrome ver. 71.0.3578.98.

Table 6 shows the result of the performance tests that were carried out over the different scenarios that were considered. As expected, the Internet bandwidth has a great impact on the performance of the application. If this value is duplicated, the time that is spent to load the scene the first time (no cache) is reduced by a factor of approximately two. This relation is not observed in the case of loading the models from the web cache. Apart from this, it was possible to observe that the asynchronous loading scheme has an impact on the overall response of the viewer, obtaining higher ratios for scenarios C and D. Both scenes have the highest number of models with asynchronous loads (seven and five, respectively).

Table 5. Scenarios considered during the performance assessment.

Scenario	3D Models	Size of the OBJ Models (Mb)			Total Size (Mb)
		Geometry	Texture	Total	
A	Pavement	0.1	0.3	0.4	2.2
	Real Alcázar	1.7	0.1	1.8	
B	Pavement	0.1	0.3	0.4	6.8
	Fort Alcázar	0.9	0.4	1.3	
	Alhondiga house	2.4	2.7	5.1	
C	Pavement	0.1	0.3	0.4	30.7
	Fort Alcázar	0.9	0.4	1.3	
	Alhondiga house	2.4	2.7	5.1	
	House I	0.1	5.8	5.9	
	House II	2.3	1.3	3.6	
	House III	3.0	1.4	4.4	
	House IV	9.1	0.9	10.0	
D	Pavement	0.1	0.3	0.4	55.0
	Medieval Wall_1	2.2	19.5	21.7	
	Medieval Wall_2	0.3	0.3	0.6	
	House IV	9.1	0.9	10.0	
	Cadaster and park	3.7	18.6	22.3	

Table 6. Web rendering statistics obtained for the different scenarios proposed. The ratio between the size of the model (in Mb) and the time spent to load the model (in s) is indicated in parentheses.

Bandwidth	Loading Time (s)			
	Scenario A	Scenario B	Scenario C	Scenario D
10.0 Mb/s	3.0 (0.7)/0.9 (2.4)	8.5 (0.8)/1.2 (5.8)	20.7 (1.5)/1.3 (23.3)	39.2 (1.4)/1.2 (46.2)
20.0 Mb/s	1.9 (1.2)/0.9 (2.5)	4.6 (1.5)/1.1 (6.1)	10.0 (2.9)/1.3 (23.8)	20.0 (2.7)/1.2 (45.5)
40.0 Mb/s	1.2 (1.9)/0.9 (2.6)	2.7 (2.5)/1.0 (6.8)	5.8 (5.3)/1.3 (24.2)	11.0 (5.0)/1.2 (45.3)

Regarding the influence of the number of hotspots in the overall response of the viewer, several loading tests were carried out with the considerations that were previously shown. During these tests, each scenario was loaded with a different number of hotspots (5/10/20). All of these elements, including the hotspots, were made up of the same amount of information: i) an image of about 1.8 Mb, and ii) a text with 250 characters. From these tests, it was possible to conclude that the number of hotspots did not have an influence on the time that was needed to load the whole scenario. This is due to the protocol implemented in the viewer (see Section 2.4), which uses an AJAX protocol to load the information contained into the modal page in an asynchronous way. Thus, each time that the user consults a modal window, the information (image and text) is loaded immediately.

4. Discussion and Conclusions

The contribution of the present paper is twofold: on the one hand, to merge heterogeneous information and expertise to deliver enhanced 4D digital products of CH assets/sites, and on the other hand, to provide qualitative and quantitative assessments of the results. The derived products can be the basis for quantitative analyses about architectural changes, visualization purposes, preservation policies, future planning, etc.

Through the experimental results, the different difficulties that face 4D modeling were highlighted, including: the lack of coherence among the available graphical material with regard to the current remains, the inconsistency among historical sources, and the need to add elements based on hypotheses, among others. The main difficulty was that the element to be reconstructed had disappeared and the current remains were limited, and in some cases, unreliable, since they were modified, such as the displacement of west wall, or the reconstruction of the south towers. Therefore, it was necessary to use constructive interpretations in some parts of the CH elements, with the consequent implications in the derived analysis. The human dependence for the reconstruction operation entails an increase of the amount of time involved. This issue is aggravated in the case of the non-parametric elements presented in CH elements that avoid the use of the semi-automatic tools of reverse modeling suites. Although the modeling process is not completely traceable, the workflow and twofold assessment enable evaluating the feasibility of 4D reconstruction.

The assessment of all of the historical sources based on spatial invariants enables providing a qualitative evaluation of the fidelity of each source. This will be of interest for future studies, since it will allow weighing them and/or the information represented in a less subjective manner. Regarding quantitative evaluation against current remains, it allows analyzing the results' accuracy, concluding that the CAD model has a centimetric accuracy that is suitable for reverse modeling applications in heritage and architecture. The improvement of the modeling software tools will increase the precision, but only up to a threshold, due to the requirement of the aforementioned constructive hypotheses.

To allow the dissemination of the 4D model, a web-based 4D viewer was presented. The main advantages of the viewer are its capacity to deal with complex 3D environments, the progressive scene loading, the 4D management, and the complementary information (by means of hotspots). As pointed out in Section 3.5, the 4D viewer shows good performance (with visualization almost in real-time) when the models have been loaded previously and they are stored in the cache. The employment of an asynchronous loading scheme allows the visualization of complex scenarios from the very first moments. For example, in the scenario of *Alcázar* after the mid-18th century, the loading of the full scene is delayed by the loading of *Alhóndiga* house and the annexed houses; however, the *Alcázar* is loaded for the first moments and accessible to the users. The hotspots were designed with the purpose of showing non-geometrical data such as historical photographs or written testimonies. Regarding the 4D visualization, in the web viewer, some possible issues were identified with the 3D models that require an additional edition stage for a better user experience, such as artifacts and inverted normals.

Future research will involve an upgrade of 4D modeling in order to improve the traceability of 4D reconstruction and obtain consistent and constructively plausible historical reconstructions. A very compelling and significant research line is devoted to the capacity of shaping informal knowledge and adding our intangible heritage to the anastylosis process.

Author Contributions: All of the authors conceived and designed the study. P.R.-G., A.L.M.-N. and D.G.-A. acquired the data and processed it. P.R.-G. and A.G.C. implemented the methodology and analyzed the results L.J.S.-A. implemented the diachronic visualization. P.R.-G. and D.G.-A. wrote the manuscript and all authors read an approved the final version.

Funding: This research was funded by European Union through the Joint Programming Initiative on Cultural Heritage (JPI-CH), for the project Cultural Heritage Through Time (CHT^2), as well as the Ministry of Economy and Competitiveness of Spain, grant number PCIN-2015-071.

Acknowledgments: The authors wish to thanks Simón Cardozo Mamani for his support in the 3D reconstruction of extramural buildings; Martín Rodríguez Vales for the viewer implementation; and Professor José Luis Pajares, Professor Serafín de Tapia, Local Archive and Provincial Archive of Ávila for their support and advice in historical documentation gathering.

Conflicts of Interest: The authors declare no conflict of interest. The funders had no role in the design of the study; in the collection, analyses, or interpretation of data; in the writing of the manuscript, or in the decision to publish the results.

References

1. Rodríguez-Gonzálvez, P.; Cardozo Mamani, S.; Guerra Campo, A.; Sánchez-Aparicio, L.J.; del Pozo, S.; Muñoz-Nieto, A.; González-Aguilera, D. Diachronic reconstruction of lost cultural heritage sites. Study case of the medieval wall of Avila (Spain). *Int. Arch. Photogramm. Remote Sens. Spat. Inf. Sci.* **2018**, *XLII-2*, 975–981. [CrossRef]
2. Fieber, K.D.; Mills, J.P.; Peppa, M.V.; Haynes, I.; Turner, S.; Turner, A.; Douglas, M.; Bryan, P.G. Cultural Heritage Through Time: A Case Study at Hadrian's Wall, United Kingdom. *Int. Arch. Photogramm. Remote Sens. Spat. Inf. Sci.* **2017**, *XLII-2/W3*, 297–302. [CrossRef]
3. Nocerino, E.; Menna, F.; Morabito, D.; Remondino, F.; Toschi, I.; Abate, D.; Ebolese, D.; Farella, E.; Fiorillo, F.; Minto, S.; et al. The Vast Project: Valorisation of History and Landscape for Promoting the Memory of WWI. *ISPRS Ann. Photogramm. Remote Sens. Spat. Inf. Sci.* **2017**, *IV-2/W2*, 179–186. [CrossRef]
4. Georgopoulos, A. 3D virtual reconstruction of archaeological monuments. *Mediterr. Archaeol. Archaeom.* **2014**, *14*, 155–164.
5. Torres-Martínez, J.; Seddaiu, M.; Rodríguez-Gonzálvez, P.; Hernández-López, D.; González-Aguilera, D. A Multi-Data Source and Multi-Sensor Approach for the 3D Reconstruction and Web Visualization of a Complex Archaelogical Site: The Case Study of "Tolmo De Minateda". *Remote Sens.* **2016**, *8*, 550. [CrossRef]
6. Guidi, G.; Russo, M.; Ercoli, S.; Remondino, F.; Rizzi, A.; Menna, F. A multi-resolution methodology for the 3D modeling of large and complex archeological areas. *Int. J. Archit. Comput.* **2009**, *7*, 39–55. [CrossRef]
7. Remondino, F.; Gruen, A.; von Schwerin, J.; Eisenbeiss, H.; Rizzi, A.; Sauerbier, M.; Richards-Rissetto, H. Multi-sensor 3D documentation of the Maya site of Copan. In Proceedings of the 22nd CIPA Symposium, Kyoto, Japan, 11–15 October 2009; pp. 241–248.
8. Guidi, G.; Russo, M.; Angheleddu, D. Digital Reconstruction of an Archaeological Site Based on the Integration of 3D Data and Historical Sources. *Int. Arch. Photogramm. Remote Sens. Spat. Inf. Sci.* **2013**, *XL-5/W1*, 99–105. [CrossRef]
9. Hanke, K.; Moser, M.; Rampold, R. Historic photos and TLS data fusion for the 3D reconstruction of a monastery altar ensemble. *Int. Arch. Photogramm. Remote Sens. Spat. Inf. Sci.* **2015**, *XL-5/W7*, 201–206. [CrossRef]
10. Ramos, M.M.; Remondino, F. Data fusion in Cultural Heritage—A Review. *Int. Arch. Photogramm. Remote Sens. Spat. Inf. Sci.* **2015**, *XL-5/W7*, 359–363. [CrossRef]
11. Chen, B.; Huang, B.; Xu, B. Multi-source remotely sensed data fusion for improving land cover classification. *ISPRS J. Photogramm. Remote Sens.* **2017**, *124*, 27–39. [CrossRef]
12. Hess, M.; Petrovic, V.; Meyer, D.; Rissolo, D.; Kuester, F. Fusion of multimodal three-dimensional data for comprehensive digital documentation of cultural heritage sites. In Proceedings of the 2015 Digital Heritage International Congress (DigitalHeritage), Granada, Spain, 28 September–2 October 2015; pp. 595–602. [CrossRef]
13. Frohlich, R.; Kato, Z.; Tremeau, A.; Tamas, L.; Shabo, S.; Waksman, Y. Region based fusion of 3D and 2D visual data for Cultural Heritage objects. In Proceedings of the 23rd International Conference on Pattern Recognition (ICPR), Cancun, Mexico, 4–8 December 2016; pp. 2404–2409. [CrossRef]
14. Dore, C.; Murphy, M. Integration of Historic Building Information Modeling (HBIM) and 3D GIS for recording and managing cultural heritage sites. In Proceedings of the 18th International Conference on Virtual Systems and Multimedia, Milan, Italy, 2–5 September 2012; pp. 369–376.
15. Gonzalez-Aguilera, D.; Del Pozo, S.; Lopez, G.; Rodriguez-Gonzalvez, P. From point cloud to CAD models: Laser and optics geotechnology for the design of electrical substations. *Opt. Laser Technol.* **2012**, *44*, 1384–1392. [CrossRef]
16. Reilly, P. Towards a virtual archaeology. In *Computer Applications and Quantitative Methods in Archaeology 1990*; BAR International Series 565; Tempus Reparatum: Oxford, UK, 1991; pp. 133–139.
17. Remondino, F. Heritage Recording and 3D Modeling with Photogrammetry and 3D Scanning. *Remote Sens.* **2011**, *3*, 1104–1138. [CrossRef]
18. Nocerino, E.; Menna, F.; Toschi, I.; Morabito, D.; Remondino, F.; Rodríguez-Gonzálvez, P. Valorisation of history and landscape for promoting the memory of WWI. *J. Cult. Herit.* **2018**, *29*, 113–122. [CrossRef]

19. Micoli, L.; Guidi, G.; Angheleddu, D.; Russo, M. A multidisciplinary approach to 3D survey and reconstruction of historical buildings. In Proceedings of the 2013 Digital Heritage International Congress (DigitalHeritage), Marseille, France, 28 October–1 November 2013; pp. 241–248.
20. Münster, S.; Kröber, C.; Hegel, W.; Pfarr-Harfst, M.; Prechtel, N.; Uhlemann, R.; Henze, F. First experiences of applying a model classification for digital 3D reconstruction in the context of humanities research. In Proceedings of the Euro-Mediterranean Conference (EuroMed 2016), Nicosia, Cyprus, 18–21 December 2016; pp. 477–490. [CrossRef]
21. Guidi, G.; Russo, M. Diachronic 3D reconstruction for lost cultural heritage. *Int. Arch. Photogramm. Remote Sens. Spat. Inf. Sci.* **2011**, *XXXVIII-5/W16*, 371–376. [CrossRef]
22. Cipriani, L.; Fantini, F. Digitalization Culture VS Archaeological Visualization: Integration of Pipelines and Open Issues. *Int. Arch. Photogramm. Remote Sens. Spat. Inf. Sci.* **2017**, *XLII-2/W3*, 195–202. [CrossRef]
23. Rodríguez-Gonzálvez, P.; Muñoz-Nieto, A.L.; del Pozo, S.; Sanchez-Aparicio, L.J.; Gonzalez-Aguilera, D.; Micoli, L.; Gonizzi Barsanti, S.; Guidi, G.; Mills, J.; Fieber, K.; et al. 4D Reconstruction and Visualization of Cultural Heritage: Analyzing our Legacy Through Time. *Int. Arch. Photogramm. Remote Sens. Spat. Inf. Sci.* **2017**, *XLII-2/W3*, 609–616. [CrossRef]
24. Aguilera, D.G.; Lahoz, J.G. sv3DVision: Didactical photogrammetric software for single image-based modeling. *Int. Arch. Photogramm. Remote Sens. Spat. Inf. Sci.* **2006**, *XXXVI-Part 6*, 171–178.
25. Garcia-Gago, J.; Gomez-Lahoz, J.; Rodríguez-Méndez, J.; González-Aguilera, D. Historical single image-based modeling: The case of Gobierna Tower, Zamora (Spain). *Remote Sens.* **2014**, *6*, 1085–1101. [CrossRef]
26. Levoy, M.; Pulli, K.; Curless, B.; Rusinkiewicz, S.; Koller, D.; Pereira, L.; Ginzton, M.; Anderson, S.; Davis, J.; Ginsberg, J.; et al. The digital Michelangelo project: 3D scanning of large statues. In Proceedings of the 27th Annual Conference on Computer Graphics and Interactive Techniques, New Orleans, LA, USA, 25–27 July 2000; pp. 131–144. [CrossRef]
27. Barber, D.M.; Dallas, R.W.; Mills, J.P. Laser scanning for architectural conservation. *J. Archit. Conserv.* **2006**, *12*, 35–52. [CrossRef]
28. Remondino, F.; Rizzi, A. Reality-based 3D documentation of natural and cultural heritage sites—Techniques, problems, and examples. *Appl. Geomat.* **2010**, *2*, 85–100. [CrossRef]
29. Xiao, W.; Mills, J.; Guidi, G.; Rodríguez-Gonzálvez, P.; Gonizzi Barsanti, S.; González-Aguilera, D. Geoinformatics for the conservation and promotion of cultural heritage in support of the UN Sustainable Development Goals. *ISPRS J. Photogramm. Remote Sens.* **2018**, *142*, 389–406. [CrossRef]
30. González-Aguilera, D.; Rodríguez-Gonzálvez, P. Drones—An Open Access Journal. *Drones* **2017**, *1*, 1. [CrossRef]
31. Stylianidis, E.; Remondino, F. *3D Recording, Documentation and Management of Cultural Heritage*; Whittles Publishing: Dunbeath, Scotland, UK, 2016.
32. Toschi, I.; Rodríguez-Gonzálvez, P.; Remondino, F.; Minto, S.; Orlandini, S.; Fuller, A. Accuracy evaluation of a mobile mapping system with advanced statistical methods. *Int. Arch. Photogramm. Remote Sens. Spat. Inf. Sci.* **2015**, *XL-5/W4*, 245–253. [CrossRef]
33. Rodríguez-Gonzálvez, P.; Jiménez Fernández-Palacios, B.; Muñoz-Nieto, Á.; Arias-Sanchez, P.; Gonzalez-Aguilera, D. Mobile LiDAR System: New Possibilities for the Documentation and Dissemination of Large Cultural Heritage Sites. *Remote Sens.* **2017**, *9*, 189. [CrossRef]
34. Chu, C.H.; Chiang, K.W. The performance of a tight INS/GNSS/photogrammetric integration scheme for land based MMS applications in GNSS denied environments. *Int. Arch. Photogramm. Remote Sens. Spat. Inf. Sci.* **2012**, *XXXIX-B1*, 479–484. [CrossRef]
35. Nocerino, E.; Menna, F.; Remondino, F.; Toschi, I.; Rodríguez-Gonzálvez, P. Investigation of indoor and outdoor performance of two portable mobile mapping systems. In Proceedings of the SPIE Optical Metrology, Munich, Germany, 25–29 June 2017. [CrossRef]
36. Cabo, C.; Del Pozo, S.; Rodríguez-Gonzálvez, P.; Ordóñez, C.; González-Aguilera, D. Comparing Terrestrial Laser Scanning (TLS) and Wearable Laser Scanning (WLS) for Individual Tree Modeling at Plot Level. *Remote Sens.* **2018**, *10*, 540. [CrossRef]
37. Rodríguez-Gonzálvez, P.; González-Aguilera, D.; Gómez-Lahoz, J. From Point Cloud to Surface: Modeling Structures in Laser Scanner Point Clouds. In Proceedings of the ISPRS Workshop on Laser Scanning 2007 and Silvilaser 2007, Espoo, Finland, 12–14 September 2007; pp. 338–343.
38. Cultural Heritage through Time. Available online: http://cht2-project.eu/ (accessed on 3 November 2018).

39. Tejeda-Sánchez, C.; Muñoz-Nieto, A.; Rodríguez-Gonzálvez, P. Geomatic Archaeological Reconstruction and a Hybrid Viewer for the Archaelogical Site of Cáparra (Spain). *Int. Arch. Photogramm. Remote Sens. Spat. Inf. Sci.* **2018**, *XLII-2*, 1105–1111. [CrossRef]
40. Evans, A.; Romeio, M.; Bahrehmand, A.; Agenjo, J.; Blat, J. 3D graphics on the web: A survey. *Comput. Graph.* **2014**, *41*, 43–61. [CrossRef]
41. Potenziani, M.; Callieri, M.; Dellepiane, M.; Corsini, M.; Ponchio, F.; Scopigno, R. 3DHOP: 3D heritage online presenter. *Comput. Graph.* **2015**, *52*, 129–141. [CrossRef]
42. Schütz, M. *Potree: Rendering Large Point Clouds in Web Browsers*; Technische Universität Wien: Wien, Austria, 2016.
43. Esri CityEngine. Available online: https://www.esri.com/es-pa/store/city-engine (accessed on 3 November 2018).
44. Nocerino, E.; Menna, F.; Remondino, F. GNSS/INS aided precise re-photographing. In Proceedings of the 18th International Conference on Virtual Systems and Multimedia, Milan, Italy, 2–5 September 2012; pp. 235–242. [CrossRef]
45. CloudCompare. Version 2.9.1. [GPL Software]. Available online: http://www.cloudcompare.org/ (accessed on 3 November 2018).
46. Rodríguez-Gonzálvez, P.; Garcia-Gago, J.; Gomez-Lahoz, J.; González-Aguilera, D. Confronting Passive and Active Sensors with Non-Gaussian Statistics. *Sensors* **2014**, *14*, 13759–13777. [CrossRef] [PubMed]
47. Höhle, J.; Höhle, M. Accuracy assessment of digital elevation models by means of robust statistical methods. *ISPRS J. Photogramm. Remote Sens.* **2009**, *64*, 398–406. [CrossRef]
48. Rodríguez-Gonzálvez, P.; Nocerino, E.; Menna, F.; Minto, S.; Remondino, F. 3D Surveying & modeling of underground passages in WWI fortifications. *Int. Arch. Photogramm. Remote Sens. Spat. Inf. Sci.* **2015**, *XL-5/W4*, 17–24. [CrossRef]
49. De Naeyer, A.; Arroyo, S.P.; Blanco, J.R. *Krakow Charter 2000: Principles for Conservation and Restoration of Built Heritage*; Bureau Krakow: Krakow, Poland, 2000.
50. The Nara Document on Authenticity. Available online: https://whc.unesco.org/document/116018 (accessed on 30 December 2018).

International Journal of
Geo-Information

Article

Application of Open-Source Software in Community Heritage Resources Management

Jihn-Fa Jan

Department of Land Economics, Taiwan Institute for Governance and Communication Research,
National Chengchi University, No. 64, Section 2, Zhi-Nan Road, Taipei 11605, Taiwan; jfjan@nccu.edu.tw

Received: 19 July 2018; Accepted: 28 October 2018; Published: 31 October 2018

Abstract: In this paper, we present a case study of community heritage resources investigation and management, which was a collaborative project conducted by researchers and participants from rural communities. Geotagged photos were obtained using smart phones, and 360-degree panoramas were acquired using a robotic camera system. These images were then uploaded to a web-based GIS (WebGIS) developed using Arches-Heritage Inventory Package (HIP), an open-source geospatial software system for cultural heritage inventory and management. By providing various tools for resources annotation, data exploration, mapping, geovisualization, and spatial analysis, the WebGIS not only serves as a platform for heritage resources database management, but also empowers the community residents to acquire, share, interpret, and analyze the data. The results show that this type of collaborative working model between researcher and community can promote public awareness of the importance of heritage conservation and achieve the research goal more effectively and efficiently.

Keywords: community; heritage resources management; open-source software; web-based GIS; Arches-HIP

1. Introduction

The World Heritage Convention (1972) is a globally recognized legal instrument in heritage conservation, which has been implemented by many countries to protect lots of sites all over the world. Over the years, a growing awareness of the importance of the involvement of local communities and indigenous people has caused paradigmatic shift in practice and strategic planning for heritage conservation. Local communities not only contribute to the protection of biodiversity, specifically agrodiversity through sustainable land use, but also play an important role in maintaining and managing heritage, whether natural or cultural [1]. Previous researches in various countries, including Zimbabwe, Tanzania, South Africa, and other parts of the world reveal that active involvement of local people are essential for understanding the dynamics surrounding heritage sites and their management [2–4].

Many cultural heritage sites had been destroyed or damaged accidentally, deliberately, or by natural disaster in the past. To preserve cultural heritages for future generations, effective measures are needed for documenting them. Digital representations of cultural heritage can be archived, disseminated, and can serve as digital surrogates of the "real world". They not only can be used for personal enjoyment but also enable visualization and scientific study without the need to have physical experience of the object or place. Emerging technologies such as digital photography, photogrammetry, panoramas, laser scanning, Global Positioning System (GPS), 3D modeling, distributed database, Web 2.0, and computer software for generating 2D and 3D information provide heritage professionals with tools that can acquire, process, and distribute heritage data more effectively and efficiently than decades ago [5]. However, CIPA Heritage Documentation, one of the oldest International Scientific Committees (ISC) of the International Council on Monuments and Sites (ICOMOS) and the International Society of Photogrammetry and Remote Sensing (ISPRS), pointed out some key problems that need further development, including duplication efforts, limited resources, and lack of uniformity. In particular, they suggested that the following issues merit more attention: open-source applications for heritage recording, use of mobile recording technologies to support documentation, usage of the CIDOC-CRM (International Committee for Documentation Conceptual Reference Model) and other ontology concepts for integration with heritage documentation activities, and use of controlled vocabularies for ensuring validity and consistency of inventory data [6,7].

In recent years, a wide range of sophisticated and freely available open-source software (OSS) have emerged, which are not only cost-saving but also ideal for developing cross-platform applications [8–10]. OSS is provided with free license that permits developers and users to study, modify, improve, and distribute the software. Many previous works have shown that this software is of great importance for various applications of the heritage field. Built with OSS, the Open Video Digital Library Toolkit (OVDLT) and CollectionSpace provide museums, libraries, and other institutions with web-based collections management software for archiving digital videos and digital replicas of cultural heritage objects. Constructed using OVDLT, the Open Video repository provides thousands of video clips from a variety of sources, including the U.S. Records and Archives Administration, National Aeronautics and Space Administration (NASA), and several universities and organizations [11–15]. Meyer et al. developed a web-based information system for managing and distributing cultural heritage data. The system allows management of diverse types of data and provides users with opportunity to explore and analyze data at spatial and temporal levels [16]. In a research project aiming to study the building techniques of the 13th–18th centuries in the Sardinia region (Italy), for managing enormous amount of data produced during the research, a spatial information system (SIS) was developed using a variety of OSS components, including PostgreSQL and PostGIS for building geodatabase, PHP and HTML for creating web-based data entry form, Geoserver for publishing geodatabase on the WWW through Open Geospatial Consortium (OGC) Web Map Service (WMS), Quantum GIS (QGIS) for accessing the database, and Leaflet JavaScript libraries for establishing the WebGIS [17,18].

Recent advancement of geographic information systems (GIS) technology offered powerful instruments for geodatabase management, spatial analysis, data exploration, geostatistical analysis, 2D and 3D visualization, and web mapping. More and more cultural heritage experts and authorities responsible for managing cultural heritage were actively engaged in the development of information system having GIS as the primary infrastructure component [19]. Mainly designed as integrated system for heritage inventory and management, two well-recognized systems exemplify the advantages of OSS, i.e., the South African Heritage Resources Information System (SAHRIS) launched by the South African Heritage Resources Agency (SAHRA) since 2012, and the Arches system jointly developed by Getty Conservation Institute (GCI) and World Monuments Fund (WMF) since 2013 [20–24].

SAHRIS was developed using Drupal, a very popular content management system (CMS) based on the LAMP stack, which is a combination of the Linux operating system, Apache web server, MySQL database, and PHP scripting language. Integrated with GIS tools and Geoserver, the system provides user-friendly heritage management tools to provincial heritage authorities, local planning authorities, and all bodies and custodians of heritage. SAHRIS consists of four main functions, namely, (1) to serve as an integrated heritage management system that facilitates online development applications, decision making, and planning process; (2) to serve as a national heritage sites repository populated with detailed information of identified sites; (3) to serve as a national collections management system that provides museums and heritage managers with software tools for working with digital inventories of their collections; and (4) to serve as a centralized platform for management of a comprehensive database of heritage objects for reporting and tracking heritage crime. By the end of December 2015, major achievements of the SAHRIS included: (1) a total of 8020 heritage cases had been processed, with 9462 heritage reports captured; (2) totally 42,456 sites had been processed by the system; (3) a total of 26,195 objects had been registered to the system; and (4) there were 260 stolen objects recorded on the system [20,25–28].

First released in October 2013, Arches is an OSS platform in the heritage conservation field providing a suite of powerful tools for compiling and managing inventories of heritages of various scales (national, regional, city or site). In addition to the professionals of Farallon Geographics, the primary developer of the software, many experts of related world-class organizations such as the English Heritage, WMF, GCI are heavily involved in its development [22]. Based on Django [29], a Python web framework that encourages rapid development of web-based applications, Arches is a flexible platform that can be tailored to meet different needs at low cost. Since its release, organizations worldwide are using Arches to build heritage resources management system, such as the HistoricPlacesLA deployed by the City of Los Angeles, the Philippine Heritage Map created by a Manila-based nonprofit organization, the Endangered Archaeology in the Middle East and North Africa project based at Oxford University, the Hong Kong Heritage Platform developed at the University of Hong Kong (HKU), and the Armed Forces Retirement Home (AFRH) Information and Resource Inventory System (IRIS) established by AFRH, a federal agency of the USA [22,30].

The current implementations of SAHRIS and Arches have proven that these two platforms are both very suitable for developing heritage resources management system. In comparison to SAHRIS and other implementations, one major advantage of Arches is that it provides source codes and documentations for developers, and it can be customized to meet individual needs. Moreover, to ensure data integrity and consistency, the Arches system adopts two CIDOC standards: (1) the revised International Core Data Standard for Archaeological and Architectural Heritage (CDS); and (2) the CIDOC Conceptual Reference Model. The former has been used as a basis to define the data fields in the generic version of the system while the latter has been used to provide the semantic framework [23,24,31]. The objective of this research was to develop a heritage inventory and management system for local communities with minimum cost. Considering the establishment of a repository of digital cultural resources that can be in line with CIDOC-CRM standard, we decided to implement Arches for a rural community in eastern Taiwan, in which the local residents are very proactive in preserving cultural and natural resources within their community.

2. Materials and Methods

2.1. Arches and Arches-HIP

The Arches project originated from MEGA (Middle Eastern Geodatabase for Antiquities), a web-based geospatial information system built with open-source tools [32,33]. Mainly sponsored by WMF and GCI, MEGA had demonstrated numerous superior features for cultural heritage management since its initial release in 2010. In addition to the support from several heritage authorities in the United States and Europe on researching the best practice and standards relevant to the development of the system, the project accepted substantial technical assistance from the English Heritage and Flanders Heritage Agency.

The version of Arches used in this study is v3.0, which was released in April 2015. However, Arches v4.0 had been released recently, and both versions can be downloaded from the official website of the Arches project [22]. Moreover, the Arches-HIP was launched simultaneously with Arches v3.0, which is an officially developed application of Arches. This application is designed to let users manage many types of heritage data and these types are mainly divided into six different resources in Arches-HIP [22,34]:

- Heritage Resources: objects that are culturally significant, for example, temple, shrines, artifact, tombstones, buildings, and monuments.
- Heritage Resource Groups: gathering of individual heritage resources that are best regarded as a group such as region, landscapes, historic zones, and graveyards.
- Activity: resources that occur over a period of time, which include religious celebrations, investigations, excavations, and restoration.

- Historic Event: culturally significant occurrences including battles, protests, natural disasters such as volcanic eruptions, earthquakes, and flooding.
- Actor: resources that are persons, groups of people, or organizations, e.g., architects, research teams, surveying firms, and association.
- Information Objects: objects that encode information such as a photo, documents, audio and video recordings, signs, drawings, and inscriptions.

Each resource provides a list of attributes that can be used to describe a heritage feature. Based on CIDOC-CRM, Arches employ ontology concept for naming and defining data types, properties, and the relationships between data entities that describe a resource. Arches-HIP requires the user to decide how to describe the resource information that will be entered into the system. An authority file is associated with an attribute such that its value is constrained by a list of controlled vocabularies. By using authority files to define the controlled vocabularies, it can be ensured that heritage data are created using just a list of predetermined terms. Table 1 shows a list of authority files included in Arches-HIP. For example, as shown in Table 2, to describe the condition of a building, controlled vocabularies can be a list of simple words such as "Good, Fair, Poor", or thesauri in multi-lingual environment. Table 3 shows the content of measurement types authority file, which contains a list of predefined words that are used to describe measurement types including: length, depth, breadth, weight, tonnage, and height [34,35]. Defined in CRM, E55 class consists of concepts represented by terms from thesauri and controlled vocabularies used to describe and classify instances of CRM classes, and E57 class may denote properties of matter. Details of the class can be found at: http://www.cidoc-crm.org/Entity/e55-type/version-6.2 and http://www.cidoc-crm.org/Entity/e57-material/version-6.2.

Table 1. A partial list of authority files included in Arches-HIP, which contain controlled vocabularies.

Entity Type	Authority Document
CONDITION_TYPE.E55	CONDITION_AUTHORITY_DOCUMENT.csv
CULTURAL_PERIOD.E55	CULTURAL_PERIOD_AUTHORITY_DOCUMENT.csv
DISTURDANCE_TYPE.E55	DISTURDANCE_TYPE_AUTHORITY_DOCUMENT.csv
LANGUAGE.E55	LANGUAGE_AUTHORITY_DOCUMENT.csv
MATERIAL.E57	MATERIAL_AUTHORITY_DOCUMENT.csv
MEASUREMENT_TYPE.E55	MEASUREMENT_TYPE_AUTHORITY_DOCUMENT.csv
MODIFICATION_TYPE.E55	MODIFICATION_TYPE_AUTHORITY_DOCUMENT.csv
NAME_TYPE.E55	NAME_TYPE_AUTHORITY_DOCUMENT.csv
SETTING_TYPE.E55	SETTING_TYPE_AUTHORITY_DOCUMENT.csv
STYLE.E55	STYLE_AUTHORITY_DOCUMENT.csv
STATUS.E55	STATUS_AUTHORITY_DOCUMENT.csv
THREAT_TYPE.E55	THREAT_TYPE_AUTHORITY_DOCUMENT.csv

Table 2. A list of values for conditions defined in "CONDITION_AUTHORITY_DOCUMENT.csv".

Conceptid	PrefLabel	AltLabels	Parent Conceptid	Concept Type	Provider
CONDITION	1	good	CONDITION_AUTHORITY_DOCUMENT.csv	Index	GCI
CONDITION	2	fair	CONDITION_AUTHORITY_DOCUMENT.csv	Index	GCI
CONDITION	3	poor	CONDITION_AUTHORITY_DOCUMENT.csv	Index	GCI
CONDITION	4	very bad	CONDITION_AUTHORITY_DOCUMENT.csv	Index	GCI

Table 3. The "MEASUREMENT_TYPE_AUTHORITY_DOCUMENT.csv" contains a list of values for measurement types.

Conceptid	PrefLabel	AltLabels	Parent Conceptid	Concept Type	Provider
MEASUREMENT_TYPE	1	length	MEASUREMENT_TYPE_AUTHORITY_DOCUMENT.csv	Index	EH
MEASUREMENT_TYPE	2	depth	MEASUREMENT_TYPE_AUTHORITY_DOCUMENT.csv	Index	EH
MEASUREMENT_TYPE	3	breadth	MEASUREMENT_TYPE_AUTHORITY_DOCUMENT.csv	Index	EH
MEASUREMENT_TYPE	4	weight	MEASUREMENT_TYPE_AUTHORITY_DOCUMENT.csv	Index	EH
MEASUREMENT_TYPE	5	tonnage	MEASUREMENT_TYPE_AUTHORITY_DOCUMENT.csv	Index	EH
MEASUREMENT_TYPE	6	height	MEASUREMENT_TYPE_AUTHORITY_DOCUMENT.csv	Index	EH

The Arches system is also designed to ensure the compliance with OGC standards and therefore is compatible with other desktop GIS applications such as ESRI ArcGIS, Google Earth, Quantum GIS, and online satellite imagery and map services including Google Maps, Microsoft Bing Maps, and OpenStreetMap [22–24]. With high compatibility, Arches is capable of displaying geographic data from different sources, which is very convenient for customizing the system to meet different needs. Since its release, Arches have been used by numerous organizations to develop various applications in heritage field. For example, the implementation of Arches at Nouli Community, Hualien County, Taiwan resulted in an online cultural heritage resource inventory system for better understanding the community and promoting public awareness of the importance of heritage conservation [22,36,37]. Figure 1 shows the map view of the system (http://arches.nccu.edu.tw/).

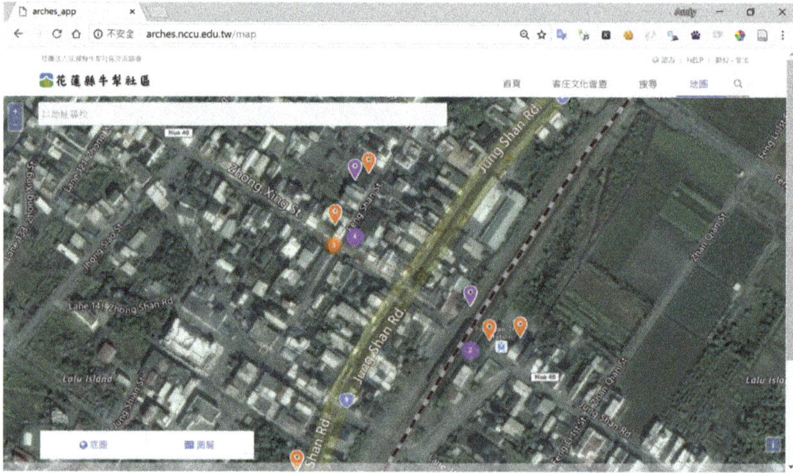

Figure 1. The MAP VIEW of Nouli Community Arches platform showing locations of heritage sites on Bing Maps.

2.2. Other Related Open-Source Tools

With many open-source geospatial software features, Arches is distributed under AGPLv3, which allows the users to copy and modify Arches without restriction and requires that derivative works to be distributed under AGPLv3. Arches is comprised of four basic components [22,34,35]:

- A Django-based web server tier, which is developed using Python programming language.
- A Bootstrap-based user interface using JavaScript and CSS (Cascading Style Sheets).
- A PostgreSQL relational database, and PostGIS, which is the spatial database extender for PostgreSQL.
- An Elasticsearch search engine, which can provide high performance full-text querying.

Developed using Python, Django is a high-level web framework that offers dozens of tools for creating web-based applications, including user authentication, content management, RSS feeds, and many more out of the box [29]. Owing to Django's flexibility, we can create applications to seamlessly integrate with Arches. To produce dynamic statistical charts that can be put in web pages, we used pygal, a Python library for creating scalable vector graphics (SVG) [38].

In addition, GEOS (Geometry Engine-Open-Source) [39] is used to process geometries and provide spatial operations such as union, distance, intersection, buffer, area, and length. Given its open-source nature, Arches has flexibility of extension and customization if the user has appropriate software skills. In this research, we deployed the heritage inventory and management system using software listed in Table 4. All the software run on Linux operating system Ubuntu 14.04 LTS (Long Term Support).

Table 4. Software used in this research.

Name of Software	Version	Purpose
Ubuntu Server	14.04 LTS	The platform used for software installation and development.
Apache	2.4.7	The web server software.
Python	2.7.6	Used for developing data processing scripts.
Django	1.6.2	A web development framework written in Python.
Bootstrap	3.3.5	An open-source toolkit for developing responsive websites.
Elasticsearch	1.4.1	Library for providing full-text searching.
PostgreSQL	9.3	The relational database management system.
PostGIS	2.2.0	Supports spatial database extension for PostgreSQL.
GEOS	3.4.2	The most popular geospatial library that is used by PostGIS.
pygal	2.1.1	A Python library for creating statistical charts in SVG format.

2.3. Heritage Resource Data

The island of Taiwan is situated in East Asia with a population of more than 23 million. It was named Formosa when the Portuguese sailors sighted the island in the 16th century. Taiwanese aborigines of Austronesian race were the native inhabitants since prehistoric period. Most Han Chinese, mainly from southeastern provinces of China, immigrated to the island between the 16th and 18th century and dispersed in fluvial plains along the west coast. A few European settlements appeared on the southwestern coast, and northern areas in the 16th and 17th century. Between 1895 and 1945 the Japanese colonized Taiwan after defeating China in a war. Moreover, many mainlanders from various regions of China immigrated to Taiwan after World War II and the Chinese Civil War, a war fought between forces loyal to KMT-led government and the Communist Party of China. The influences of these different cultures are reflected in many aspects, including architecture, clothing, languages, artifacts, foods, agriculture, religion, and customs.

In the past few decades, Taiwan has experienced very rapid economic growth and large-scale development occurred in numerous industrial parks and urban areas as well as infrastructures such as high-speed railway, highways, and roads; however, rural villages remain largely undeveloped. Therefore, even though Taiwan has become a highly developed country in the world, cultural heritage resources are still quite abundant in rural communities such as agricultural and aquatic farm lands, tribal villages in uphill areas, and coastal settlements.

Similar to more economically developed countries, population aging has caused prominent impacts on Taiwan's economy, education, health care, and social welfare policy [40]. Especially in rural communities, lack of working-age people not only affects the economic income of residents but also has adverse influences on social stability, and preservation of valuable cultural heritages. Consequently, in Taiwan, many community associations were established in recent years with an aim to help local inhabitants preserve heritage resources through inventory, protection, and educational programs. Heritages resources not only are helpful in learning local culture and landscape of a community, but also can preserve the sense of belonging and identity of people that can be passed on to the following generations. In many cases, preserving heritage resources even help generate more income to the community residents through well-developed tourism programs. However, due to limited available resources, only few of these community associations gained successful results. To address the needs of effective tools for community-oriented heritage resources management, this research aims to use Arches-HIP, a new open-source geospatial software system for cultural heritage inventory and management, to develop a spatial database and platform that can be applied to various rural communities.

The study site, Nouli Community, of this research is part of Hualien County located in eastern Taiwan with a total area of 24.6 km^2 (Figure 2). Nouli Community is comprised of three villages, including Fengping, Fengli and Fengshan. During Japanese occupation (1895–1945), this area was named Toyota in 1911, therefore this name also appears in several places and documents ever since. Although the Japanese had withdrawn from Taiwan after World War II, a great number of Japanese architecture and construction were well preserved. Since Japanese established Toyota Village in 1913, Hakka, Holo, aborigines, Chinese mainlanders, and new immigrants (mostly from southeastern Asia countries) had moved to this area in different time periods spanning over a century. Hence, Nouli Community is a multi-ethnic inhabited village filled with diverse cultural characteristics. According to the statistics of July 2016, the total population of the Nouli Community is 4883. Similar to a typical rural area in Taiwan, the elderly population in the Nouli Community makes up a large proportion, with about one-sixth of the population being over the age of 65. The residents of Nouli Community mainly belong to three ethnic groups, including Hakka (65%), Holo Taiwanese (22%), and aborigines (11%). Chinese mainlanders and new immigrants totally amount to only 2% of the total population. Starting in 1996, in response to common problems in rural communities such as farmland fallow, population aging and outflows, members of several families formed an organization to launch a few programs aiming to boost economic development of the community. The organization later became the Nouli Community Development Association (NCDA) [41–43].

Figure 2. The study site of this research. The overview image of Taiwan is from Google Earth, and the base map is from a WMS server created by the National Land Surveying and Mapping Center, Ministry of the Interior, Taiwan.

In the study site, most of the areas are less-developed agricultural lands, representing distinct cultural characteristics of various ethnics and eras, heritage resources such as monuments, shrines, artifacts, sites, and buildings dispersed in different places. To collect data about the heritage resources in the Nouli Community, this research used GPS logger and a mobile application, My Footprint, to record GPS coordinates and geotagged photos acquired using smart phones. My Footprint is a product of Madtec Solutions Limited and can run on various mobile devices installed with Android 2.2 and above. Moreover, a robotic camera system, LizardQ, was used to obtain high-resolution panoramic imageries.

3. Results

3.1. Heritage Resources Database

The heritage resources database resulted from collaborative efforts between the research team and the NCDA [43]. The main objective of this research is to inventory all types of resources that are of culturally significant values to the community such as natural and cultural landscape, historic sites, buildings, architectures, monuments, stelae, artifacts, and documents. For each resource, we recorded as much detail information as possible.

First, by using LizardQ we took photos of each site in seven viewing directions. Equipped with a Canon EOS 5D camera, LizardQ is capable of capturing high dynamic range (HDR) images. With different exposure times and ISO settings, six photos for each viewing direction were taken, with a total of 42 photos for each site. Panoramic images can then be created by stitching all photos together using Stitcher software provided by the vendor of LizardQ. The resulting images have very wide contrast range, therefore, the darkest and the brightest objects are all conspicuous in the same panoramic HDR image [44,45]. Figure 3 is a panorama of the Hakka Culture Exhibition Center in the community, with a resolution of 8000 × 4000 pixels. Panoramic images are placed on Google Cloud, and can be viewed from different perspectives using regular browsers running on smart phones, tablets, and desktop computers. Furthermore, 360° videos can be produced from these panoramas using image processing software, e.g., Photoshop, and GIMP (GNU Image Manipulation Program), a free and open-source image editor [46]. YouTube supports 360° videos with 24, 25, 30, 48, 50, or 60 frames per second. The video should be in equirectangular format with a 2:1 aspect ratio at a resolution of 7168 × 3584 or higher, up to 8192 × 4096. These video files must have certain metadata for 360° playback to be enabled. Provided by Google, a tool called "Spatial Media Metadata Injector" was used to insert metadata into these files, and then these videos were uploaded to YouTube [47]. A 360° video can be rotated by pressing WASD keys if it is played on computers in browsers such as Chrome, Firefox, Safari, MS Edge, or Opera. More interestingly, these videos can be viewed with head-mounted displays (HMDs), also referred to as "virtual reality (VR) headsets", or "VR glasses". For mobile device, users can wear a VR headset, such as Google Daydream View and Google Cardboard, to watch 360° videos in YouTube app [48]. Thus, the 360° videos can provide viewers with immersive virtual reality experience without the need to physically visit the sites. Figure 4 shows a 360° video viewed in YouTube app in VR mode.

Figure 3. Panoramic image of the of Hakka Culture Exhibition Center, which was obtained by using LizardQ camera system. The image is on Google Cloud so that viewers can observe the scene from every viewing angle. https://bit.ly/2PYxE7U.

Figure 4. A screen capture of a 360° video derived from panorama of the Hakka Culture Exhibition Center viewed in YouTube app. The video is on YouTube so that viewers can use VR headset to experience immersive virtual reality without the need to visit the site. https://youtu.be/x4PnMW1FG9U.

Second, several field investigations were conducted by the researchers and NCDA to collect data about the heritage sites. The location of each site was recorded using My Footprint mobile application and then exported as Google KML (Keyhole Markup Language) format. Based on XML (Extensible Markup Language) standard, KML uses tag-based structure to describe features, including placemarks, descriptions, ground overlays, paths, and polygons [49]. For example, Figure 5 depicts a placemark described in tags, including "name", "TimeStamp", "when", "styleUrl", "Point", and "coordinates". Through interviews with local people and assistance of NCDA, we also collected maps, photographs, documents, stories, reports, and other related data downloaded from official websites of government agencies. For this research we documented a total of 188 heritage resources in Nouli Community, and entering all the attribute data that describe each heritage would take much time and caution to assure data quality. Therefore, a more effective and reliable measure was needed to facilitate construction of heritage database.

```
166    <Placemark>
167        <name>2016/01/26 16:13 (起點)</name>
168        <TimeStamp><when>2016-01-26T08:13:57.369Z</when></TimeStamp>
169        <styleUrl>#start</styleUrl>
170        <Point>
171            <coordinates>121.491200,23.860372,54.90000152587891</coordinates>
172        </Point>
173    </Placemark>
```

Figure 5. Definition of a placemark in a KML file.

Third, to ensure data quality, we used Microsoft Excel to input field data into spreadsheets, and collaborated with NCDA to validate the data before entering them into the Arches-HIP system. By using Python scripts developed by the researcher, the GPS coordinates were extracted from KML files and then integrated with the resources investigation report provided by NCDA. For areal features, we used Quantum GIS to digitize the boundaries of sites using orthoimages. Figure 6 shows a spreadsheet used for entering heritage resource data, including site IDs, names, town IDs, description, and geometries such as points and polygons. Several worksheets were created for entering pertinent data, e.g., photographs, panoramas, ethnic groups, township, and population. In addition, relationships between resources were indicated in these worksheets.

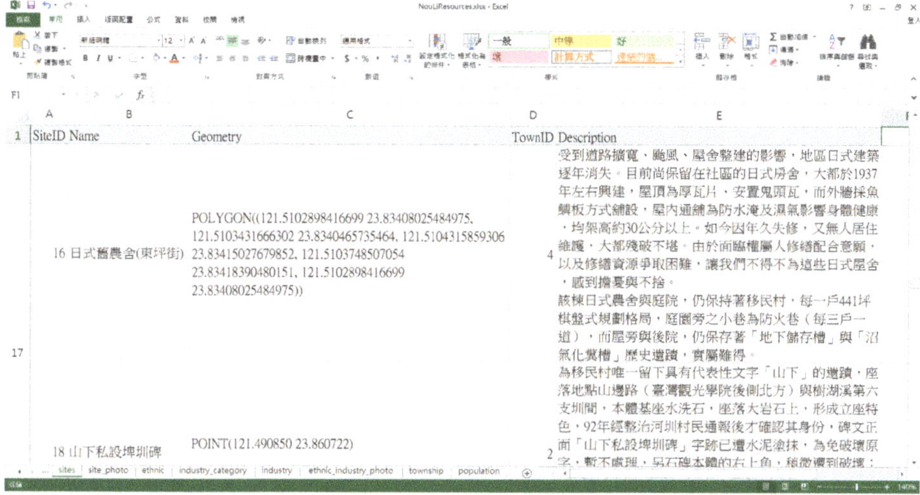

Figure 6. Microsoft Excel spreadsheets were used to enter heritage resource data, including site IDs, names, geometries, town IDs, and description.

The last step of data processing involved entering heritage resource data into the Arches-HIP system. Adopting the CIDOC-CRM ontological model, Arches-HIP uses a set of controlled vocabularies, defined in authority files, to ensure data consistency. It also provides Reference Data Manager for the users to interactively define the terms to be used for controlled vocabularies, and Resource Data Manager to let user enter resource data into the system manually [23,34,35]. However, entering attribute data of 188 heritage resources manually not only takes much time but also is prone to errors.

Arches-HIP allows user to create inventory database in three ways: (a) manually enter data using data entry form of Resource Data Manager: it is easy to use; however, it is inefficient and prone to errors if data volume is large; (b) import from a GIS shapefile: can import geometry and attribute; however, it is not suitable if the attributes are complicated, such as multiple names for a heritage resource; (c) import from text file in "arches" format: the attribute data is defined in plain text, and allows to import very sophisticated datasets, e.g., heritage data with complex attributes such as multiple geometry types, names. More details about how to create heritage inventory database can be found in Arches-HIP Documentation [34]. For this research, we decided to develop software tools to automate creation of "arches" file for importing data into Arches-HIP so that our work can be applied to future implementation for other communities more easily.

To define domain values that are suitable for describing heritage resources of this research, we created authority files by following the format of authority files that came with Arches-HIP (Tables 1–3). Before loading data into the Arches-HIP, we need to prepare two text files with the same name but different suffixes: an ".arches" file and a ".relations" file. The former contains the content of heritage data, and the latter defines the relationships among various resources. For example, we may relate many resources to one resource district, or many actors to a historical event. The ".arches" file contains attribute data of each heritage resource, which include RESOURCEID, RESOURCETYPE, ATTRIBUTENAME, ATTRIBUTEVALUE, and GROUPID. For example, the "Toyota Culture and History Exhibition Hall" (with RESOURCEID of SITE_7) in Nouli Community has several entries in the ".arches" file as shown in Figure 7. These entries are used to specifically describe the heritage resource, including "heritage resource type", "heritage resource use type", "name type", "name", "spatial coordinates of geometry", "description type", "description", "administrative subdivision type", and "administrative subdivision".

```
RESOURCEID|RESOURCETYPE|ATTRIBUTENAME|ATTRIBUTEVALUE|GROUPID
SITE_7|HERITAGE_RESOURCE.E18|HERITAGE_RESOURCE_TYPE.E55|HISTRES_
TYPE_ID:4|7
SITE_7|HERITAGE_RESOURCE.E18|HERITAGE_RESOURCE_USE_TYPE.E55|USE_
TYPE:2|7
SITE_7|HERITAGE_RESOURCE.E18|NAME_TYPE.E55|NAME_TYPE:1|7
SITE_7|HERITAGE_RESOURCE.E18|NAME.E41|Toyota Culture and His-
tory Exhibition Hall|7
SITE_7|HERITAGE_RESOURCE.E18|SPATIAL_COORDINATES_GEOMETRY.E47|PO
LYGON((121.5030117183072 23.84526956165155,121.5031067930137
23.84521614373416,121.503155699835
23.84528198284967,121.5030578975389
23.84533415994341,121.5030117183072 23.84526956165155 ))|7
SITE_7|HERITAGE_RESOURCE.E18|DESCRIPTION_TYPE.E55|DESCRIPTION_TY
PE:5|7
SITE_7|HERITAGE_RESOURCE.E18|DESCRIPTION.E62|(Description of the
Toyota Culture and History Exhibition Hall)|7
SITE_7|HERITAGE_RESOURCE.E18|ADMINISTRATIVE_SUBDIVISION_TYPE.E55
|ADMINISTRATIVE_SUBDIVISION:2|7
SITE_7|HERITAGE_RESOURCE.E18|ADMINISTRATIVE_SUBDIVISION.E48|Feng
li Village|7
```

Figure 7. An example of ".arches" file, which contains attributes and geometry of the "Toyota Culture and History Exhibition Hall".

As shown in the ".arches" file, both attributes data and geometry of a heritage resource can be defined in it. The geometry must be defined by values using WKT (well-known text) with coordinates set to EPSG (European Petroleum Survey Group) 4326 or latitude/longitude WGS84 [50]. WKT is a standard format for representing vector geometry objects on a map in human-readable text [51]. Moreover, the ".relations" file contains entries that define relationships among heritage resources. For example, the relationships between "Toyota Culture and History Exhibition Hall" and the other heritage resources in Nouli Community are defined as shown in Figure 8. The entries in the file relate this particular heritage resource to "historic district", "ethnic group", and several "image objects". One important thing worth noting is that the only valid values for ATTRIBUTENAME in the ".arches" file are entity types from the resource graph nodes file indicated in the RESOURCETYPE column. The resource graphs and nodes are listed in the online documentation of Arches-HIP (http://arches-hip.readthedocs.org/en/stable/appendix/).

From these spreadsheets, the heritage resources inventory data were converted to relational tables of PosgreSQL database, which were then used to create geospatial database of heritage resources in subsequent steps. To attain a more efficient data processing procedure, we designed Python scripts to extract data from PostgreSQL database and created "arches" files by following the format shown in Figures 7 and 8. Once we have created the ".arches" files and ".relations" for the heritage resources, we can load data into the Arches-HIP database using tools provided by Arches-HIP [34].

```
RESOURCEID_FROM|RESOURCEID_TO|START_DATE|END_DATE|RELATION_TYPE|
NOTES
SITE_7|HIST_DIS_32|||RELATIONSHIP_TYPE:1|
SITE_7|ETHNIC_1|||RELATIONSHIP_TYPE:1|
SITE_PH_91|SITE_7|||RELATIONSHIP_TYPE:26|
SITE_PH_11|SITE_7|||RELATIONSHIP_TYPE:26|
SITE_PH_12|SITE_7|||RELATIONSHIP_TYPE:26|
SITE_PH_13|SITE_7|||RELATIONSHIP_TYPE:26|
SITE_PH_14|SITE_7|||RELATIONSHIP_TYPE:26|
```

Figure 8. An example of ".relations" file, which defines relationships among heritage resources.

3.2. Arches Web Applications

Although the Arches-HIP can be installed on a variety of different platforms, the deployment of Arches-HIP for this research was under the Ubuntu Linux 14.04 LTS environment. The system setup in this research followed the installation procedures described in the online documentation of Arches and Arches-HIP. After completion of the installation process, the system contains a PostgreSQL database with a series of attribute tables that are ready for use. Built with many OSS tools, Arches-HIP provides powerful management tools out of the box for heritage resource managers. The overall architecture of the entire system is shown in Figure 9 [34,35,42].

Figure 9. The overall architecture of the Arches heritage management system.

The Arches-HIP provides ease of use web-based interface for managing heritage resources data. For example, with Resource Data Manager (Figure 10), the resource manager can do the following tasks on historic resources:

- Maintaining resource data: to update descriptions, location, classifications, components, measurements, condition assessment, images and files, related resources, and external system references of the resource.
- Evaluating Resource: to update the designation and evaluation criteria for the resource.
- Managing Resource: to review the edit history of the resource or delete the resource.

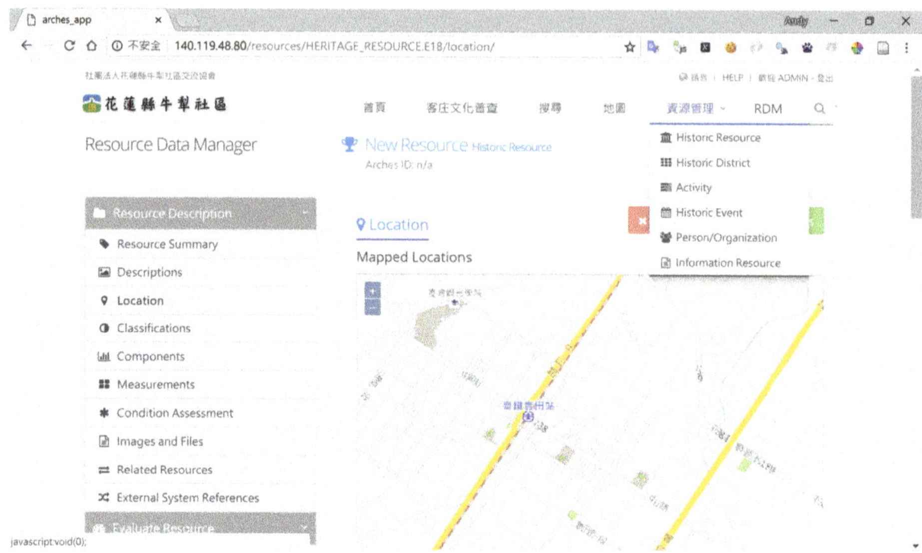

Figure 10. The Resource Data Manager of Arches-HIP.

All the data about a heritage resource can be viewed and updated by using Resource Data Manager, including modifying the geometry of a resource whether it is a point or a polygon feature. As shown in Figure 10, the resource manager can choose from six types of heritage resources, and the elements associated with the selected resource type will be available for editing. Table 5 shows the resource types and associated elements for each type.

Table 5. Resource types and associated elements for editing in Resource Data Manager of Arches-HIP.

Resource Type	Associated Elements
Historic Resource	Resource Summary, Descriptions, Location, Classifications, Components, Measurements, Condition Assessment, Images and Files, Related Resources, External System References, Designation, Evaluation Criteria
Historic District	Resource Summary, Descriptions, Location, Classifications, Measurements, Condition Assessment, Evaluation Criteria, Designation, Related Resources, and External System References
Activity	Resource Summary, Descriptions, Location, Actions, Related Resources, and External System References
Historic Event	Resource Summary, Descriptions, Location, Phase, Related Resources, and External System References
Person/Organization	Actor Summary, Descriptions, Location, Role, Related Resources, and External System References
Information Resource	Resource Summary, Creation and Publication, Coverage, Descriptions, File Upload, Related Resources, and External System References

With coordinates information filled, the location of heritage resource can be overlaid on a variety of basemaps, including streets map, satellite images, and a combination of both maps and images. Moreover, additional map layers can be added to the map view, for example Figure 11 shows the heritage resources on a 1:200,000 historic map produced by Japanese in 1897.

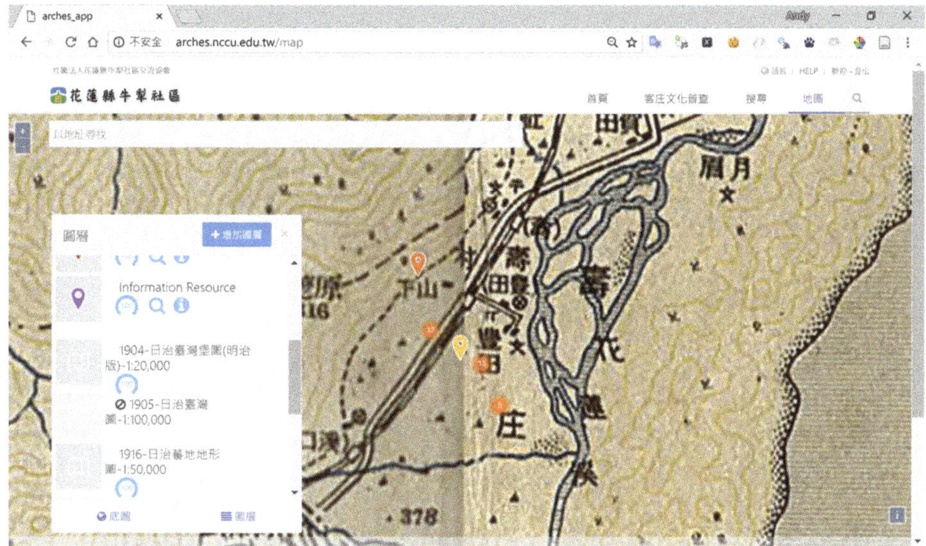

Figure 11. Heritage resources overlaid on historic map produced in 1897.

The search function allows users to query database using keywords, and then the Elasticsearch running in the background will perform a full-text search in all fields from all tables of the database. Location filter and time filter can also be used to alter selection if the related information is available. In addition to the location and description, Arches-HIP enables users to check the other related resources and show their relationships by using a dynamic graph. Figure 12 shows the search result of Culture and History Exhibition Hall at Shoufeng Township, and the relationships of this particular heritage resource with the other heritage resources, such as image files, district names, and ethnic information.

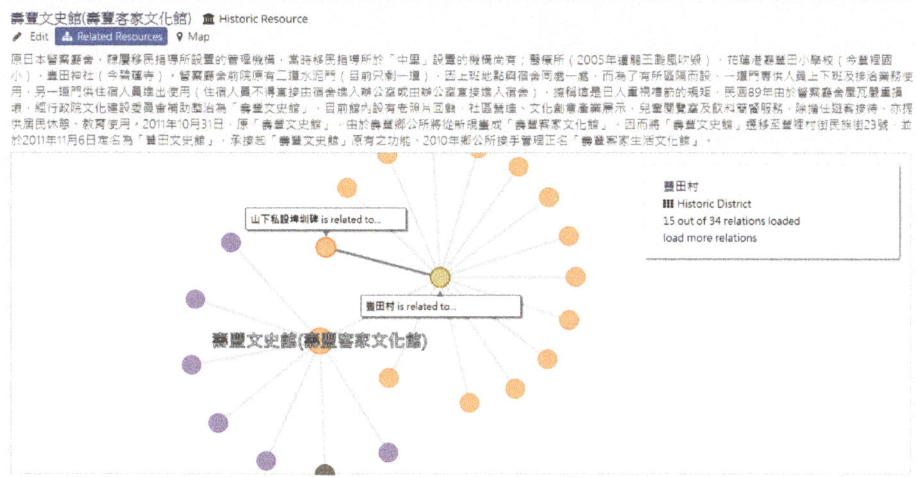

Figure 12. The detail information of a heritage resource and the relation graph.

Arches-HIP is built on top of Django, which has many powerful features for rapid development of web-based applications and can be easily customized to meet individual needs. In addition to the functionalities offered by Arches-HIP, we can also use Django as a full-featured web CMS, and create applications that can be integrated with the system. The majority of the people within the community are of Hakka ethnic group, therefore, the cultural characteristics of Hakka are prominent in various aspects. Based on open data downloaded from government agencies and statistics provided by NCDA, we used pygal to generate dynamic statistical charts in SVG format for some web pages. Figure 13 depicts the immigration of Hakka people to the Nouli Community spanning a period of more than a century. More dynamic statistical charts derived from the census of Hakka cultural resources can be found at the website.

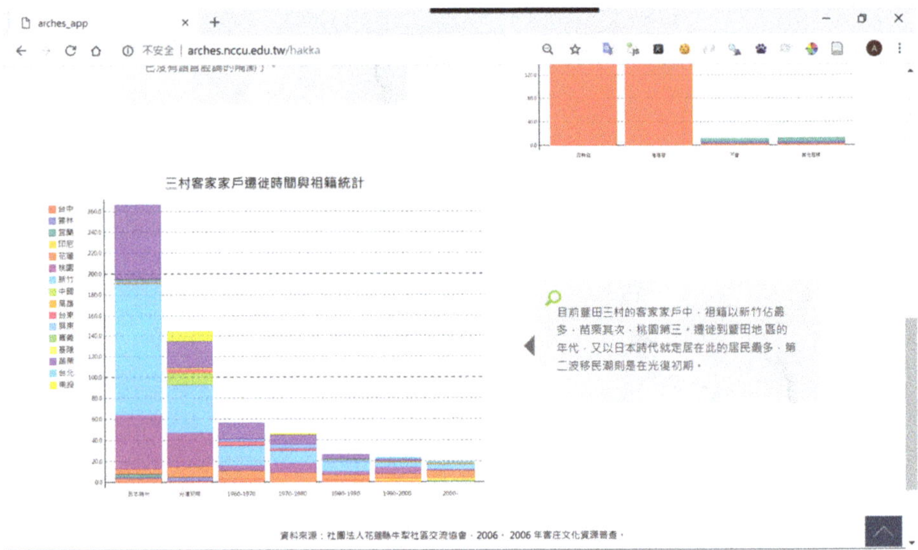

Figure 13. The website uses scalable vector graphics (SVG) to show dynamic statistical chart for open data downloaded from government agencies. This figure depicts the immigration of Hakka people to the Nouli Community spanning a period of more than a century. http://arches.nccu.edu.tw/hakka.

4. Discussion

Tang and Tang [52] presented a comparative study on conservation practices of two indigenous communities in Taiwan, which are located in an isolated island, and a mountainous tribal village, respectively. The research results indicate that traditional institutional support on conservation of local resources may collapse as a result of external influences. In contrast, active involvement from local people is an important factor for adapting traditional institutions to confront the challenges from the outside world while preserving natural resources within a local community.

Aiming to establish collaborative relationships between local communities and the government on conservation of natural and cultural resources, the Taiwan Forestry Bureau (TFB) started "Community Forestry Initiative" in 2002. Since then TFB had sponsored more than 1000 projects proposed by hundreds of communities dispersed in rural areas, coastal villages, mountainous settlements, and aboriginal tribes. Most of these community forestry projects focused on developing ecotourism programs that were meant to preserve significant natural and cultural values of the communities while trying to improve economic income of the local residents. Among these communities, Nouli Community is one of the most prominent case that demonstrates the importance of local involvement in cultural and natural conservation [37].

Attempting to develop a system that can be used by many community associations in Taiwan, this research established a community-oriented heritage resources inventory and management system using open-source geospatial software, particularly the Arches-HIP application. Based on the implementation results of this current study, some notable advantages of using the Arches framework to develop a WebGIS (web-based GIS) are described as follows.

To encourage local community to participate in cultural heritage conservation, through Arches-HIP we use the website to provide heritage information to the public, including descriptions of each heritage resource, photographs, panorama, 360-degree videos, and various types of data. The Arches-HIP system implements internationally adopted standards (CIDOC Conceptual Reference Model) for heritage data definition and management, which is essential for data sharing and international collaboration on heritage conservation. Following these standards not only assure data integrity, but also are beneficial to customize or extend the system to meet different needs.

The Arches-HIP provides an array of powerful tools for heritage resources inventory and management. To fully use its functionalities, however, the resource manager must plan the implementation with caution, especially preparing authority files with proper controlled vocabularies is an important step for creating the database. Well-organized authority files can ensure data consistency and facilitate maintaining high quality of the database.

The Arches-HIP supports geospatial technology standards, such as OGC standards on spatial data and web services. This is very important to develop a system that can be interoperable both on retrieving images and maps from the other servers, and publishing heritage data as standard web service that may be consumed by the other GIS applications. For example, we are able to add historic map layers and satellite imageries to the map view, which are very helpful to understand the landscape and historic changes of the study site.

The fact that Arches-HIP supports web application standards is also very important for building a cross-platform system. Unlike some web-based applications that require special browsers and plugins to work on the system, the heritage inventory and management system established in this research works well on different platforms and web browsers. The desktop platforms we have tested include Mac OS, Linux, and Microsoft Windows 7/8/10, using modern web browsers such as Chrome, Firefox, Safari, Internet Explorer, and Edge. The Bootstrap framework is adopted by Arches for responsive website development with CSS styles that can easily adjust to desktops, tablets, and mobiles. For mobile and smart devices running Apple iOS and Google Android, the embedded browser can automatically fit the web page to the screen, whether it is a smart phone, or a tablet computer. This feature is very helpful for field investigation and promoting tourism program of the community.

Although Arches-HIP provides user-friendly management tools for maintaining the database, it is still a challenge for the researcher to enter resource data into the database if the data volume is large. Arches-HIP provides a convenient tool for importing shapefile to the database, including attribute data and geometries of features. However, if the relationships among heritage resources are more complicated, loading data into the database by using text file in "arches" format is more preferable [34]. Based on our experience, we find that spreadsheet software is an effective tool for inputting field data because of its availability and ease of use. We used spreadsheets to store raw data, and in some instances, we used them to communicate with local participants for data validation. Once the data is ready for entering to the Arches-HIP database, the Python scripts developed for this study allow us to import field data to the database in a more efficient and effective way.

Unlike some implementations of heritage management system that seem to be successful but do not provide required software and documentation, Arches-HIP is an integrated OSS platform specifically built for heritage field. The official release of the system provides source codes, detailed documentation, and sample data, which are very helpful for understanding how CIDOC-CRM can be implemented for heritage resources inventory and management. Arches-HIP comprises many OSS tools. Therefore, it can be implemented for different communities or countries with low cost if the application developer has appropriate software skills. Arches-HIP depends on several OSS software tools with certain versions. This software may need to be updated for enhancement or security problems; however, we had experienced some problems due to software dependency issues if we installed newer versions of these required software. As a result, we had to install the software in an isolated Python environment [53] such that multiple versions could coexist without interfering each other. For future development, we suggest that the software should be packaged using Docker container technology [54] because it works much better especially if non-Python software are required. Although Arches 4.0 had been released, it does not provide software package such as HIP, which we think can be useful for future implementations.

5. Conclusions

Community involvement has been proven to be an important element for strategical planning and practice of cultural and ecological conservation. In this research, the researcher collaborated with NCDA in collecting heritage resources data and creating a heritage resources inventory and management system. The results of this study show that, using open-source geospatial software, we were able to build the system with very low cost. Moreover, the implementation is scalable, and can be easily adapted to different communities without needing to purchase expensive proprietary software. This can be very helpful to develop applications for a wide array of local communities and indigenous people and encourage public participation in heritage conservation.

Funding: This research received no external funding. The research was supported by grants from the National Science Council of Taiwan (NSC 102-2621-M-004-008-MY3).

Acknowledgments: The author is grateful to the Nouli Community Development Association for sharing their data with us, and the research team for field data collection and manipulation, particularly Wan-Hsin Mao for her help in implementing the Arches-HIP application. The author also would like to thank GCI and many developers of the OSS that were used in this research.

Conflicts of Interest: The author declares no conflict of interest.

References

1. Rössler, M. Partners in site management. A shift in focus: Heritage and community involvement. In *Community Development through World Heritage*; Albert, M.T., Richon, M., Vnals, M.J., Witcomb, A., Eds.; UNESCO: Paris, France, 2012; pp. 27–31, ISBN 978-92-3-001024-9.
2. Chirikure, S.; Pwiti, G. Community involvement in archaeology and cultural heritage management: An assessment from case studies in Southern Africa and elsewhere. *Curr. Anthropol.* **2008**, *49*, 467–485. [CrossRef]
3. Mawere, M.; Sagiya, M.E.; Mubaya, T.R. Conservation conversations and community participation in the management of heritage sites in Zimbabwe. *Greener J. Environ. Manag. Public Saf.* **2012**, *2*, 7–16.
4. Bushozi, P.M. Towards sustainable cultural heritage management in Tanzania: A case study of Kalenga and Mlambalasi sites in Iringa, Southern Tanzania. *S. Afr. Archaeol. Bull.* **2014**, *69*, 136–141.
5. Mudge, M.; Ashley, M.; Schroer, C. A Digital Future for Cultural Heritage. *Int. Arch. Photogram. Remote Sens. Spat Inf. Sci.* **2007**, *36–5/C53*. Available online: http://www.isprs.org/proceedings/XXXVI/5-C53/papers/FP104.pdf (accessed on 29 October 2018).
6. Santana Quintero, M.; Georgopoulos, A.; Stylianidis, E.; Lerma, J.L.; Remondino, F. CIPA's Mission: Digitally Documenting Cultural Heritage. *APT Bull. J. Preserv. Technol.* **2017**, *48*, 51–54.
7. Crofts, N.; Doerr, M.; Gill, T.; Stead, S.; Stiff, M. *Definition of the CIDOC Conceptual Reference Model*; ICOM/CIDOC CRM Special Interest Group; ICOM Deutschland: Berlin, Germany, 2011.
8. Steiniger, S.; Bocher, E. An overview on current free and open source desktop GIS developments. *Int. J. Geogr. Inf. Sci.* **2009**, *23*, 1345–1370. [CrossRef]
9. Steiniger, S.; Hunter, A.J. Free and open source GIS software for building a spatial data infrastructure. In *Geospatial Free and Open Source Software in the 21st Century. Lecture Notes in Geoinformation and Cartography*; Bocher, E., Neteler, M., Eds.; Springer: Berlin/Heidelberg, Germany, 2012; pp. 247–261, ISBN 978-3-642-10594-4.
10. Brovelli, M.A.; Minghini, M.; Moreno-Sanchez, R.; Oliveira, R. Free and open source software for geospatial applications (FOSS4G) to support Future Earth. *Int. J. Digit. Earth* **2016**, *10*, 386–404. [CrossRef]
11. Gkoumas, G.; Lazarinis, F. Preserving Cultural Heritage Using Open Source Collection Management Tools. *Dig. Present. Preserv. Cult. Sci. Herit.* **2013**, *3*, 169–175.

12. Open Video Digital Library Toolkit. Available online: https://github.com/ovdlt/open-video-digital-library-toolkit (accessed on 10 September 2018).

13. Marchionini, G.; Geisler, G. The Open Video Digital Library. *D-Lib Mag.* **2002**, *8*, 12. [CrossRef]

14. CollectionSpace. Available online: http://www.collectionspace.org (accessed on 10 September 2018).

15. The Open Video Project: A Shared Digital Video Collection. Available online: https://open-video.org/index.php (accessed on 10 September 2018).

16. Meyer, E.; Grussenmeyer, P.; Perrin, J.P.; Durand, A.; Drap, P. A web information system for the management and the dissemination of Cultural Heritage data. *J. Cult. Herit.* **2007**, *8*, 396–411. [CrossRef]

17. Vacca, G.; Fiorino, D.R.; Pili, D. A WebGIS for the knowledge and conservation of the historical buildings in Sardinia (Italy). *Int. Arch. Photogram. Remote Sens. Spat. Inf. Sci.* **2017**. [CrossRef]

18. Vacca, G.; Fiorino, D.R.; Pili, D. A spatial information system (SIS) for the architectural and cultural heritage of Sardinia (Italy). *ISPRS Int. J. Geo-Inf.* **2018**, *7*, 49. [CrossRef]

19. Petrescu, F. The Use of GIS Technology in Cultural Heritage. *Int. Arch. Photogram. Remote Sens. Spat. Inf. Sci.* **2007**, *XXXVI–5/C53*. Available online: http://www.isprs.org/proceedings/XXXVI/5-C53/papers/FP114.pdf (accessed on 29 October 2018).

20. Smuts, K.; Mlungwana, N.; Wiltshire, N. SAHRIS: South Africa's integrated, web-based heritage management system. *J. Cult. Herit. Manag. Sustain. Dev.* **2016**, *6*, 38–152. [CrossRef]

21. South African Heritage Resources Agency. South African Heritage Resources Information System (SAHRIS). Available online: https://www.sahra.org.za/sahris/about/sahris (accessed on 10 September 2018).

22. Paul Getty Trust and World Monuments Fund. Arches: Fact Sheet, Standards & Interoperability, Project Background, Implementations of Arches, and Downloads. Available online: http://www.archesproject.org (accessed on 10 September 2018).

23. Carlisle, P.K.; Avramides, I.; Dalgity, A.; Myers, D. The Arches Heritage Inventory and Management System: A Standards-Based Approach to the Management of Cultural Heritage Information. In Proceedings of the CIDOC (International Committee for Documentation of the International Council of Museums) Conference Access and Understanding—Networking in the Digital Era, Dresden, Germany, 6–11 September 2014.

24. Myers, D.; Dalgity, A.; Avramides, I.; Wuthrich, D. Arches: An open source GIS for the inventory and management of immovable cultural heritage. In *Progress in Cultural Heritage Preservation. EuroMed 2012. Lecture Notes in Computer Science*; Ioannides, M., Fritsch, D., Leissner, J., Davies, R., Remondino, F., Caffo, R., Eds.; Springer: Berlin/Heidelberg, Germany, 2012; Volume 7616, pp. 817–824.

25. Wiltshire, N.G. The use of SAHRIS as a state sponsored digital heritage repository and management system in South Africa. *Int. Ann. Photogram. Remote Sens. Spat. Inf. Sci.* **2013**, *II–5/W1*, 325–330. [CrossRef]

26. Mlungwana, N. Using SAHRIS a web-based application for creating heritage cases and permit applications. *Int. Arch. Photogram. Remote Sens. Spat. Inf. Sci.* **2015**, *XL–5/W7*, 337–341. [CrossRef]

27. Smuts, K. SAHRIS: Using the South African Heritage Register to report, track and monitor heritage crime. *Int. Arch. Photogram. Remote Sens. Spat. Inf. Sci.* **2015**, *XL–5/W7*, 395–402. [CrossRef]

28. Drupal. Documentation. Available online: https://www.drupal.org/docs/user_guide/en/understanding-drupal.html (accessed on 10 September 2018).

29. Django. Overview of Django. Available online: https://www.djangoproject.com/start/overview/ (accessed on 10 September 2018).

30. Barton, C.; Cox, A.; Cruz, S.D.; Hansen, J. Cultural-heritage inventory implementations: The versatility of the Arches system. *APT Bull. J. Preserv. Technol.* **2017**, *48*, 19–28.

31. International Core Data Standard for Archaeological Sites and Monuments. In *Documenting the Cultural Heritage*; Thornes, R., Bold, J., Eds.; The Getty Information Institute: Los Angeles, CA, USA, 1998.

32. The Getty Conservation Institute and World Monuments Fund. MEGA-Jordan: The National Heritage Documentation and Management System. Available online: http://megajordan.org/ (accessed on 10 September 2018).

33. Myers, D.; Dalgity, A. The Middle Eastern Geodatabase for Antiquities (MEGA): An Open Source GIS-Based Heritage Site Inventory and Management System. Available online: http://www.getty.edu/conservation/our_projects/field_projects/jordan/mega_jordan_cot_article.pdf (accessed on 10 September 2018).

34. Arches-HIP Documentation. Available online: https://arches-hip.readthedocs.io/ (accessed on 10 September 2018).

35. Arches Documentation v3.1.2. Available online: https://arches.readthedocs.io/en/3.1.2/ (accessed on 10 September 2018).

36. Arches Project: Customization of Arches for City of Los Angeles. Available online: http://www.getty.edu/conservation/our_projects/field_projects/arches/arches_city_la.html (accessed on 10 September 2018).
37. Nouli Community Association. Available online: http://www.nlica.org.tw/ (accessed on 10 September 2018).
38. Pygal. Available online: http://www.pygal.org/en/latest/ (accessed on 10 September 2018).
39. GEOS—Geometry Engine, Open Sources. Available online: http://trac.osgeo.org/geos (accessed on 10 September 2018).
40. Chen, L.K.; Inoue, H.; Won, C.W.; Lin, C.H.; Lin, K.F.; Tsay, S.F.; Lin, P.F.; Li, S.H. Challenges of urban aging in Taiwan: Summary of urban aging forum. *J. Clin. Gerontol. Geriat.* **2013**, *4*, 97–101, doi.org/10.1016/j.jcgg.2013.05.001. [CrossRef]
41. Jan, J.F. Digital heritage inventory using open source geospatial software. In Proceedings of the 22nd International Conference on Virtual Systems & Multimedia (VSMM), Kuala Lumpur, Malaysia, 17–21 October 2016.
42. Mao, W.H. Application of Open Source GIS in Digital Conservation of Heritage: A Case Study of Arches. Master's Thesis, National Chengchi University, Taipei, Taiwan, 2016.
43. Nouli Community Association (Built with Arches). Available online: http://arches.nccu.edu.tw/ (accessed on 10 September 2018).
44. The LizardQ Stitcher. Available online: https://www.lizardq.com/en/features-software/ (accessed on 10 September 2018).
45. LizardQ User Manual. Available online: https://www.lizardq.com/cdn/docs/LizardQUserManual.pdf (accessed on 10 September 2018).
46. GNU Image Manipulation Program (GIMP). Available online: https://www.gimp.org/ (accessed on 10 September 2018).
47. Spatial Media Metadata Injector. Available online: https://github.com/google/spatial-media/blob/master/spatialmedia/README.md (accessed on 10 September 2018).
48. YouTube Help. Upload 360-Degree Videos. Available online: https://support.google.com/youtube/answer/6178631?hl=en (accessed on 10 September 2018).
49. Keyhole Markup Language. KML Tutorial. Available online: https://developers.google.com/kml/documentation/kml_tut (accessed on 10 September 2018).
50. EPSG (European Petroleum Survey Group):4326. Available online: http://spatialreference.org/ref/epsg/4326/ (accessed on 10 September 2018).
51. Wikipedia. Well-Known Text. Available online: https://en.wikipedia.org/wiki/Well-known_text (accessed on 10 September 2018).
52. Tang, C.P.; Tang, S.Y. Institutional Adaptation and Community-Based Conservation of Natural Resources: The Cases of the Tao and Atayal in Taiwan. *Hum. Ecol.* **2010**, *38*, 101–111. [CrossRef]
53. Virtualenv. Available online: https://virtualenv.pypa.io/en/stable/ (accessed on 10 September 2018).
54. Docker. Available online: https://www.docker.com/ (accessed on 10 September 2018).

International Journal of
Geo-Information

Article

CATCHA: Real-Time Camera Tracking Method for Augmented Reality Applications in Cultural Heritage Interiors

Piotr Siekański [1,*], Jakub Michoński [1], Eryk Bunsch [2] and Robert Sitnik [1]

[1] Faculty of Mechatronics, Warsaw University of Technology, ul. św. Andrzeja Boboli 8, 02-525 Warszawa, Poland; j.michonski@mchtr.pw.edu.pl (J.M.); r.sitnik@mchtr.pw.edu.pl (R.S.)

[2] Museum of King Jan III's Palace, ul. Stanisława Kostki Potockiego 10/16, 02-958 Warszawa, Poland; jerzywart@yahoo.de

* Correspondence: p.siekanski@mchtr.pw.edu.pl

Received: 6 November 2018; Accepted: 13 December 2018; Published: 15 December 2018

Abstract: Camera pose tracking is a fundamental task in Augmented Reality (AR) applications. In this paper, we present CATCHA, a method to achieve camera pose tracking in cultural heritage interiors with rigorous conservatory policies. Our solution is real-time model-based camera tracking according to textured point cloud, regardless of its registration technique. We achieve this solution using orthographic model rendering that allows us to achieve real-time performance, regardless of point cloud density. Our developed algorithm is used to create a novel tool to help both cultural heritage restorers and individual visitors visually compare the actual state of a culture heritage location with its previously scanned state from the same point of view in real time. The provided application can directly achieve a frame rate of over 15 Hz on VGA frames on a mobile device and over 40 Hz using remote processing. The performance of our approach is evaluated using a model of the King's Chinese Cabinet (Museum of King Jan III's Palace at Wilanów, Warsaw, Poland) that was scanned in 2009 using the structured light technique and renovated and scanned again in 2015. Additional tests are performed on a model of the Al Fresco Cabinet in the same museum, scanned using a time-of-flight laser scanner.

Keywords: CATCHA; camera tracking; cultural heritage; real-time; point clouds

---

1. Introduction

In Augmented Reality (AR), virtual objects are superimposed on a real-world view observed by a user. The main requirement for an immersive AR application is correct spatial alignment between the real-world objects and virtual ones, which means that the virtual objects must be placed in the same coordinate system as the real objects. If the camera moves, their 2D projections on the screen must move accordingly. To achieve this, a precise camera pose, relative to the world coordinate system, must be known. This process is often called camera pose estimation or camera tracking.

Multiple camera tracking methods are available, such as magnetic, inertial, GPS-based, vision, and other methods [1]. Inasmuch as our interest is in cultural heritage interiors, usually guarded by rigorous conservatory policies, we focus on a vision-based camera tracking algorithm, which does not require physical integration into the interior geometry, as compared with alternative solutions that employ external tracking devices. Moreover, we consider only the visible-light spectrum because the use of infrared light may be dangerous for pigments used to paint the walls [2].

In recent years, 3D scanning has become an increasingly popular technique in cultural heritage documentation. Until now, many cultural heritage sites have been digitized using different techniques such as laser scanning [3], Digital Photogrammetry (DP) in which the Structure from Motion

(SfM) algorithm is widely used [4], or structured light [2] resulting in precise 3D documentation. These high-quality scans may be used in model-based camera tracking algorithms to create specialized AR applications.

The aim of our work is to develop an AR system called Camera Tracking for Cultural Heritage Applications (CATCHA) with the following functionalities.

1. It should work on any interior model, both planar and non-planar, regardless of its registration technique. The solution should depend only on point cloud data.
2. It should work in a wide range of object distances from 0.5 to 2 m.
3. It should be as simple as possible. Neither input data/parameters nor any configuration and initialization in a known initial position should be required from the user.
4. It should work in real-time, i.e., it must have an interactive frame rate of at least 15 Hz on a mobile device.
5. It should not use any artificial markers.
6. It should exhibit stable visualization with negligible drift and jitter.

To our knowledge, no system has been developed that satisfies all these requirements. CATCHA was implemented and tested in two real cultural heritage interiors scanned using structured light and laser techniques.

Our paper is organized as follows. Related work is presented in the next section, and Section 3 presents the materials used in the investigation. The details of our method are provided in Section 4. Verification of the algorithm on real data and the results are presented in Section 5 and discussed in Section 6, respectively. The last section contains our conclusions.

2. Related Work

Markerless camera tracking approaches can be divided into two main categories [5]: relative (image based) and tracking-by-detection (model based). A summary of these approaches is listed in Table 1. The first category exploits frame-to-frame spatiotemporal information to determine the camera motion between frames. This approach is very sensitive to occlusions and rapid camera movements, and suffers from drift because of the accumulation of tracking errors. It does not require a model of the tracked object to be known beforehand. On the other hand, the tracking-by-detection approach requires a model and relies on extracted features (points, lines, etc.) from both the model and current camera frames. They are independently matched in every frame, and thus, this method does not suffer from the drift effect and can handle partial occlusions and rapid camera movements at the cost of higher computational complexity.

Table 1. Comparison of the relative and tracking-by-detection approaches.

Method	Relative Tracking	Tracking-by-Detection
Tracking	Image based	Model based
Model of a tracked object	May be constructed on the fly	Must be known in advance
Robustness to partial occlusions	Low	High
Robustness to rapid camera movement	Low	High
Drift	Yes	No
Computational complexity	Lower	Higher

Combined approaches are also available, such as in [6], that employ relative tracking to track the camera motion between consecutive frames and the tracking-by-detection approach to recover from tracking failures and compensate the drift.

We are also aware of the SLAM (simultaneous localization and mapping) [7] based methods that simultaneously track the camera pose and compute the geometry of the surrounding environment. Therefore, it does not depend on a predefined model. However, extracting the real-world scale information from a SLAM-based system may be challenging. Further, displaying a specific object content is not directly possible. Thus, model-based tracking is more suitable.

As mentioned earlier, three main techniques are available in capturing point clouds in cultural heritage applications, but many model-based camera tracking algorithms require the model to be built using one particular method. For example, in the last decade, an enormous progress in SfM-based location recognition was achieved [8,9]. In addition, some approaches are available that estimate the camera pose based on laser scans. Although very accurate, they are not suitable for real-time applications [10,11].

With regard to general point cloud based camera pose estimation algorithms, Wu et al. [12] proposed a method to extract SIFT (Scale Invariant Feature Transform) [13] descriptors out of orthonormalized patches that exploit the 3D normal direction of a single patch. However, their approach relies on fronto-parallel surfaces for an initial match, and features are then extracted relative to this direction. This algorithm was originally designed to interactively compute the camera pose in a single query image. However, large errors may be produced when the camera is close to the object surface, as presented in [14], and thus fail to fulfill Requirement No. 2 presented earlier.

Jaramillo et al. [15] proposed a framework that uses a point cloud constructed by an RGBD sensor and thereafter only uses a monocular camera to estimate the pose. In contrast to the SLAM-based approaches, this solution does not track points from frame to frame. Instead, previous poses are only used to limit the search space. Their solution was reported to work at only approximately 3 Hz. Therefore, Requirement No. 4 is not satisfied.

Crombez et al. [16] used laser scanners to create a dense point cloud model of an interior and then used a fisheye camera and synthetic views to localize its position. However, their method requires a manual initialization step. This solution does not fulfill Requirement No. 3.

Probably, a solution most similar to that presented in this paper is the work from Rambach et al. [17]. They rendered approximately 10000 training images of a model and then chose the best matching set. Subsequently, they used relative tracking combined with the tracking-by-detection approach to recover from tracking failure. In the event of a tracking failure, their initialization procedure takes more than 100 ms, which may be visible as a frame drop; thus, their approach may not satisfy Requirement No. 6 in some circumstances.

With respect to markerless camera tracking algorithms in cultural heritage, many solutions rely on available libraries and track only the planar structures in the scene. To name a few, Vanoni et al. [18] created a visual see-through AR application that maps multimodal data on a renascence fresco in Florence, Italy. Their solution relies on the Vuforia [19] library. Gîrbacia et al. [20] used AR to recreate sculptures in the Black Church in Brasov, Romania. In their project, the Instant Player library [21] was used. Verykokou et al. [22] used specially designed algorithm to track the feature points on a planar wall to recreate an ancient stoa in Athens, Greece. All mentioned solutions track only planar structures; thus, they do not satisfy Requirement No. 1. Battini and Laudi [23] tracked a statue of Leonardo da Vinci in Milan, Italy. In their project, ARmedia SDK [24] was used. Compared with the approaches mentioned earlier, this library allows 3D camera tracking based on object texture and non-planar geometry but requires time-consuming manual data preparation and algorithm parameters tuning; thus, Requirement No. 3 is not satisfied.

3. Materials

Our algorithm was used to track the camera pose in the King's Chinese Cabinet in the Museum of King Jan III's Palace at Wilanów, Warsaw, Poland. This interior was scanned using structured light technique in 2009 shortly before renovation, and scanned again in 2015 [2]. Figure 1 shows the visual comparison between these two scans. The point cloud registered in 2009 contains 4 billion points and was scanned using SLR cameras, and that registered in 2015 contains 35 billion points and was scanned using industrial cameras with color calibration [25]. In both scans, every point contains its position, color, normal vector, and quality value that represents the certainty of a registered point. Based on these two scans, we designed an AR application that allows a user to see the previous state of the Cabinet. Additional tests were performed on the Al Fresco Cabinet in the same museum. This interior was scanned once using the FARO Focus3D X130 HDR time-of-flight (TOF) laser scanner. The cloud contains 172 million points and has the same internal structure.

(a)   (b)

Figure 1. Visual comparison between two registered point clouds of the King's Chinese Cabinet: (**a**) from 2009 and (**b**) from 2015.

In our camera tracking solution, we used a single Logitech B910 camera with the resolution reduced from 1920 × 1080 pixels to 640 × 480 pixels (Figure 2 shows a sample frame). We connected it to a Lenovo Yoga convertible laptop with Intel i5-5200U 2,20 GHz processor and Nvidia GT 840M graphics card. We implemented our solution using Microsoft Visual Studio 2015 in C++ and OpenCV library on the Windows 8.1 Operating System. We also tested the streaming version of our approach, in which only the image acquisition and final model rendering were performed on the device, and all processing algorithms were remotely executed on a PC with Intel i7-6700K 4.0-GHz processor and Nvidia GeForce 960 (see Section 3 for details). For streaming, we decided to separately send every frame with optional JPEG-LS compression realized using the CharLS library [26]. The application based on CATCHA is shown in Figure 3.

Figure 2. Sample frame taken using a Logitech B910 camera.

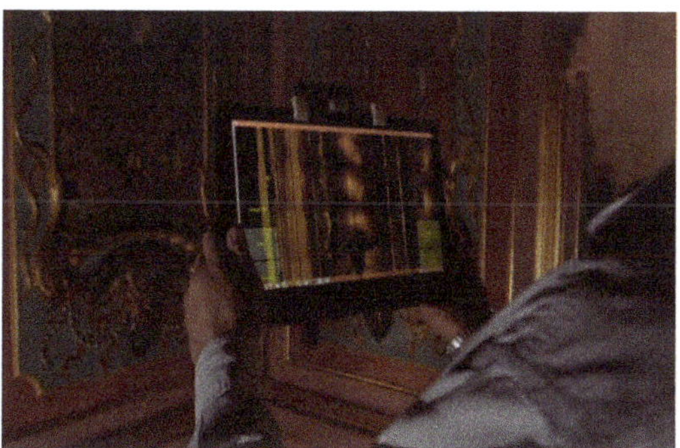

Figure 3. CATCHA-based application in use.

4. Method

To track the camera pose relative to the model, we must determine rotation matrix R and translation vector t of the camera in the model coordinate system in each frame. Given camera intrinsic matrix K from the calibration and a set of correspondences between 3D model points P and image points p, both in homogeneous coordinates, we must determine the camera extrinsic matrix that minimizes the reprojection error between p and projected model points p'.

$$p' = K\cdot[R|t]\cdot P \tag{1}$$

Therefore, to determine the camera pose, valid 3D–2D correspondences between the 2D points in the current camera image and the 3D model points must be known.

We assume that CATCHA can work on any point cloud-based data without prior knowledge of their origin. We consider three main sources of the clouds of points: structured light, TOF, and DP. To achieve model origin invariance, we decide to generate high-resolution orthographic renders (Figure 4) with the associated 3D coordinates in every pixel using an approach similar to that in [12].

We segment the point cloud using normal vectors to search for suitable nearly planar surfaces. Then we place the virtual orthographic camera in front of each planar area and render it. In the case of two analyzed interiors, five orthomaps are generated for each interior. If an interior has more complicated geometry, more orthomaps have to be generated. On each render, we extract the keypoints and their descriptors. In our case, the SIFT descriptor extracted by the GPU is used, but we can exploit any descriptor by taking advantage of considerably faster binary descriptors (e.g., ORB [27]) if necessary. Because we store the 3D coordinates of each pixel, we can easily obtain the 3D–2D correspondences between each point found in the orthomap and the associated model point.

This approach leads us to a sparse interior model regardless of the initial point cloud density. In our case, we only extract a few thousand keypoints to cover the entire interior geometry. Further, reduction in the model complexity can be achieved by creating masks on the orthomaps (Figure 5). Mask is a simple binary image with the same size as the orthomap. The black areas on the mask indicate that these areas are rejected from further processing. The purpose of this process is to filter out points and areas of noise and the noise near the borders between the model and its background, as well as the reflective or transparent fragments of a surface. The masks have to be created by the user. We used a semi-automated method based on hue to filter the areas.

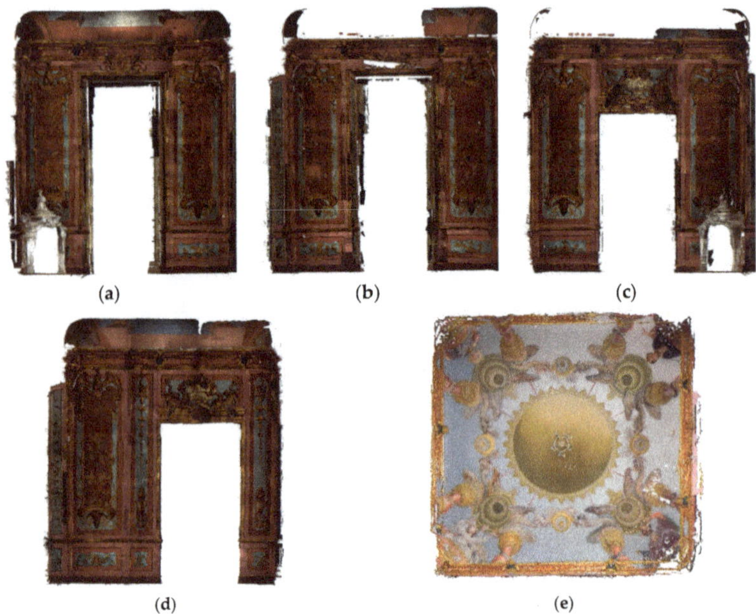

Figure 4. Generated orthographic renders. (**a**–**d**) Four walls of the Cabinet. (**e**) The ceiling.

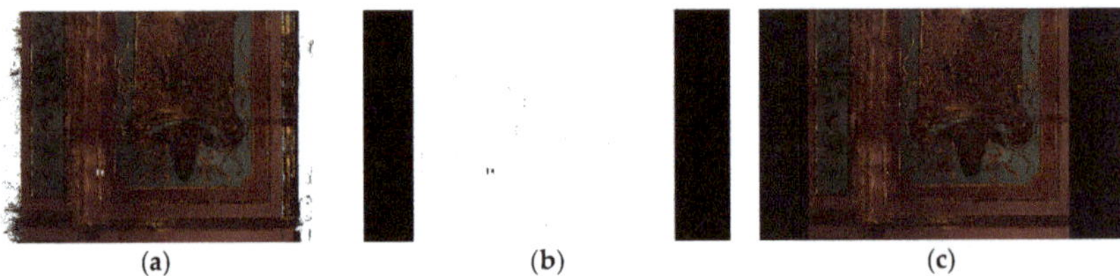

Figure 5. Illustration of the masks: (**a**) raw orthomap, (**b**) mask, and (**c**) result of filtration.

Depending on the type of interior, a user can be interested in bigger or smaller details. The optimal orthomap resolution must be computed depending on the assumed average distance between the camera and observed wall [we call it later the optimal viewing distance (OVD)]. In our tests, the application smoothly works from approximately 50% to 200% of the OVD. In our case, we set this parameter to 1 m. Desired orthomap resolution depends on OVD, camera resolution, its FOV and the real-world size of the interior.

From Figure 6, we obtain:

$$x = \tan \frac{\varphi}{2} \cdot 2d \tag{2}$$

where φ indicates the camera field of view (FOV), d is distance to the wall, and x denotes the observed fragment dimension.

We want to maintain the scales of the keypoints as closely as possible. Thus, we intend to keep the ratio between camera resolution res_c (in pixels) and real-world dimension of observed fragment x (in meters) equal to the ratio between orthomap resolution res_o (in pixels) and real-world dimension of the interior w (in meters) (Figure 7).

$$\frac{x}{res_c} = \frac{w}{res_o} \tag{3}$$

From Equations (2) and (3), we obtain:

$$res_o = \frac{w \cdot res_c}{\tan \frac{\varphi}{2} \cdot 2d} \tag{4}$$

Therefore, we propose equation (4) to compute the desired resolution. If we substitute our values, namely, $\varphi = 78°$, which is the FOV of a Logitech B910 camera, $res_c = 640$ pixels, and $w = 6$ m, we can draw a plot of d against res_o (Figure 8). In our case, we set OVD to 1 m, which allows the application mentioned in Section 3 to run smoothly from approximately 0.5 to 2 m. Therefore, we decide to render orthomaps with a 2048-px resolution because it has the closest power of two to the chosen OVD (1 m).

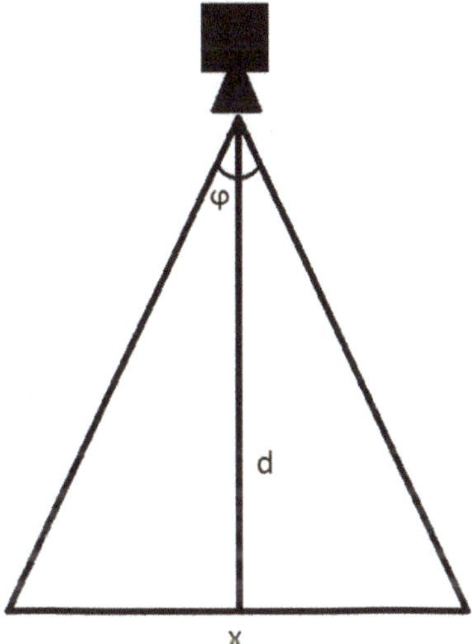

Figure 6. Camera FOV.

Figure 7. Real-world size of interior w and the size of observed fragment x.

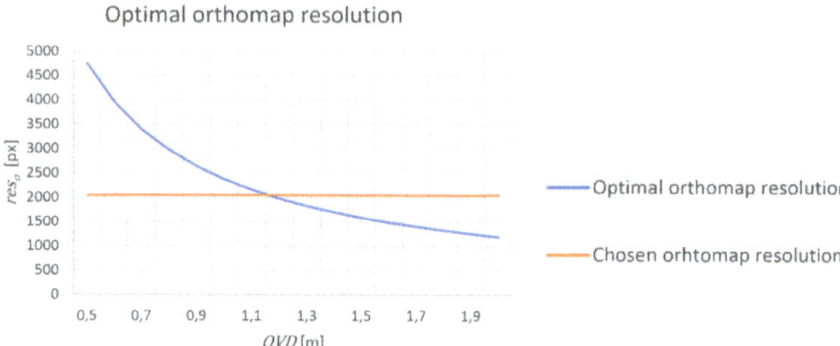

Figure 8. Orthomap resolution res_o and OVD: optimal resolution with respect to optimal viewing distance (blue) and closest power of 2 for viewing distance equal to 1 m (orange).

The model distinctiveness strongly depends on the quality of the renders. The lesser the noise in the rendered image is, the more reliable and stable are the keypoints. Therefore, we decide to render the orthomaps using 4x antialiasing. We render them with four times bigger resolution and downscale them four times using bicubic interpolation to reduce the noise (Figure 9). A detailed review of the different point cloud rendering methods with focus on descriptor extraction can be found in [28]. As the paper suggests, we do not need splatting because our data is dense enough to render the point cloud without gaps, even with resolution four times higher in each dimension.

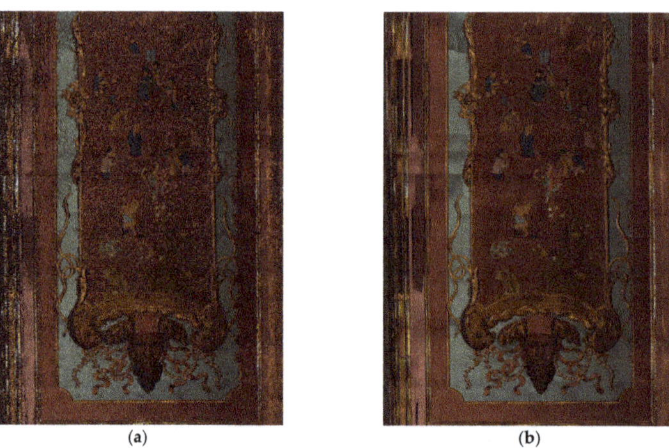

(a) (b)

Figure 9. Comparison between orthomaps rendered in different resolutions: (**a**) native resolution and (**b**) with four times of antialiasing.

Our runtime diagram (Figure 10) is similar to the state-of-the-art solutions. For every frame captured by the camera, we extract the keypoints and compute the descriptors. Because our model is very sparse (a few thousand keypoints), we can very effectively match it using a brute-force method on the GPU. We are aware that more methods are available to much more effectively match the descriptors on a CPU [29], but matching on the GPU allows us to avoid costly data transfer between the descriptors extracted on the GPU and matches on the CPU side. Later, the matched keypoints are filtered using the ratio test proposed by Lowe [13]. Then, the camera pose is computed using the EPnP (Efficient Perspective-n-Point) method [30] supported by RANSAC (RANdom SAmple Consensus) sampling in which the P3P algorithm [31] is used. The minimum required number of correspondences is 4, but we reject all the estimations that have less than 7 inliers to ensure that the camera pose is estimated correctly. The camera pose is then transferred to the visualization module. We take advantage of the fact that the camera is tracked inside a room; thus, if the computed camera pose is outside it, we reject the pose and do not send it to the visualization module. If the frame is rejected, the previous frame is rendered. We separately compute the camera pose in each frame and later smooth the camera trajectory using Gaussian filtration [32] to reduce jitter in the camera movement.

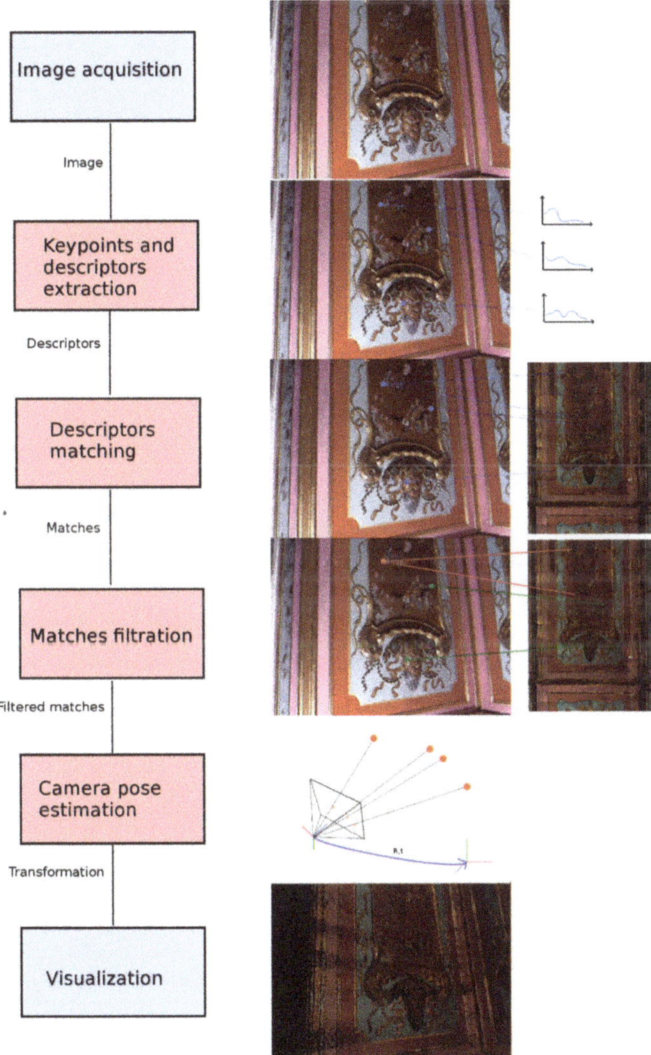

Figure 10. Runtime diagram: steps indicated with violet are performed on the mobile device; those in pink can be remotely computed.

Because our approach relies on orthomaps generated parallel to the walls, processing of the frames with high angular difference from this direction could be challenging. This problem in descriptor matching is called angular invariance. Standard wide-baseline descriptors such as SIFT or SURF (Speeded Up Robust Features) [33] have limited angular invariance of up to 30° [34]. Thus, a very low number of correct matches (if any) are present if the angle between the camera axis and normal to the wall is large. This problem is clearly apparent when the ceiling of the room is very high and the user is looking at the elements located just under the ceiling (Figure 11). To overcome this type of problems, the ASIFT (Affine SIFT) [35] algorithm may be used, but the computational complexity is much too high for a real-time application. We differently deal with this problem by dividing the interiors into two main categories based on the ceiling altitude.

1. The average altitude in which the details located just under the ceiling do not require extreme viewing angle and camera tracking is based on the walls.
2. High ceiling in which the problem described above may occur. To overcome this problem, we exploit the fact that the ceiling may also have a rich texture (for instance, it may be covered by frescoes). We rotate the physical camera to point up to the ceiling so that it is always nearly parallel to the camera sensor plane. We estimate the camera position based on the ceiling and rotate it by 90° to visualize the view in front of the user. This solution also reduces the computational complexity by decreasing the necessary orthomaps from one per wall to one from the ceiling. However, this approach does not allow the user to view the ceiling. If the view of the ceiling is also important, we can render additional orthomaps for the walls and track the ceiling using the planar walls. In this case, the performance gain mentioned earlier is no longer available.

Figure 11. Example of the details under a high ceiling.

5. Results

To evaluate the accuracy of the proposed approach, we tested it on two printed posters: on a real-size fragment of a scanned Chinese Cabinet (Figure 12a) and actual size of the Al Fresco Cabinet (Figure 13a). Both posters are surrounded by ARUCO [36] fiducial markers. We recorded six different sequences that simulated the user behavior in front of the posters. In each frame, we estimated the camera pose twice, i.e., using keypoints from the poster and the corners of the extracted ARUCO markers. Both camera poses were then compared to evaluate the accuracy of our method. We considered the camera pose estimated from the ARUCO markers as our ground truth. Figures 12b–d and 13b–d show the estimated camera trajectories in each sequence. The quantitative results are listed in Tables 2 and 3. To measure the translation error, we calculated the Euclidean distance between the estimated camera poses from both methods and divided it by the L2 norm of the camera position estimated by ARUCO (Equation (5)).

$$e_t = \frac{\sqrt{(x_1 - x_2)^2 + (y_1 - y_2)^2 + (z_1 - z_2)^2}}{\sqrt{x_1^2 + y_1^2 + z_1^2}} \cdot 100\% \tag{5}$$

where:

e_t—translation error

x_1, y_1, z_1—coordinates of the ground-truth camera position

x_2, y_2, z_2—coordinates of the computed camera position

We decided to normalize the translation error using the Euclidean distance from the marker to the camera because if the ratio between the error and distance remains the same, the camera image would also stay the same, regardless of the distance to the marker. To estimate the rotation error, we first converted the rotation matrices into quaternions and then computed the angle difference between them. We considered the angle obtained from ARUCO as the ground truth. Let us define two quaternions (q_1 and q_2) to represent the camera rotations from both methods.

$$q_1 = (x_1, x_2, x_3, x_4), q_2 = (y_1, y_2, y_3, y_4)$$

Then, let us compute:

$$q_3(z_1, z_2, z_3, z_4) = q_1 \cdot conj(q_2). \tag{6}$$

Finally, let us compute the rotation error in radians as:

$$e_r = 2 \cdot \text{acos}(z_1). \tag{7}$$

To examine the quality of the pose-estimation algorithm used in CATCHA, we decided to use statistical analysis. In each sequence, we calculated the four quartiles of the errors (Q1–Q4) and computed the interquartile range (*IQR*) of our errors to determine how many pose estimations we could consider as gross errors.

$$IQR = Q3 - Q1 \tag{8}$$

To determine *IQR*, we computed both the upper (*UF*) and lower (*LF*) fences with $k = 1.5$ and marked the pose estimation errors that exceeded the upper limit as gross errors.

$$UF = Q3 + k \cdot IQR \tag{9}$$

$$LF = Q1 - k \cdot IQR \tag{10}$$

The mean error was below 2.5% in the translation, and it was below 1.5° in the rotation. The error values in each sequence are shown in Figure 14. In each case, the level of outliers was approximately 5% for the King's Chinese Cabinet and approximately 1% for the Al Fresco Cabinet. We note that the erroneous single-pose estimations were rejected and filtered using the visualization module so as not to affect the quality of the user experience. In our test procedure, we skipped the frames where only the markers in one row or one column were visible to avoid a situation where the reference pose was computed with a high rotation error.

Table 2. Statistical accuracy evaluation of the three analyzed sequences of the King's Chinese Cabinet.

Sequence	1		2		3	
Number of Frames	182		620		483	
	Translation [%]	**Rotation [°]**	**Translation [%]**	**Rotation [°]**	**Translation [%]**	**Rotation [°]**
Median	0.71	0.45	0.95	0.61	0.99	0.69
Average	0.83	0.54	1.19	0.80	1.26	0.90
Min	0.14	0.03	0.05	0.03	0.08	0.06
Max	3.50	2.20	9.03	6.54	6.93	5.43
Q1	0.44	0.28	0.57	0.35	0.62	0.41
Q3	1.02	0.69	1.50	1.00	1.55	1.08
IQR	0.58	0.41	0.94	0.65	0.93	0.67
Lower fence ($k = 1.5$)	−0.44	−0.34	−0.84	−0.62	−0.78	−0.59
Upper fence ($k = 1.5$)	1.90	1.31	2.90	1.97	2.94	2.08
Outliers	8	8	26	28	23	30
% of outliers	4.4	4.4	4.2	4.5	4.8	6.2

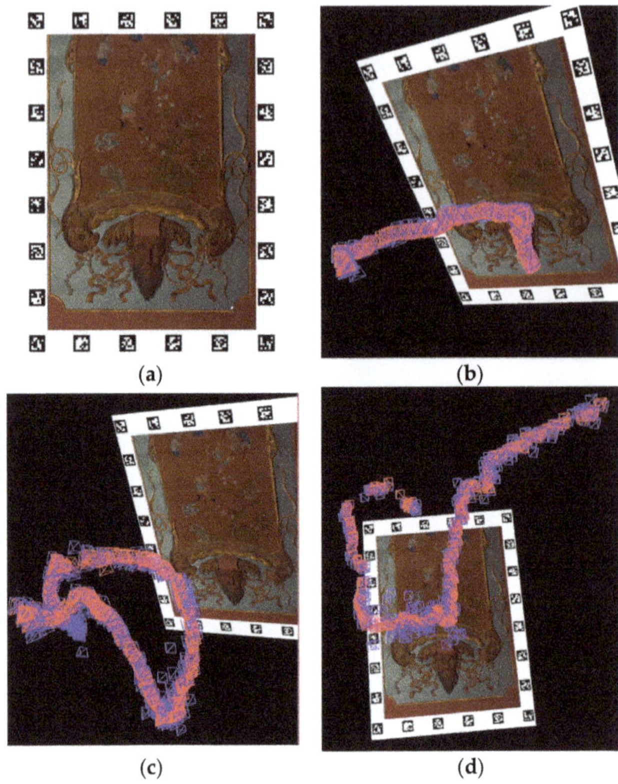

Figure 12. Algorithm evaluation of the King's Chinese Cabinet. (**a**) Poster used to test the accuracy. (**b–d**) Camera poses in each sequence (red—ground truth; blue—our method).

Table 3. Statistical accuracy evaluation of the three analyzed sequences of the Al Fresco Cabinet.

Sequence	4		5		6	
Number of Frames	353		391		209	
	Translation [%]	Rotation [°]	Translation [%]	Rotation [°]	Translation [%]	Rotation [°]
Median	2.18	1.33	1.86	1.28	2.04	1.35
Average	2.28	1.48	2.04	1.46	2.26	1.43
Min	0.15	0.13	0.10	0.04	0.20	0.09
Max	5.65	4.89	7.66	4.88	7.04	3.95
Q1	1.38	0.92	0.93	0.66	1.19	0.80
Q3	3.03	1.95	2.79	2.04	3.12	2.00
IQR	1.65	1.04	1.86	1.38	1.92	1.20
Lower fence ($k = 1.5$)	−1.10	−0.64	−1.87	−1.41	−1.69	−1.01
Upper fence ($k = 1.5$)	5.51	3.51	5.59	4.11	6.00	3.81
Outliers	1	6	4	2	2	1
% of outliers	0.3	1.7	1.0	0.5	1.0	0.5

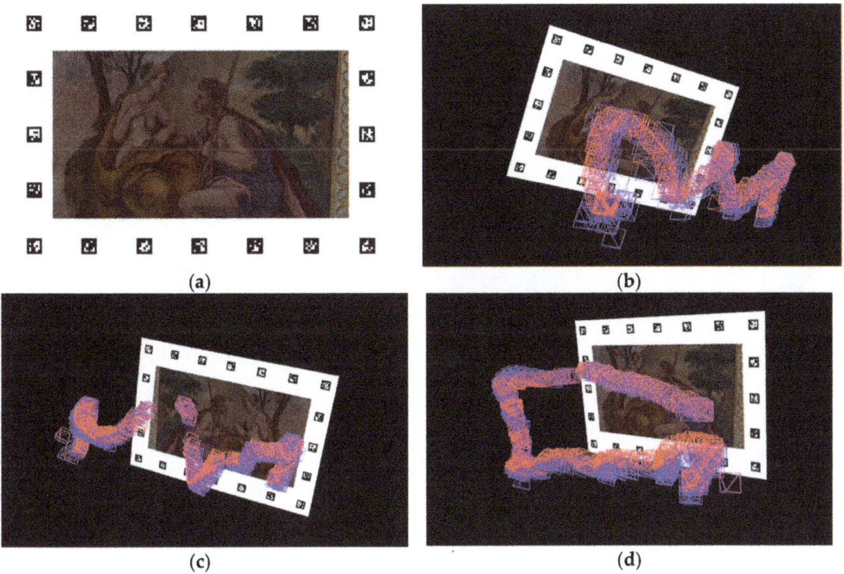

Figure 13. Algorithm evaluation of the Al Fresco Cabinet. (**a**) Poster used to test the accuracy. (**b–d**) Camera poses in each sequence (red—ground truth; blue—our method).

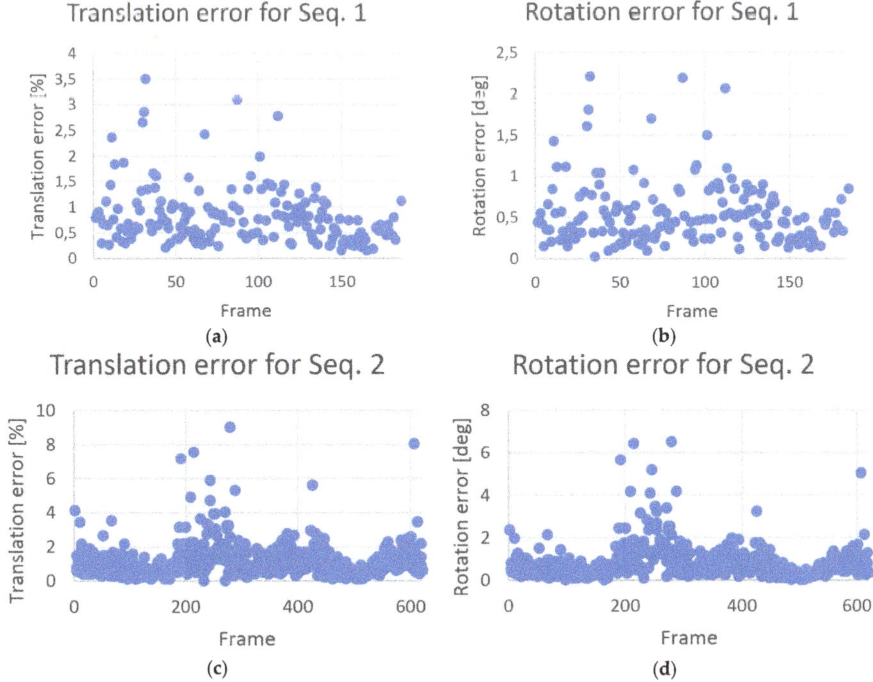

Figure 14. *Cont.*

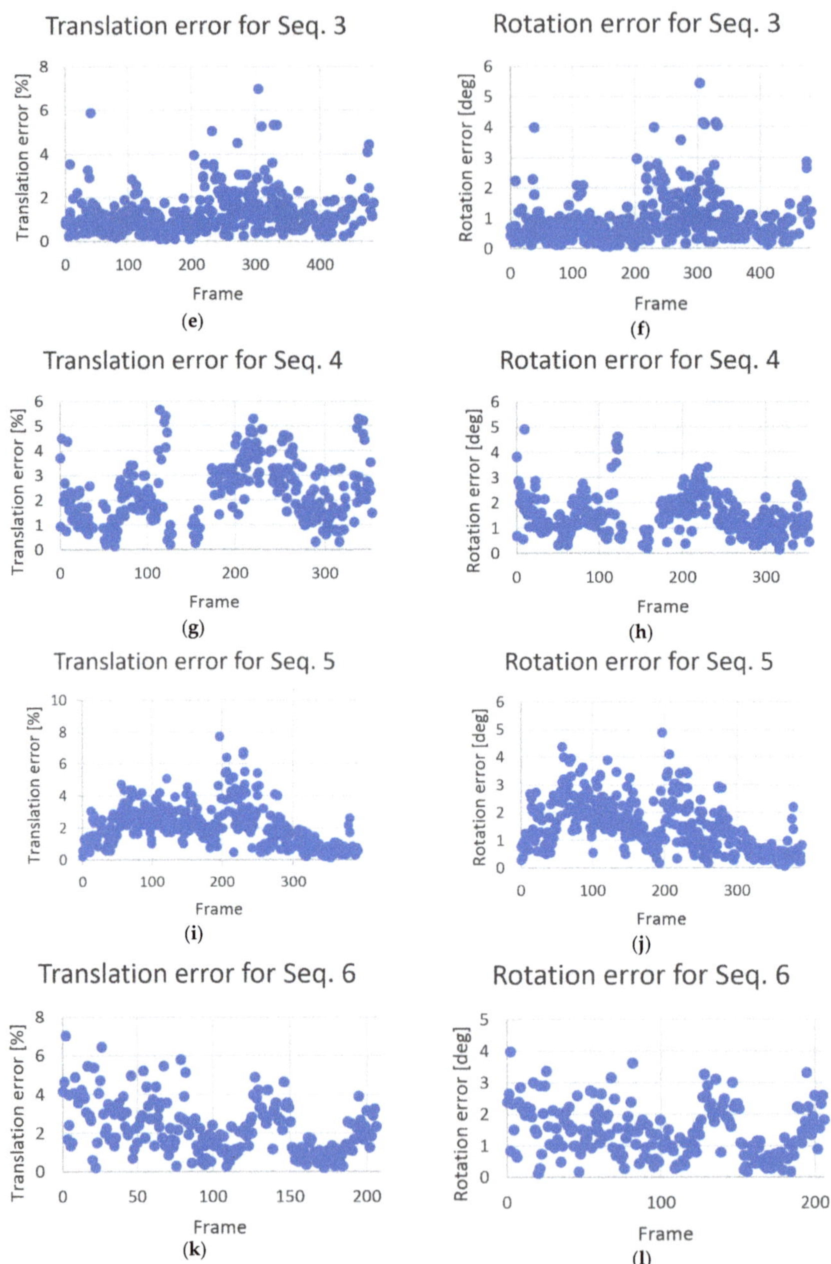

Figure 14. Error values in each test sequence. **Left** (**a**), (**c**), (**e**), (**g**), (**i**), and (**k**). Translation errors for sequences 1–6, respectively. **Right** (**b**), (**d**), (**f**), (**h**), (**j**), and (**l**). Rotation errors for sequences 1–6, respectively.

With respect to the application performance in the museum, by implementing the setup mentioned in Section 3, we achieved a mean frame rate of over 15 Hz directly using the SIFT descriptors on a mobile device. The mean frame rate increases up to 40 Hz using remote processing. This value is sufficient for smooth AR experience because many cameras on commercial devices and webcams have a maximum frame rate of up to 30 Hz.

6. Discussion

Cultural heritage objects are more or less under strict protection, and their limitations must be taken into consideration while designing AR solutions. The King's Chinese Cabinet is an example of an interior under a very strict protection because of the very fragile pigments on the walls. The paintings may be damaged if illuminated by an infrared or ultraviolet light. In addition, the maximum permissible level of visible light is limited and the room is protected from any illumination sources, including sunlight. The presented restrictions forbid the use of RGBD cameras that emit infrared light. In contrast to the solutions in which light can be fully adjusted, such as those in shopping galleries, we are forced to limit the amount of light. To overcome the problem of darkness in the Cabinet, we could increase the exposure time of the camera or increase the gain. The former results in motion blur, and the latter increases noise. We decided to adjust the gain of the camera, but this also decreased the quality of the extracted and matched keypoints. We believe that this decision may have influenced the results.

Compared with most of the state-of-the-art approaches that depend on the SfM or DP models and images used to build them, CATCHA can be used on any textured point cloud without prior knowledge of its origin or the data used to construct it. In addition, the developed method can be used on point clouds built using any method, including SfM or DP. This factor indicates that CATCHA is substantially more versatile than many existing approaches. CATCHA is obviously limited to interiors with rich matte textures on the walls because its camera pose estimation algorithm relies on the keypoints extracted from the images. Line-based approach may be used if no textures are available or the surfaces are very glossy. CATCHA builds model from orthomaps that can be rendered from any model, regardless of its registration method and works well with interiors with perpendicular walls and ceilings. With regard to more sophisticated interior geometry (e.g., dome viewed from the inside), a much higher number of orthomaps may be required to cover the entire geometry and possible viewing angles. In this case, the model size and matching time increases. For larger models, more sophisticated matching strategies should be adopted. For those cases, the k-means-based descriptor matching, proposed by Nister and Stewenius [29], or matching using only a part of the original model based on the previous frames with a region-growing may be used.

Because CATCHA relies on a scanned 3D model of the tracked interior, if a significant change occurs in the interior's appearance between the current state and the registered state (e.g. additional furniture), the 3D model has to be updated. Minor changes can be handled effectively, either in the matches-filtration step or by the RANSAC camera pose estimation algorithm. In the first case, unknown points should fail the ratio test due to a high distance to known points from the orthomap, and the ratio between the first and second nearest point would be close to 1. One of the possible solutions would be to integrate a SLAM-based tracking module to detect potential significant changes in model geometry or to track the model in areas that are poorly covered on the renders.

We partially overcome the problem of high-viewing angles by dividing the interiors into two groups, as mentioned in the previous section. However, this solution does not ensure that a camera pose would be established, regardless of the viewing angle. The first approach is limited by the viewing angle (Figure 11), and the second approach is limited by the fact that the camera has to point in the direction of the ceiling. In the Cabinet example, where the ceiling is approximately 6 m high, we examined both strategies mentioned in Section 3. The second strategy resulted in better user experience because the ceiling was always nearly parallel to the camera sensor plane. Additionally, keypoints were extracted and more robustly matched. Further, in the case of this particular interior, the walls contain much gilding that caused reflections in the camera frame, which, in combination with high viewing angles, resulted in less stable keypoints.

Because CATCHA uses RANSAC to eliminate outliers in the matches between the model and current camera frame, its performance depends on the number of points after filtration of the camera image. If this number is extraordinarily high, the frame rate may significantly drop. To eliminate this problem, different outlier rejection strategies such as that in [37] may be applied. The problem of a high outlier ratio can be partially handled by higher orthomap quality because the lesser the noise is, the more accurate is the keypoint matching and thus the fewer are the outliers. The experimental results presented in the previous section showed that the shapes of the translation and rotation errors (Figure 14) were almost identical because translation and rotation were simultaneously estimated and one compensated the other. The mean and median error values obtained in the King's Chinese Cabinet were much lower than those obtained in the Al Fresco Cabinet because the quality of the registered texture of the latter model was lower. Despite this condition, the outlier ratios were higher in the King's Chinese Cabinet because much gilding and many other shiny surfaces were present that could cause unstable keypoints. The reason for the high error values on certain frames was that the matched keypoints populated only part of the image (Figure 15). Because of the non-uniform distribution of points used as an input to the PnP algorithm, the solution may fall in the local minimum, instead of the global minimum. The error distribution does not depend on a specific camera layout but rather on the configuration of matched keypoints found in the image. Single-error camera-pose estimations did not affect the animation quality because the camera path was smoothed using Gaussian filtration. In both analyzed interiors, the outlier ratios and error values were sufficiently low to allow smooth and precise visualization of the details.

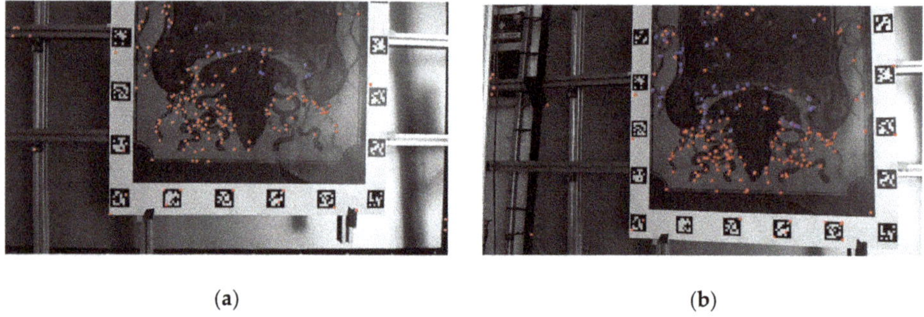

<table>
<tr><td align="center">(a)</td><td align="center">(b)</td></tr>
</table>

Figure 15. Point distribution: blue—inliers; red—outliers. (**a**) Non-uniform inliers are located only on a part of an image. (**b**) Uniform point distribution.

Beside the developed AR application allows the user to visually compare two states of the interior, the presented algorithm may be a base of many potential AR applications for professionals such as interactive model analysis (e.g. real-time curvature visualization and multispectral data mapping), presentations of interior design or interactive annotations for cultural heritage restorers. Ordinary visitors may be interested in simple augmentation of the interior with digital content, such as virtual guides who could walk around the interior and tell the story of the location. That would make the sightseeing more immersive.

7. Conclusions

We propose CATCHA—a novel AR camera tracking method inside cultural heritage interiors that satisfies very strict preservation requirements. The main novelty of this method is its orthographic-based data preparation that reduces the model complexity and does not depend on a point cloud registration method. We also propose a method that estimates the optimal orthographic resolution, which depends on real-world interior dimensions, camera FOV, and predicted optimal viewing distance. CATCHA was tested in two cultural heritage interiors in the Museum of King Jan III's

Palace at Wilanów, Warsaw, namely, the King's Chinese Cabinet and Al Fresco Cabinet. In our test, we achieved a mean error of below 2.5% in translation and below 1.5° in rotation. The accuracy of the camera-pose estimation algorithm is sufficient to provide visual feedback to the user. In our future work, we plan to examine different matching and data preparation strategies to improve robustness. One of the possible extensions may be to simultaneously track lines and keypoints. This approach may be helpful in handling glossy or textureless surfaces. Another direction of development can be automation of the process of mask generation using, for instance, machine learning methods. Finally, we plan to test CATCHA in different interiors and reduce the computational complexity to achieve a 60 Hz frame rate.

Author Contributions: Funding acquisition, R.S.; Investigation, P.S. and J.M.; Resources, E.B.; Software, P.S.; Supervision, R.S.; Validation, P.S. and E.B.; Visualization, J.M.; Writing—original draft, P.S.; Writing—review & editing, R.S.

Funding: This work was partially supported by the Ministry of Culture and National Heritage (Poland), by the KULTURA+ framework, and by the project "Revitalization and digitalization Wilanów, the only Baroque royal residence in Poland," (grant No. POIS.11.01.00-00.068/14) which is co-financed by the European Union within the Infrastructure and Environment Operational Programme and Statutory Work of the Warsaw University of Technology.

Acknowledgments: We would like to thank Marta Nowak for providing the implementation of network transmission used for remote tracking.

Conflicts of Interest: The authors declare no conflict of interest.

References

1. Bekele, M.K.; Pierdicca, R.; Frontoni, E.; Malinverni, E.S.; Gain, J. A Survey of Augmented, Virtual, and Mixed Reality for Cultural Heritage. *J. Comput. Cult. Herit.* **2018**, *11*, 1–36. [CrossRef]
2. Sitnik, R.; Bunsch, E.; Mączkowski, G.; Załuski, W. Towards automated, high resolution 3D scanning of large surfaces for cultural heritage documentation. In *IS&T/SPIE Symposium on Electronic Imaging: Science &Technology, IS&T*; Society for Imaging Science and Technology: Springfield, VA, USA, 2016.
3. Costantino, D.; Angelini, M.G.; Caprino, G. Laser Scanner Survey of An Archaeological Site: Scala Di Furno (Lecce, Italy). *Int. Arch. Photogramm. Remote Sens. Spat. Inf. Sci.* **2010**, *38*, 178–183.
4. Aicardi, I.; Chiabrando, F.; Maria Lingua, A.; Noardo, F. Recent trends in cultural heritage 3D survey: The photogrammetric computer vision approach. *J. Cult. Herit.* **2018**, *32*, 257–266. [CrossRef]
5. Marchand, E.; Uchiyama, H.; Spindler, F. Pose estimation for augmented reality: A hands-on survey. *IEEE Trans. Vis. Comput. Graph.* **2016**, *22*, 2633–2651. [CrossRef] [PubMed]
6. Pauwels, K.; Rubio, L.; Diaz, J.; Ros, E. Real-time model-based rigid object pose estimation and tracking combining dense and sparse visual cues. In Proceedings of the IEEE Computer Society Conference on Computer Vision and Pattern Recognition, Portland, OR, USA, 23–28 June 2013; pp. 2347–2354.
7. Fuentes-Pacheco, J.; Ruiz-Ascencio, J.; Rendón-Mancha, J.M. Visual simultaneous localization and mapping: A survey. *Artif. Intell. Rev.* **2012**, *43*, 55–81. [CrossRef]
8. Zeisl, B.; Sattler, T.; Pollefeys, M. Camera Pose Voting for Large-Scale Image-Based Localization. In Proceedings of the 2015 IEEE International Conference on Computer Vision (ICCV), Santiago, Chile, 7–13 December 2015; pp. 2704–2712.
9. Rubio, A.; Villamizar, M.; Ferraz, L.; Ramisa, A.; Sanfeliu, A. Efficient Monocular Pose Estimation for Complex 3D Models. In Proceedings of the 2015 IEEE International Conference on Robotics and Automation (ICRA), Seattle, WA, USA, 26–30 May 2015; Volume 2, pp. 1397–1402.
10. Yang, G.; Becker, J.; Stewart, C.V. Estimating the location of a camera with respect to a 3D model. In Proceedings of the 3DIM 2007 6th International Conference on 3-D Digital Imaging and Modeling, Montreal, QC, Canada, 21–23 August 2007; pp. 159–166.
11. Guan, W.; You, S.; Pang, G. Estimation of camera pose with respect to terrestrial LiDAR data. In Proceedings of the IEEE Workshop on Applications of Computer Vision, Tampa, FL, USA, 15–17 January 2013; pp. 391–398.
12. Wu, C.; Fraundorfer, F.; Frahm, J.-M.; Pollefeys, M. 3D model search and pose estimation from single images using VIP features. In Proceedings of the 2008 IEEE Computer Society Conference on Computer Vision and Pattern Recognition Workshops, Anchorage, AK, USA, 23–28 June 2008; pp. 1–8. [CrossRef]

13. Lowe, D.G. Distinctive image features from scale-invariant keypoints. *Int. J. Comput. Vis.* **2004**, *60*, 91–110. [CrossRef]
14. Taketomi, T.; Sato, T.; Yokoya, N. Real-time and accurate extrinsic camera parameter estimation using feature landmark database for augmented reality. *Comput. Graph.* **2011**, *35*, 768–777. [CrossRef]
15. Jaramillo, C.; Dryanovski, I.; Valenti, R.G.; Xiao, J. 6-DoF pose localization in 3D point-cloud dense maps using a monocular camera. In Proceedings of the 2013 IEEE International Conference on Robotics and Biomimetics, ROBIO 2013, Shenzhen, China, 12–14 December 2013; pp. 1747–1752.
16. Crombez, N.; Caron, G.; Mouaddib, E.M. Using dense point clouds as environment model for visual localization of mobile robot. In Proceedings of the 2015 12th International Conference on Ubiquitous Robots and Ambient Intelligence (URAI), Goyang, Korea, 28–30 October 2015; pp. 40–45. [CrossRef]
17. Rambach, J.; Pagani, A.; Schneider, M.; Artemenko, O.; Stricker, D. 6DoF Object Tracking based on 3D Scans for Augmented Reality Remote Live Support. *Computers* **2018**, *7*, 6. [CrossRef]
18. Vanoni, D.; Seracini, M.; Kuester, F. ARtifact: Tablet-Based Augmented Reality for Interactive Analysis of Cultural Artifacts. In Proceedings of the 2012 IEEE International Symposium on Multimedia, Irvine, CA, USA, 10–12 December 2012; pp. 44–49.
19. PTC Inc. Vuforia Library. Available online: https://www.vuforia.com/ (accessed on 3 October 2018).
20. Gîrbacia, F.; Butnariu, S.; Orman, A.P.; Postelnicu, C.C. Virtual restoration of deteriorated religious heritage objects using augmented reality technologies. *Eur. J. Sci. Theol.* **2013**, *9*, 223–231.
21. The Fraunhofer Institute for Computer Graphics Research IGD Instant Player. Available online: http://www.instantreality.org/ (accessed on 3 October 2018).
22. Verykokou, S.; Ioannidis, C.; Kontogianni, G. 3D visualization via augmented reality: The case of the middle stoa in the ancient agora of Athens. In *Lecture Notes in Computer Science, Proceedings of the Euro-Mediterranean Conference, Limassol, Cyprus, 3–8 November 2014*; Springer: Berlin, Germany, 2014; Volume 8740, pp. 279–289101007978.
23. Battini, C.; Landi, G. 3D tracking based augmented reality for cultural heritage data management. *Int. Arch. Photogramm. Remote Sens. Spat. Inf. Sci.* **2015**, *40*, 375. [CrossRef]
24. Inglobe Technologies ARmedia 3D SDK. Available online: http://dev.inglobetechnologies.com/index.php (accessed on 3 October 2018).
25. Lech, K.; Mączkowski, G.; Bunsch, E. Color Correction in 3D Digital Documentation: Case Study. In *Image and Signal Processing*; Mansouri, A., Nouboud, F., Chalifour, A., Mammass, D., Meunier, J., Elmoataz, A., Eds.; Springer International Publishing: Cham, Switzerland, 2016; pp. 388–397.
26. CharLS. CharLS Library. Available online: https://github.com/team-charls/charls (accessed on 3 October 2018).
27. Rublee, E.; Rabaud, V.; Konolige, K.; Bradski, G. ORB: An efficient alternative to SIFT or SURF. In Proceedings of the IEEE International Conference on Computer Vision, Barcelona, Spain, 6–13 November 2011; pp. 2564–2571.
28. Sibbing, D.; Sattler, T.; Leibe, B.; Kobbelt, L. SIFT-realistic rendering. In Proceedings of the 2013 International Conference on 3D Vision, 3DV 2013, Seattle, WA, USA, 29 June–1 July 2013; pp. 56–63.
29. Nister, D.; Stewenius, H. Scalable recognition with a vocabulary tree. In Proceedings of the IEEE Computer Society Conference on Computer Vision and Pattern Recognition, New York, NY, USA, 17–22 June 2006; Volume 2, pp. 2161–2168.
30. Lepetit, V.; Moreno-Noguer, F.; Fua, P. EPnP: An Accurate O(n) Solution to the PnP Problem. *Int. J. Comput. Vis.* **2008**, *81*, 155. [CrossRef]
31. Gao, X.S.; Hou, X.R.; Tang, J.; Cheng, H.F. Complete solution classification for the perspective-three-point problem. *IEEE Trans. Pattern Anal. Mach. Intell.* **2003**, *25*, 930–943. [CrossRef]
32. Siekański, P.; Bunsch, E.; Sitnik, R. Seeing the past: An augmented reality application for visualization the previous state of cultural heritage locations. In *Electronic Imaging*; Society for Imaging Science and Technology: Springfield, VA, USA, 2018; Volume 2018, pp. 4521–4524.
33. Bay, H.; Tuytelaars, T.; Van Gool, L. SURF: Speeded Up Robust Features. In *Lecture Notes in Computer Science, European Conference on Computer Vision, Graz, Austria, 7–13 May 2006*; Springer: Berlin, Germany, 2006; pp. 404–417.

34. Irschara, A.; Zach, C.; Frahm, J.M.; Bischof, H. From structure-from-motion point clouds to fast location recognition. In Proceedings of the 2009 IEEE Computer Society Conference on Computer Vision and Pattern Recognition Workshops, CVPR Workshops 2009, Miami, FL, USA, 20–25 June 2009; pp. 2599–2606.

35. Morel, J.-M.; Yu, G. ASIFT: A New Framework for Fully Affine Invariant Image Comparison. *SIAM J. Imaging Sci.* **2009**, *2*, 438–469. [CrossRef]

36. Garrido-Jurado, S.; Muñoz-Salinas, R.; Madrid-Cuevas, F.J.; Marín-Jiménez, M.J. Automatic generation and detection of highly reliable fiducial markers under occlusion. *Pattern Recognit.* **2014**, *47*, 2280–2292. [CrossRef]

37. Ferraz, L.; Binefa, X.; Moreno-Noguer, F. Very fast solution to the PnP problem with algebraic outlier rejection. In Proceedings of the IEEE Computer Society Conference on Computer Vision and Pattern Recognition, Columbus, OH, USA, 23–28 June 2014; pp. 501–508.

International Journal of
Geo-Information

Article

An Architecture for Mobile Outdoors Augmented Reality for Cultural Heritage

Chris Panou [1], Lemonia Ragia [2,*], Despoina Dimelli [2] and Katerina Mania [1]

[1]   Department of Electrical and Computer Engineering, Technical University of Crete, Kounoupidiana, 73100 Chania, Greece; cpanou@isc.tuc.gr (C.P.); k.mania@ced.tuc.gr (K.M.)
[2]   School of Architectural Engineering, Technical University of Crete, Kounoupidiana, 73100 Chania, Greece; dimelli@arch.tuc.gr
*   Correspondence: lragia@isc.tuc.gr

Received: 10 October 2018; Accepted: 26 November 2018; Published: 30 November 2018

Abstract: In this paper, we present the software architecture of a complete mobile tourist guide for cultural heritage sites located in the old town of Chania, Crete, Greece. This includes gamified components that motivate the user to traverse the suggested interest points, as well as technically challenging outdoors augmented reality (AR) visualization features. The main focus of the AR feature is to superimpose 3D models of historical buildings in their past state onto the real world, while users walk around the Venetian part of Chania's city, exploring historical information in the form of text and images. We examined and tested registration and tracking mechanisms based on commercial AR frameworks in the challenging outdoor, sunny environment of a Mediterranean town, addressing relevant technical challenges. Upon visiting one of three significant monuments, a 3D model displaying the monument in its past state is visualized onto the mobile phone's screen at the exact location of the real-world monument, while the user is exploring the area. A location-based experience was designed and integrated into the application, enveloping the 3D model with real-world information at the same time. The users are urged to explore interest areas and unlock historical information, while earning points following a gamified experience. By combining AR technologies with location-aware and gamified elements, we aim to promote the technologically enhanced public appreciation of cultural heritage sites and showcase the cultural depth of the city of Chania.

Keywords: augmented reality; navigation; instant tracking; 3D reconstruction; cultural heritage; computer graphics

---

1. Introduction

Augmented reality (AR) is the act of superimposing digital artifacts onto real environments. Across the reality–virtuality continuum [1,2], AR is part of the broader mixed reality spectrum. It enables real-time mixing of computer generated content and real content and has already been employed for medical, military, manufacturing, and robotics training applications [3]. In contrast to virtual reality (VR), where the user is completely immersed in a synthetic environment, AR aims to supplement reality [4,5]. While early research limited the definition of AR in a way that required the use of specialized head-mounted-displays (HMDs), a taxonomy introduced by the authors of [6] defined that any system that combines virtual and real imagery, accurately registering (aligning) real and virtual objects with each other, and running interactively in three dimensions and in real-time, is considered an AR system. AR software development kits (SDKs) are available to develop AR applications [7]. Current advances based on novel, commercial AR displays such as Microsoft Hololens and Magic Leap, as well as the huge investment drawn towards their development, showcase AR's potential for integration into the every-day life of the consumer.

Based on mobile phones' current technical specification enabling them to render 3D content and combine it with camera input, we developed the proposed mobile AR (MAR) application for Android mobile phones, offering a real-time, on-site, 3D depiction and visualization of historical monuments of the old town of Chania, Crete, Greece. We present a location-based AR application for Android devices that provides a sightseeing experience that aims to challenge and motivate the visitors to further explore and uncover the city's underlying history. In this paper, we present the software architecture of the system, which seamlessly incorporates AR features for the challenging, sun-wrenched outdoor environment of a busy Mediterranean town. The system proposed includes gamified scenarios and advanced AR features that enhance user experience while the visitor walks around interest points. The main functionality is incorporated in a digital map including the AR camera view. The main focus of the AR feature is to superimpose 3D models of historical buildings in their past state onto the real world, while users hold their consumer-grade mobile phones while walking on-site. Historical information in the form of text and images is available at the same time.

We are faced with significant technical, registration, and tracking challenges that relate to the accurate registration of 3D content at the exact location of the real-world monument, without the use of black and white markers, which are normally placed on surfaces. Moreover, the visualization of 3D content outdoors, taking into account extreme sunlight, visitors obscuring views, inaccuracy of geo-location based on global positioning system (GPS), as well as unstable computer vision registration approaches, poses challenging technical issues. We review potential solutions to the registration problem for outdoor AR implementations, evaluate them, and put forward a stable solution for mobile outdoor AR visualization without the need for placing markers.

Taking into consideration the historical significance of the monuments, our approach offers the opportunity to interact with them in non-intrusive ways, thus eliminating the need to interfere with the remains and on-going archaeological research. Based on a mobile, personalized, location-aware experience taking place in various areas of Chania, Crete, we aim to showcase the city's cultural wealth. The mobile AR application is based on a database that stores historical information of a large variety of monuments. The database also stores a real-time record of personalized user visits in their selected areas of interest, which is only available to each specific user after signing in.

The old Venetian part of the city of Chania was selected as the main spatial location for its rich history, enriched by influences from other cultures. It is distinguished by the presence of historical monuments from the Minoan, Roman, Arabic, Venetian, and Ottoman periods until the Greek historical period. Adverse climate conditions, modern city planning, and rapid expansion have slowly compromised the state of these sites, thus endangering their historical value over time, as well as their original beauty. For this work, a specific route around the Venetian part of the city of Chania was selected as the main route that a tourist holding a mobile phone would traverse. In relation to the AR visualization, three monuments were selected, representing three distinct historical periods and architectural interests. These are the following: (a) the Glass Mosque (Figure 1a) from the 17th century [8]. This is the first mosque built in Crete and the only one remaining in the city. It included a minaret (Figure 1, right 1b, 1c), which was demolished. (b) The Saint Rocco temple is one of many Venetian churches in Chania (Figure 2) [9]. Only part of the church is in place, while the exterior facades are painted and the original texture has been lost. (c) The Byzantine wall, which is one of the most famous monuments in the city, built around the sixth and seventh century AD (Figure 3) [10].

The Glass Mosque (Figure 1) is located in the Venetian harbor of Chania and is the first mosque that was built in Crete and the only surviving one in the city, dating from the second half of the 17th century. Erected in honor of the first garrison commander of Chania, named Küçük Hasan, it is a jewel of Islamic art in the Renaissance. The mosque is a cubic building covered by a large hemispherical cupola supported by four ornate stone arches. Its western and northern parts are surrounded by a covered arcade of six small cupolas, which are open at the top, as is customary in mosques. Around 1880, the arcade was covered with arched openings of neoclassic style. The small, but picturesque minaret was demolished in 1920 or in 1939. It was quite badly damaged by bombing during WWII. After suffering substantial damage during the war, it was finally restored and moved to Chania's archaeological museum. Later, it was used as either a warehouse, a folk art museum, the Information Office of the Hellenic Organization of Tourism, and recently as a home to seasonal events and exhibitions.

(a)   (b)   (c)

Figure 1. (a) The Ottoman Glass Mosque in its current state on the left (author's picture). Its original state in the middle (b) and on the right (c). Image (b) comes from G. Despotaki's archive and image (c) from M. Manousaka's archive.

Figure 2. Front side of the temple showing the northern (**left**) and the southern (**right**) part.

The Saint Rocco temple is a Venetian chapel on the northwest corner of Splantzia square, and consists of two different forms of vaulted roof aisles (Figure 2). Although the southernmost part is preserved in good condition, the northern and oldest one has had its exterior painted over, covering its stony façade, while a residential structure has been built on top.

Figure 3. The Byzantine wall and the part chosen to be 3D modeled in green (**left**). The demolished and built over towers of the wall (**right**).

The Byzantine wall was built over the old fortifications of the Chydonia settlement around thesixth and seventh century AD (Figure 3). Its outline is irregular with a longitudinal axle from the east to the west, where its two central gates were located. The wall consists of rectilinear parts, interrupted by small oblong or polygonal towers, many of which are now partly or completely demolished (Figure 4).

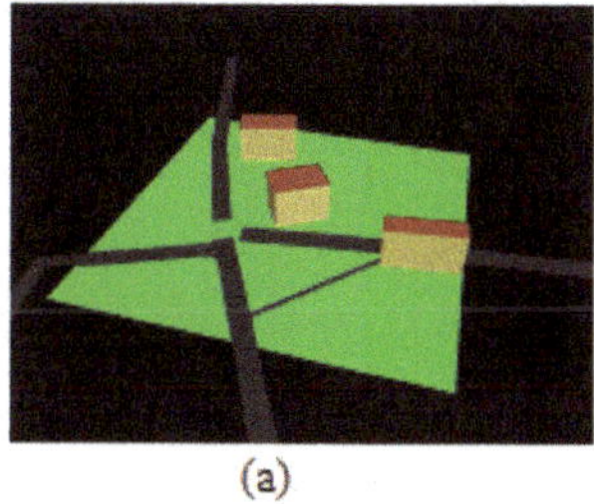 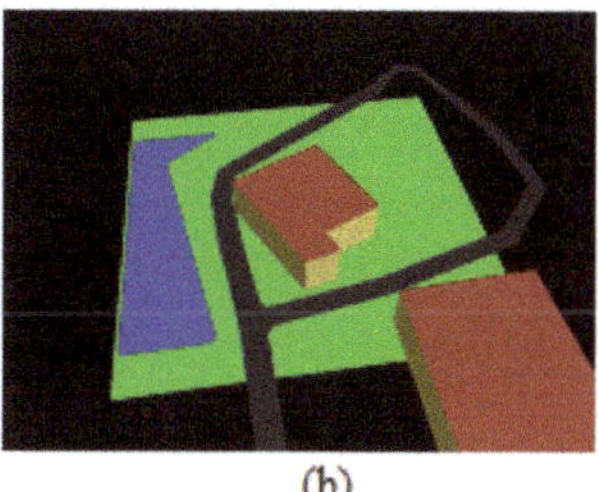 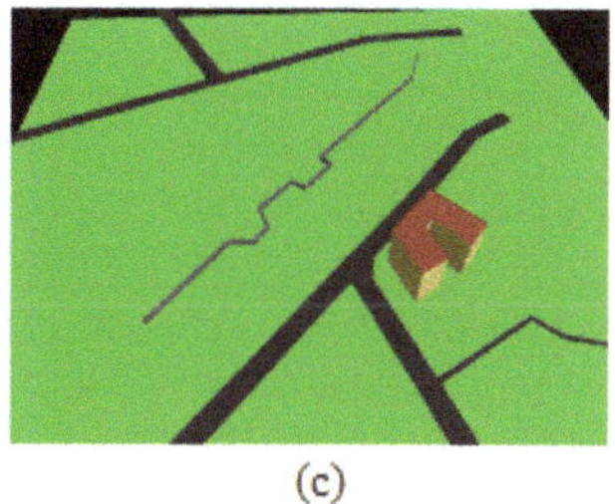

(a)   (b)   (c)

Figure 4. (**a**) Glass Mosque; (**b**) Saint Rocco temple; (**c**) Byzantine wall.

The scope of this work is to virtually restore partially or fully damaged buildings and structures on historic sites and enable visitors to see them integrated into their real environment through a mobile phone. Our aim is to design and implement a mobile AR application for Android devices that will help visitors interact with the city's monuments. We aim to deliver geo-located information to the users, through text, images, and AR, and help them document their visits. By integrating digital maps and a location-based experience, we aim to urge the users to further investigate interest areas in the city and uncover their underlining history. The AR mobile application developed aims to provide an easily extendable platform for future additions of digital content, and requires a moderate amount of development and technical expertise. The goal is to provide a complete and operational AR experience to the end-user by tackling significant AR technical challenges related to the accurate registration and positioning of the 3D content integrated into a mobile application.

Previous Work

In comparison with older see through AR displays, which were head-mounted, based on cumbersome hardware and complicated software modules, a recent emergence in mobile technology has led to an integrated platform, ideal for the development of AR experiences, often referred to as mobile AR (MAR). MAR is a concept first conceived in the late 1990s, producing early AR systems for cultural and archaeological sites, mainly indoors. However, today's modern smartphones have brought AR to a wider audience. The presence of high processing power, cameras, and inertial and global positioning system (GPS) sensors provides the necessary components of an AR system in an ergonomic hand-held device.

In past years, AR has been utilized for a number of applications in cultural heritage. *Archeoguide* was first presented using AR for personalized tours in cultural heritage sites [11,12]. The system allowed users to experience a VR world featuring computer generated 3D reconstructions of ruined sites without isolation from the real world. Two years later, an extended version presented a personalized mobile guide for outdoor archaeological sites [13], employing the site of Ancient Olympia as a test case [14]. The system provided on-site help and AR reconstructions of ancient ruins. It made use of a compass, a DGPS (differential global positioning system) receiver integrating images streamed from a webcam, and the users' location and orientation. Visitors carried a backpack computer that performed the calculations, and wore a see-through head-mounted-display (HMD) to display the digital content. By today's standards, the system was cumbersome and heavy. Despite the ergonomic restrictions, it was very well-received by the visitors as it provided a unique site-seeing experience [14].

A MAR application representing a historical tour guide including old photographs and information about a historical street has been reported in the work of [15]. An early overview of AR in cultural heritage highlighted the technical problems of AR development, mostly centered on registration and rendering issues, as well as having to rely on black and white markers for correct alignment [16]. Other AR approaches aimed to connect the excavation site with the artifacts in the museum in order to enable contextual awareness of the visitor [17]. The Augmented Representation of Cultural Objects ARCO system combined VR and AR, enabling museums to display their 3D, digitized collections on the web [18]. The users could interact with sensor-enhanced physical objects while their digital reconstructions were simultaneously manipulated [19]. An AR framework for on-site visualization of archaeological data employed visual tags, overlaying 3D models on images provided in real-time by the phone camera, aiming to display accurately positioned information [20].

MARCH was a MAR application for digitally enhancing visits to prehistoric caves [21]. It was developed in Symbian C++, running on a Nokia N95. The system made use of the phone's camera to detect images of cave engravings and overlay an image of the original drawings on them, on-site. This was the first attempt of a real-time MAR application without the use of grey-scale markers. Instead, colored patches positioned at the corners of photographs were employed, viewing these and not the real-world environment. An integrated mixed reality system for recreating virtual human actors by integrating the real world and a 3D scene populated by them was presented by the authors of [22]. This platform was employed on the real-world site of ancient Pompeii [23].

With the emergence of smart mobile devices, sophisticated AR experiences were made possible, such as the one placed on the Bergen-Belsen memorial site. This was a former WWII concentration camp in northern Germany, which was burned down after its liberation [24]. The application integrated database interaction, reconstruction modeling, and content presentation in a hand-held device. The system was developed for an iPhone. Real-time tracking was performed based on the device's GPS and orientation sensors. Navigation was conducted via either maps or the camera. The system superimposed3D reconstructed buildings on the actual site, visualizing their past and present state. Focusing on the promotion of cultural heritage in outdoor settings, VisAge aimed to transform users into authors of cultural stories in urban environments through physical space [25]. A story consisted of spatially distributed POIs (points of interest). Each POI was assigned its own digital content including images, text, or audio. A viewing tool was developed for mobile tablets in Unity3D using Vuforia's tracking library in order to overlay the digital content on the real space. The users could then follow these routes in the city and experience new stories.

Further work related to 3D reconstructions was shown in CityViewAR [26], a mobile outdoor AR application allowing the visualization of destroyed buildings after major earthquakes in Christchurch, New Zealand. Besides providing stories and pictures of the buildings, the main feature of the application was the ability to visualize 3D models of the buildings in AR. A practical solution presented by the authors of [27] guided groups of visitors in noisy indoor environments, employing analogue audio transmission and reliably trackable AR markers. Preparation of the environment with fiducials and supervision of the visits by experts was necessary in order to avoid accidents and interference with the working environment.

PRISMA combined the concept of tourist binoculars and AR technologies [28]. The real-world scene was enhanced by digital information, providing an interactive user-friendly system. AR interfaces for visiting cultural heritage sites employed multimedia sketches as a means of communication between experts [29]. An AR tourist guide was designed and implemented for mobile devices, which enabled tourists to present historical information [30]. Real-time tracking was performed using either computer vision techniques or sensors such as GPS and gyroscopes. Another approach puts forward a 3D reconstruction methodology applied to the restoration of historical monument based on tacheometry data [31]. AR technologies were developed to visualize restoration areas and their effect after restoration.

Putting forward a mobile, personalized, location-aware MAR experience taking place in the historical part of Chania, Crete, we aim to enhance user experience and interaction with cultural heritage sites and showcase the city's cultural wealth, as well as to address technical challenges related to registration techniques for AR outdoors. The mobile AR application presented features a database that holds records of historical monuments on-site. The database stores users' documentation of their visits and interactions in their areas of interest, available to each user that signs in. AR development of a city poses significant technical, as well as user interaction, challenges. Reliable position and pose tracking is paramount so that the 3D content representing the monuments in their past state is accurately superimposed on real settings at the exact position required. This is one of AR's most significant research challenges. Our system features a geo-location and sensor approach, which, compared with optical tracking techniques, allows for free user movement throughout the site, independent of changes in the buildings' structure. Moreover, we combined this implementation with a hybrid registration technique (sensor-based and vision-based), in order to showcase the capabilities of future AR technologies.

2. AR Methodology

In this work, we present the design and implementation of a MAR (mobile augmented reality) application for Android devices that provides on-site 3D visualization of historical buildings located in the historical part of Chania, Greece. These are superimposed over their real-world equivalent, as part of a smart AR tourist guide. Further to 2D images and text often presented by mobile tourist guides, we aim to enrich the sightseeing experience by providing a means to visualize the past glory of these sites in the context of their real-world surroundings. Taking into consideration the historical significance of the monuments, our approach offers the opportunity to interact with them in non-intrusive ways, without physically interfering with the remains and on-going archaeological research. In order to accurately record the geo-location of the monument and provide real-time tracking, the standard sensors of a mobile phone are used. We do not place markers or patches on the monument and the implementation of the software does not require any physical contact with the monument.

2.1. The AR Registration Technical Challenge

AR registration is the degree to which 3D information is accurately placed and integrated as part of the real environment. The objects in the real and 3D scene should be correctly aligned with respect to each other, or the illusion that the two co-exist is compromised [32]. In contrast to VR, where such errors result in visual–kinesthetic conflicts, in AR, such conflicts are driven by the visual sense alone and are easier to detect. A user wearing a VR headset raising an arm and actually viewing a virtual one at the same time that is off by a few centimeters may not detect the spatial misplacement because of the conflict between the "sensed" position of the real arm and the "seen" position of the virtual one. In the corresponding AR application, the virtual arm should completely overlap the real one; therefore, such an error would be easily detectable.

Registration errors are defined as dynamic and static. Static errors are the ones that affect the AR scene even when both user and environment are in stasis. Sources of such errors can be the inaccurate calibration of mechanical parts, incorrect tracker-to-eye ratio, field-of-view parameters, optical lens distortions, and so on. Static errors depend mostly on mechanical parts and the correct initial calibration of the system, and can be accounted for to a very satisfying degree. Dynamic errors, on the other hand, are the ones that take effect if either the viewpoint (user) or the annotated object begins moving. For MAR, this error is by far the largest contributor to the registration problem and varies depending on the implementation.

In early AR systems, the single most important factor for dynamic errors was end-to-end system delays. A tracker reports user movements, and the system should then update the digital artifacts on the screen. This computation and its delivery should precede changes in the user's pose, which proved to be a very difficult task at the time. With today's hardware, system delays have been minimized and the main source of errors in registration is pose estimation, for example, position and orientation tracking. AR tracking is a complicated task with no single best solution.

In order to register virtual content in the real world, the pose (position and orientation) of the viewer with respect to some "anchor" in the real world must be determined. Depending on the application and technologies used, the real-world anchor may be a physical object, such as a magnetic tracker source or paper image marker, or may be a defined location in space, determined using GPS or dead-reckoning from inertial tracking.

In order for an AR system to overlay the world with digital information, it is required to track its position in 6 DoF (degrees of freedom), three for position and three for orientation. Inertial sensors, GPS positioning, and optical sensor scans provide the necessary data for such computations. In this paper, we will focus on AR pose estimation for consumer-grade smartphones. The main approaches are the following: vision-based tracking, which relies on the device's camera and ways to process the live feed; and sensor-based, which combines GPS positioning and the inertial sensors of the device.

Optical and sensor-based tracking were tested on-site. Two dimensional image recognition, which relies on natural feature detection algorithms, was tested using a variety of captured images of the annotated monuments. Those images are processed to produce 2D point clouds of the detected features to be identified by the mobile device in the camera's feed. When detected, pose and position estimations are available, relative to the surfaces they reproduce. While image recognition has presented great results thanks to fast hardware and improved algorithms, we could not rely on it. Outdoor environments present challenging conditions. Building façades provided a small amount of features to be recognized and, taking into account variations in lighting conditions, the sets of features provided to the AR system differed greatly when compared with the real scene. Thus, the targets were not reliably recognized. In the rare cases where tracking was possible, simple movements of the device resulted in miscalculations of the orientation. Taking also into account the cramped environment of a touristic site compromising the visibility of the targets, image recognition was not perceived as a realistic choice for AR registration and tracking in our case.

2.2. The AR Registration Approach

There is no single best solution to AR tracking and registration. Vision-based approaches are best suited for controlled and small environments, but their performance diminishes in wide and outdoor areas, where the sensor-based approaches provide the best results. Tracking in AR systems is an open problem. Although the future seems to lie within hybrid implementations, they are currently in their infancy and most often require additional hardware components.

The most important criterion when selecting an AR tracking solution is reliability. Graphics have little meaning when tracking is not possible. In this paper, we selected the geo-location approach and the instant tracking option of the Wikitude SDK. While a case can be made for the registration difficulties of the geo-location approach due to sensor filtering and low GPS accuracy, these implementations require fewer actions by the users and ensure that the AR experience will be delivered independent of external conditions. In contrast, the instant tracking option provides greater registration accuracy and eliminates the latency, but is susceptible to occlusions and requires additional actions by the users. The decision to include both implementations was taken in order to provide users who are unaccustomed with AR applications an intuitive way of visualizing the 3D models based on geo-location, while also being able to provide a more sophisticated AR experience with instant tracking. The Wikitude JavaScript API was selected because of its robust tracking, educational licensing option, documentation, large community, and efficient customer service.

The geo-location approach employs the GPS and the inertial sensors of the device so that when the specific location of the actual monument is registered, the 3D visualization of it would be displayed. Locations containing latitude and longitude information are received by the GPS, while the accelerometer and geo-magnetic sensors are used to estimate the pose of the device in the Earth's frame. A visual reconstruction is then matched to the user's position and viewing angle, displaying the overlaid models on the mobile phone's screen. This implementation offered the most reliable registration of 3D content, accurately superimposed on the real-world site, demanding less action from the users and, therefore, ensuring a robust and intuitive experience [30].

The preparation of the 3D models required the acquisition of historical information and their accurate depiction in scale with the real world. Because of the lack of accurate plots and outlines of the buildings, LiDAR (Light Detection and Ranging) and DSM (Digital Surface Models) data were exported from Open Street Maps and used to create the final models. Developing such models for a mobile device means that the limited processing power and the requirements of the AR technologies need to be taken into account. Complex geometries can impair performance, so we followed a low-poly, high-resolution texture approach to avoid drops in frame rate.

A key concept for any location-aware application is the cognitive map held in the form of mental images by the user [33]. Digital maps and an AR camera displaying the interest areas were integrated to assist in navigation through the geo-located content. The client-server architecture ensures that we provide personalized experiences by storing information about user visits and progress. Changes in the server can be conducted without interfering with the mobile application, allowing for an easily extendable platform where new monuments could be added as visiting areas. The monuments' information and assets are stored in a database and delivered to the mobile application on a location-request basis. The application was developed for Android in Java, and employs the Wikitude Javascript API to handle the AR views. It features a local database based on SQLite cashing the downloaded content. The complete design and implementation is described in detail in the following sections.

The main screen of the proposed application is the map indicating the location of the user, the available POIs, as well as the main navigation buttons. The aim of the map is to help the user navigate the city. This navigation can also be accomplished via the camera view where the POIs are displayed in 3D space on the camera's surface. In its initial state, the user is shown the available monuments that can be augmented by 3D models as markers on the map. The path between them is shaded with polylines in order to visually integrate these points. Upon visiting the monuments, the user has access to the 3D reconstructions through the AR camera. The initial monuments act as an introduction to the application's main features. After users visit a specific monument, they are awarded a number of points. When all the available 3D models are visited, the game enters its main state. The interest areas of the system pop out as question marks on the map. The goal is to visit all interest areas and unlock them to earn more points. In order to unlock a POI, the user has to correctly classify it according to its respective historical period. Information concerning historical periods relevant to specific monuments can be acquired from the application. When a user correctly unlocks a monument, access to its historical information is available. The aim of this approach is to urge the user to closely observe these monuments, consult the information already unlocked, or even interact with the locals to get as much information as possible. The user is transformed as an active participant of the sightseeing experience instead of a passive spectator.

The historical and user-specific information is stored in a remote database connected to the mobile application via a REST (Representational State Transfer) web service. The application requests the additional interest areas based on a location request and updates the user's progress and monuments' statistics depending on the actions that took place. We can then display user-related rankings and leaderboards, creating a challenging and competitive experience that aims to further motivate the users to explore the city and subsequently promote its cultural heritage.

3. 3D Depiction of Historical Monuments

3.1. 3 D Modelling and Texturing

In order to record the past state of the selected monuments, old photographs, historical information, and estimates from experts were utilized. The 3D models visualizing their past state will be presented in real size and superimposed over the real-world monument, and must be in proportion with their surroundings. Therefore, accurate measurements of their structure are necessary. Because of the lack of schematics and plots, we relied on data derived from online mapping repositories that provide outlines and height. In order to ensure historical accuracy and avoid the communication of false information, the final models and their reconstructed parts are in abstract form, depicting only the main structural elements of each monument. Instead of presenting information to its full extent, it can be represented by a decided level of abstraction, or 'level of detail', providing the minimum information required. An abstract form is often used by architects to depict the general form of a structure, emphasizing its open-ended character function as an initiator of possibilities and potentialities. This can be related to how missing parts of a historical building could have appeared in the past, shedding light on archeological or historical uncertainty concerning its past form [34,35]. The focus of this work was not the accurate reconstruction of buildings, but mostly the AR component working as seamlessly and 'in-place' as possible in relation to the registration of 3D content with the real-world.

The outlines of the three monuments were acquired from Open Street Map (OSM). By selecting specific areas of the monuments on the map, we can then export a .osm file that contains the available information concerning that area, including building outlines and height, where available. This file is essentially an xml file comprising OSM raw data including roads, nodes, tags, and so on. The file is then imported into OSM2World, a Java application aiming to produce a 3D scene.

These representations are basic triangulated meshes of the outlines raised to reach the height value for each building. As is evident in relation to the Byzantine wall as it stands now, height data were not available to us. In relation to the two monuments shown in Figure 4, it is not clear whether the domes and roofs have been taken into consideration. These models formed an initial basis and any disproportions were to be corrected after on-site testing. The models were then exported to .obj format and imported to the Blender 3D modeling software.

On the basis of the basic structures of the buildings, the final 3D mesh to be included in the mobile AR application was created. The modeling process was focused on preserving a low vertex count as complex geometry compromises an interactive framerate in systems with low processing power such as mobile phones. During the initial stage, we only created the demolished part, to be overlaid on the real buildings (Figure 5).

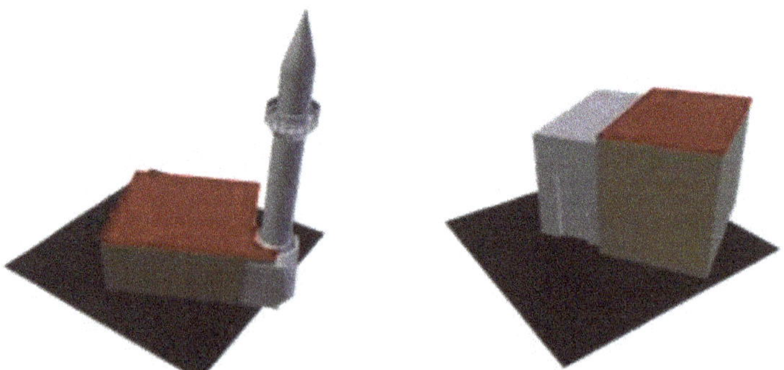

Figure 5. The 3D mesh of the 3D models in scale with the existing buildings.

However, on-site testing showed that the average accuracy of the GPS receiver of three meters and the constrained viewpoint in certain sites, such as the Saint Rocco temple, break the illusion of the 3D information coexisting as part of the real world because the 3D parts were not registered accurately with the real world. Instead, complete 3D depictions of the selected monuments were created so that they completely overlap the real ones. The most important aspect of this strategy is to keep the 3D models of the monuments in proportion in terms of size. The final scale and size were defined in Google Sketch-up, while the final model was positioned in the world coordinate system. The final models are shown in Figure 6.

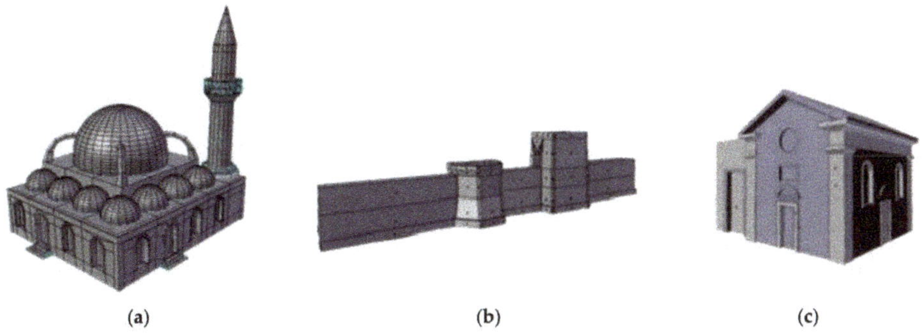

(a)   (b)   (c)

Figure 6. The final 3D meshes: (**a**) Glass Mosque; (**b**) Byzantine wall; (**c**) Saint Rocco temple [36].

These models constitute the state of the building in the respective historical periods. Reference images were used for modeling the existing parts of the buildings, while further details were added to keep them consistent with the buildings as they stand in the present time. We perceived this as a valid procedure, however, we accept that elements of historical structures visible at the present time should be validated by employing historical information. Such detailed work, especially in the domes and railings areas of the Glass Mosque (Figure 6a) and the column area in Saint Rocco (Figure 6c), unavoidably heightens the polygon count. However, the performance of displaying them on a mobile device was deemed satisfactory after testing. The Byzantine wall (Figure 6b), being the most abstract, resulted in 422 vertexes. The Glass Mosque resulted in7614 vertexes and the Saint Rocco temple in 4919 vertexes.

Texture mapping is a method of adding photorealism and detail to a flat surface without adding extra geometry. A texture map is a bitmap image that is applied to the surface of a polygon, creating a high fidelity visual result. Images captured from the real monuments on site were used as references in order to locate the surface materials. The texture mapping procedure followed a multi-material approach. A diverse range of materials is assigned to different parts of the buildings. Each material is then assigned images as texture to the surfaces. Because of the lack of information, the actual texture of the reconstructed parts is unknown. The aim is to accurately represent the compositing material rather than the actual surface, taking into account the texture of the remains. Another approach would be to use supported hypotheses of similar constructions of the same era [37]. At this point of the procedure, testing of the rendering capabilities of the AR framework was conducted utilizing the derived models. The AR framework only supports a power of two png or jpeg single material texture map. This means that bumps, normal maps, and multi-textures cannot be included for better surface representation.

The materials that compose the entire texture are "baked" into one image that will serve as the final texture employing the 3D modeling software 'Blender'. UV (ultra violet) mapping is the process of unwrapping the 3D shape of the model into a 2D map. This map contains the coordinates of each vertex of the model placed on an image. The materials that were assigned to the surfaces are then baked onto the image that forms the final texture. Taking into account that the monuments will be displayed on a mobile phone screen in real size, we needed high resolution textures. This unfortunately raises the final size of the texture files, however, the process results in a high quality visual result. The final texture resolution is 2048 × 2048 pixels (Figure 7). In order to light the scene, a simple hemi light was used, provided by 3D modelling software. The final models were exported without a light source. Although the AR framework supports lighting, having a static light in a real-world scene would impair the experience, and dynamically controlling the light source is not possible.

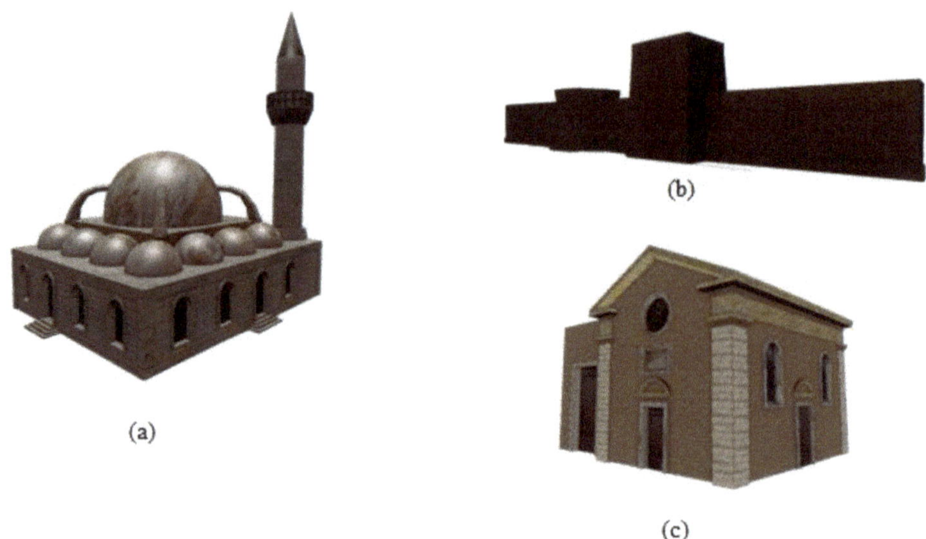

Figure 7. Final textured models, (**a**) Glass Mosque; (**b**) Byzantine wall; (**c**) Saint Rocco temple

3.2. Geo-Localization

Sensor approaches rely on the inertial sensors of the device for orientation tracking and on the GPS for positioning. For Android devices, these sensors can either be hardware components such as magnetometers, accelerometers, and gyroscopes or software components that rely on multiple physical components and sensor fusion to produce specific results, such as the linear-acceleration sensor, the orientation sensor, the gravity sensor, and others. These sensors can be categorized as motion sensors that measure acceleration forces and rotational forces along three axes of the device, such as the accelerometer, gyroscope, and positioning sensors that provide absolute measurements about the physical position of a device such as the magnetometer and the orientation sensors. The geo-location approach requires measurements relative to the Earth's frame of reference. This requires the combination of motion and positioning sensors to determine the orientation and their combination with GPS, which provides latitude and longitude measurements for position.

If the virtual objects are accurately positioned and oriented in the world coordinate system, the AR system is aware of the distance between them because that model is built into it. The digital content is associated with a geo-location that is then matched to the user's position and viewing angle. For these implementations, we used the HitlabNZ mobile AR framework and the Wikitude JavaScript API. These tracking techniques rely on cheap sensors present in every modern smartphone.

The most significant challenge is the registration errors introduced by the GPS accuracy and the latency derived from filtering sensor data. The AGPS (assisted global positioning system) present in mobile phones has an average accuracy of three meters. This error margin is proven to be impactful, especially in the case where we overlay the reconstructed parts of a monument onto the real-world. The illusion of digital and real coexistence completely breaks, even if the 3D model is registered slightly away from the real building. The orientation calculation is based on the accelerometer and the magnetic field sensors and a combination of these two with a gyroscope, if present. While the magnetic field sensor provides absolute measurements and needs no filtering, the accelerometer measures all forces that act on the device, which means that unwanted motions and mechanical noise need to be filtered out in order to isolate the force of gravity. This filtering process unavoidably introduces latency to the system following relative motions of the camera. For example, when a user moves the device to look higher at an overlaid building, the 3D model will be dragged along with the motion until it is significant enough to pass through the filter.

In order for the 3D models of the monuments to be accurately displayed in combination with real-time viewing of the real world, an initial transformation and rotation needs to be applied. The 3D models were exported in .dae format and imported into Google Sketch-up. The final models are then processed through Google Sketch-up to geo-reference and position them onto the real world while looking at the screen of a mobile phone (Figure 8).

The area of the monuments provided by Google maps was projected onto a ground plane. We then positioned the monument on its counterpart on the map. Given that the proportions of the monuments are in line, the final model was scaled to fit on the outlines. The location of the monument was then added to the file and provided to the framework.

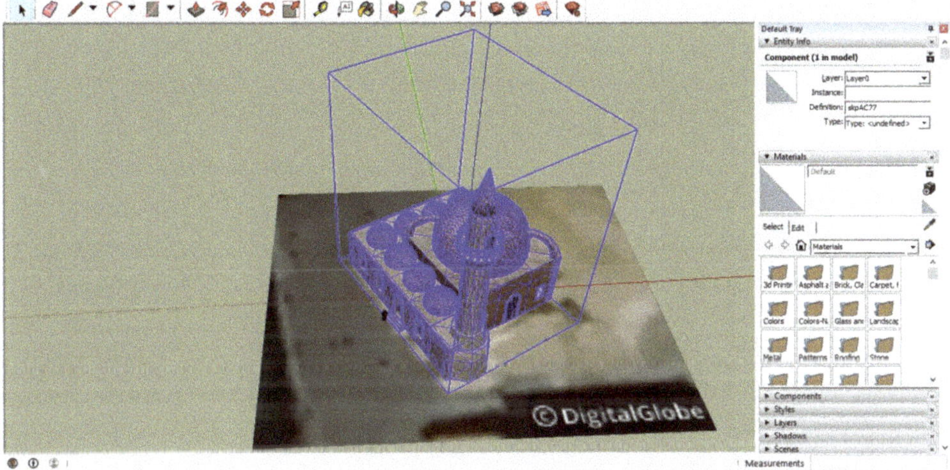

Figure 8. Geo-locating the model in Google Sketch-up.

In order to include the model in the AR framework, we exported it and used the provided 3D encoder to make a packaged version of the file together with the textures in the custom wt3 format. This file was then channeled to the MAR application.

4. System Architecture

4.1. Client-Server Implementation

One of the main challenges of MAR is the lack of established guidelines in the application and integration of AR technologies in outdoor heritage sites. The overall design of our system consists of two main parts, that is, the mobile client and the server. The server facilitates a database developed in MySQL, which holds the monument records (name, description, latitude, longitude, etc.) and user-specific information. While polygon geometry would be more accurate in defining and differentiating the areas of interest, circles centered on latitude and longitude points make calculations based on location far easier. Should the current model be expanded, polygon geometry would be a major consideration. The information is delivered to the mobile unit on a location request basis. The database is exposed to the users via a REST-full web service. A basic registration to the system is required. The functions provided to the users include storing data about visits, marked places, and the overall progress in an area, while all the monument information is delivered to the mobile device's views. The system architecture is shown in Figure 9.

4.2. Mobile Application

The mobile application's architecture has been designed to be extendable and to respond easily to changes in the underlying model of the server. It is based on three main layers (Figure 9). The *Views layer* is where the interactions with the users take place. Together with the background location service, they act as the main input points to the system. The events that take place are forwarded to the *Handling layer*, which consists of two modules following the singleton pattern. The data handler is responsible for interaction with the local content and communicating with the views, while the REST client is responsible for communicating with the web service. The *Model layer* consists of basic helper modules to parse the obtained JSON (JavaScript object notation) files and interact with the local database. The actions flow from the *Views layer* to the lower components. Responding to a user event or a location update, a call is made to the handling layer, which will access the model to return the requested data.

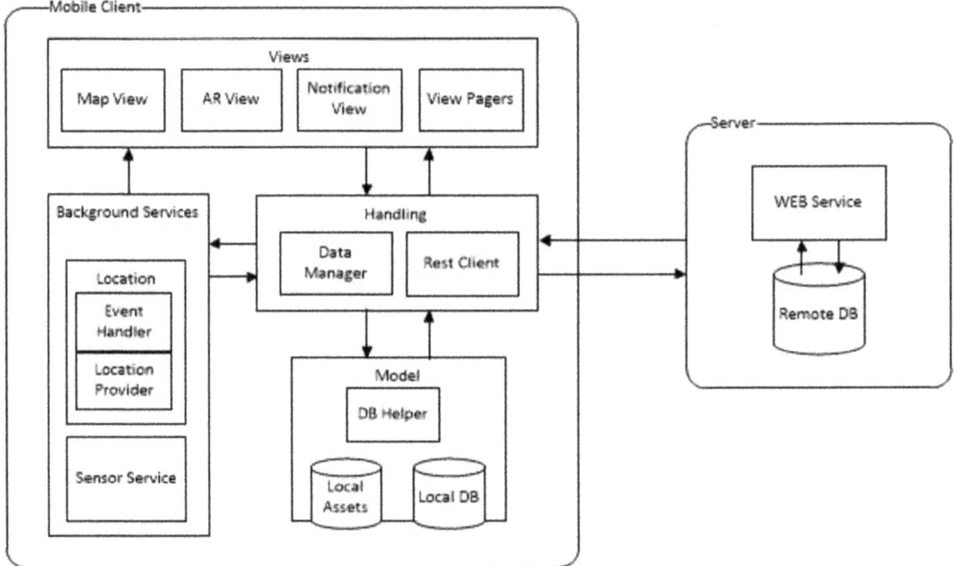

Figure 9. Architecture of the system. AR—Augmented Reality; DB—database.

The *Handling layer* is the most important of the three because interactions, exchange of information, and synchronization pass through this layer. The REST client provides an interface for receiving and sending information to the remote database as requested by the other layers. It is responsible for sending data about user visits, saved places, updates in progress, and personal information. It also provides functions for receiving data from the server concerning POIs' information and images. It also allows for synchronizing and queuing the requests. While the REST client is responsible for interactions with the server, the data handler is responsible for managing the communication of the local content. Information received is parsed and stored in the local database. The handler responds to events from the views and background service and handles the business logic for the other components. It forms and serves the available information based on the state of the application.

The views are basic user interface components facilitating the possible interactions with the users. The *Map View* is a fragment containing a 2D map developed with the Google Maps API. It displays the user's location as obtained by the background service and the POIs as markers on the map. The *AR View* is based on the Wikitude JavaScript API and this is where the AR experiences take place. It is a *Web View* with a transparent background overlaid on top of a camera surface. It displays the 3D models of the historical monuments while receiving location updates from the background service and orientation updates from the underlying sensor implementation. It also contains a *Navigation View*, where the POIs are displayed as labels on the real world. Interactivity is handled in JavaScript and is independent of the native code. The view pagers are framework specific user interface elements that display lists of the POIs, details for each POI, user leaderboards, and user profiles. Finally, the *Notification View* is used when the application is in the background and aims to provide control over the location service. It is a permanent notification on the system tray, where the user can change preferences of the location strategy and start, stop, or pause the service at will.

The aim of the standalone background service is to allow users to walk freely in the city while receiving notifications about nearby POIs. It is responsible for supplying the locations obtained by the GPS to the *Map View* and *AR View*. The *Location Provider* is the component responsible for obtaining the locations. It offers the option to swap between the Google Play Services API and the Android Location API, which are two different location strategies. In order to offer control over battery life and data usage, the users can customize its frequency settings from the preferences' menu. The views requesting location updates are registered as *Listeners* to the *Location Service* and receive locations containing latitude, longitude, altitude, accuracy information, and so on. The *Location Event Controller* is the component responsible for serving the location events to the registered views. The user's location is continuously compared with that of the available POIs. If the corresponding distance is in an acceptable range, the user is able to interact with the POI. The events sent include entering and leaving the active area of a POI. If the application is in the background, a notification is issued leading to the *AR View* and *Map View*.

The *Model layer* consists of standard storing units and handlers to enable parsing JSON files obtained from the server and interfaces to interact with the local database. It stores historical information about the current local, user-specific information and additional variables needed to ensure the optimal flow of the application. The local assets, including the 3D models and the html and JavaScript files required by the Wikitude API, are stored in this layer and provided to the handling layer as requested. The SQLite helper is the component responsible for updating the local storage and offers an interface to the data handler containing available interactions.

4.3. Server Implementation

The aim of the application server is to provide an online storage unit of historical and user-specific information. It employs a REST-full web service providing information to the mobile clients. The server holds a database of historical information as well as user information for authentication and accessibility control. The mobile units request the information and store it in their local database. The data can be queried to provide personalized information to the users and useful statistics about the city. The Web API exposes its resources via unique custom defined URLs held by the mobile client. Each key entity in the database schema is mapped to a relative path from the base URL of the server and for each entity identified with the URL, the client uses a different type of HTTP request methods (GET, PUT, POST, and DELETE) (Figure 10).

By accessing these URLs and defining the HTTP method, the clients can perform CRUD (create, read, update, and delete) operations on the underlying data. Information is transferred in JSON. Below, we outline the key entities of the database schema (Figure 11) and an explanation of their relationships:

a.   Player Table

The system supports the registration of new users. Minimal requirements for this action are the email, username, and password, as well as certain demographic information. The *visits* and *places* tables are used to record the interactions of the user with the *scenes*. Each user has a collection of monuments visited or saved. The *player_plays_in_levels* table holds the score of each player for each level and is used to provide level specific leaderboards.

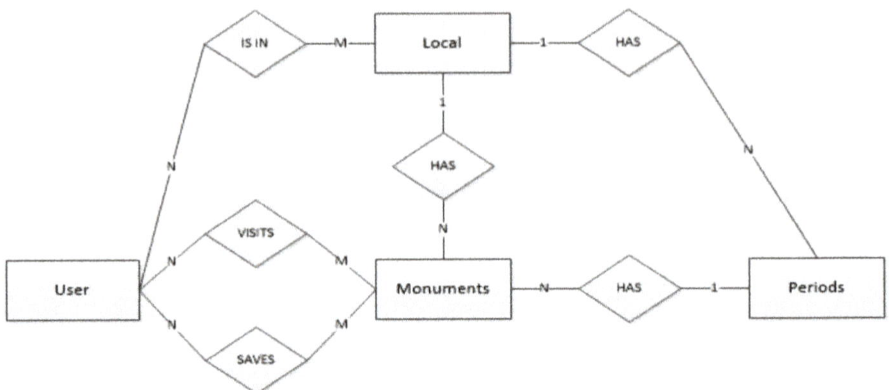

Figure 10. Entity Relationship diagram of the database.

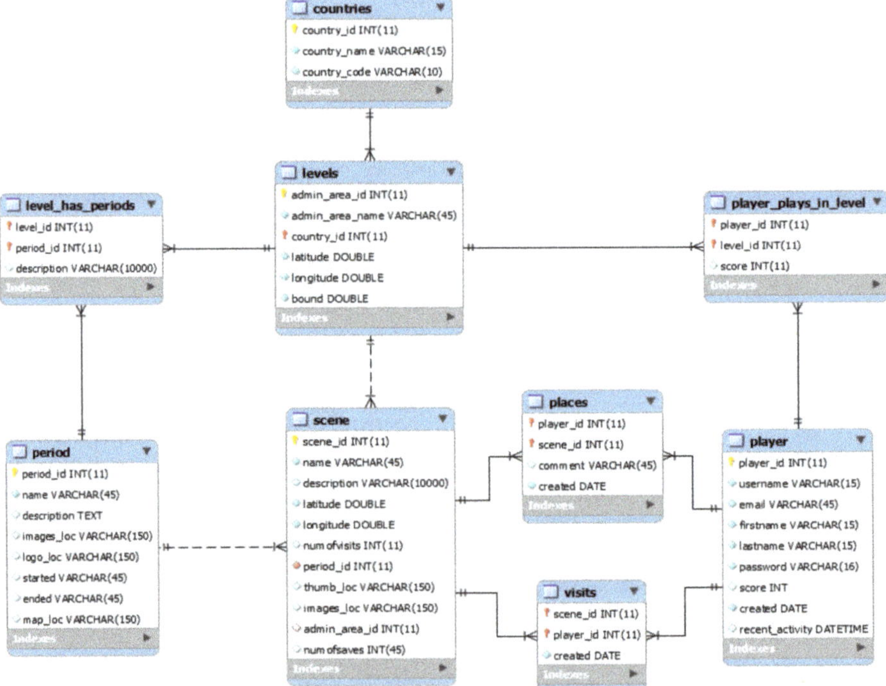

Figure 11. Database schema.

b. Scene Table

The *scene* table holds the records concerning the monuments. Each monument is uniquely identified with an auto-incremented ID. For each monument, the system records its name, description, latitude, and longitude values, as well as relative paths to its images. Similar to the players table, the *visits* and *places* entities keep track of additional statistics about each monument. Each monument can be a member of only one level (explained below) and relates to only one historical period.

c.   Period Table

The *period* table holds the information concerning the historical periods used to classify the monuments. Each period has a name, description, paths to its images, and ended and started dates. The period table is used to classify the monuments and provide additional historical information about the *levels*.

d.   Levels Table

The *levels* table is used to identify the playable areas that the system supports. Each level has a location described in latitude and longitude values, as well as a radius (*'bound'*) that sets its boundaries. Each monument in the database corresponds to a playable area and each area can relate to a number of periods defined by the *level_has_periods* table. In the current state, the database only holds information about the city of Chania, Greece, but it can be easily expanded to additional areas.

4.4. Augmented Reality Activity

The AR activity was implemented based on the Wikitude JavaScript API. The Wikitude SDK is based on web technologies (HTML, JavaScript, CSS). In order to integrate the web view with Android, the SDK provides a specific view component called *ARchitectView*, which we add to the activity's layout. Typical HTML pages are then loaded, located in the assets folder utilizing the API to create objects in AR. The first step is integrating the API with the activity's lifecycle. During the *onCreate()* call, the *ARchitectView* object is initialized and an interface is created that communicates with the pages. Information is transferred in JSON, parsing relevant arguments sent or received by the AR experiences.

The AR activity is started from the map activity with an *intent*. That intent contains a key-value pair specifying the AR experience (HTMLpage) selected to be loaded. Three AR experiences are designed, that is, the *ARNavigation*, *3DModelAtGeoLocation*, and *InstantTracking*. Each one is a separate HTML page loaded at the point of creation of the *ARActivity*. After initialization, the activity binds to the *LocationService* to receive location updates. The acquired locations are sent to the *ArchitectView* to draw the AR objects on the screen. A single native AR activity runs each AR page and is responsible for initializing each one in a different manner depending on the intent passed on by the *MapsActivity*. Creating each AR experience follows the standard web development process. The user interface (UI) includes a 3D scene containing the AR objects and standard 2D elements designed with JQuerryin order to provide additional control over the content.

The AR navigation page displays camera screen scenes at their geo-locations. After the page is loaded, a JSON array is provided including scene information derived from the native code. The array is then parsed and for each scene, a marker object is created to be displayed on the screen. The marker object provides its own logic in order to animate its changes between selected and deselected states using *AR.PropertyAnimations*.

5. Interface and User Experience

The main goal is to create an application that provides visitors of a historical city a MAR sightseeing experience including easily accessible historical information. On-site 3D depictions of monuments are provided in a MAR context, as well as information such as text and images. A complete location-aware experience promoting the exploration of a city's cultural wealth is put forward, addressing key technological challenges related to registration and tracking of AR technologies. In this section, we present the final screens of the application and explain in detail the flow of the experience.

Upon the activation of the application, the user is welcomed by a splash screen and is requested to create an account or login with an existing one. After the login process, the application checks the user's location and requests the POIs and the historical information relevant to the area from the server while it transfers to the main screen. The map activity is the main screen of the application and facilitates the core of the functionality, as shown in Figure 12. POIs are displayed on the map in their corresponding geo-locations. By clicking on a marker, the user views the info window of the POI containing a thumbnail, and the distance between POIs. By interacting with the bar at the bottom of the screen, the user can navigate to the remaining pages of the application. These include the profile, leaderboards, library, and preferences, while the middle round button is used to swap between map and camera navigation.

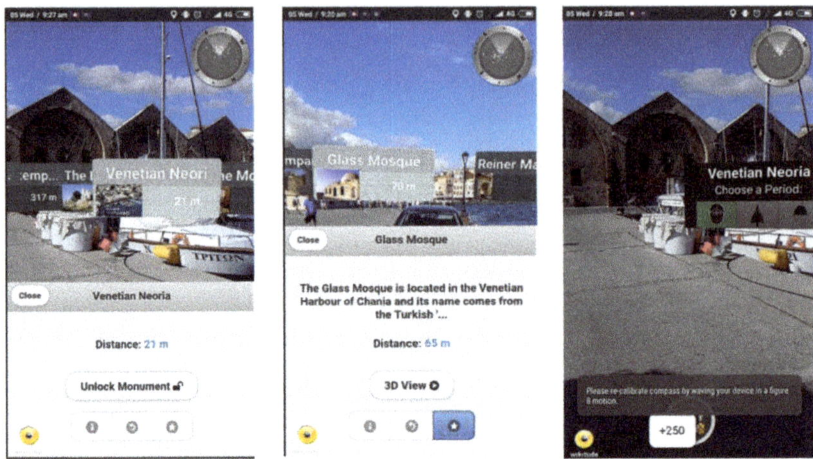

Figure 12. Navigation and classification in the camera view.

In the camera view, the content is displayed as 2D labels that contain basic information concerning the POIs. The POIs are displayed as dots on the radar to assist navigation. Users can save the POIs for later reference, access the reconstruction if available, or focus the marker and return to the map. The main goal of the experience is to reveal the available information with easy to control navigation conditions in order to facilitate efficient interaction. The UI displaying map activity is shown in detail in Figure 12.

The POI information is stored in the monuments table (Figure 13). It contains information concerning latitude, longitude, name, brief description, description, image locations, periods, and locality. The monuments are delivered on a local basis and are classified in periods to help manage the content. The locality is used to cluster the monuments in playable areas for the users. Their structure complies with ISO 3166-1 and combines countries and admin areas to ensure uniqueness. The server receives a location request by the client and then queries the database to find the POI information. The users can see the monument's specific information by selecting the items on the list, as shown in Figure 13. In this activity, the user can acquire historical information, including text and images; mark and save monuments; or get directions to specific locations.

In order to facilitate the client–server communication, a REST-full web service using is developed [38]. If such information exists, it is sent back to the server. The API exposes its resources via unique custom-defined URLs held by the mobile client. Each entity is defined as a resource and has its own URL. The service maps each resource with an implementation for varied HTTP request methods (GET, PUT, POST, and DELETE). By accessing these URLs and defining the HTTP method, the clients can then perform all CRUD (create, read, update, and delete) operations on the data. The NM relationships in the diagram are defined as sub-resources exposed by their parents links. Visits and places, for example, can be accessed from the users and monuments resources. For example, a client can make a get request to personalized visits or a get request accessing the overall amount of visits of a monument. Each response from the main resources contains links to sub-resources providing the client with the necessary information for communicating and enabling the server to evolve independently. Information is transferred in JSON.

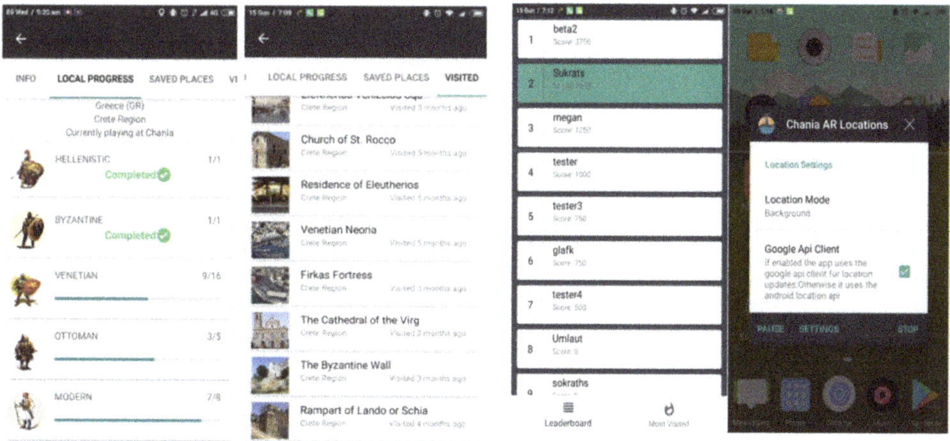

Figure 13. Local standings (**left**). Background service controls (**right**).

During the initial state of the application, the user is shown the available 3D models representing the historical monuments on the map. After visiting the monuments and viewing them in AR, the rest of the POIs' locations are unlocked and displayed on the map as question marks, shown in Figure 14. The goal is to visit them and classify them to the historical periods based on their architectural characteristics and clues obtained in the library page, as well as from the already visited monuments. The exploration of the POIs can be conducted freely by employing the background service of the application. The users can enable or disable this functionality from the settings. The aim of this approach is to urge the visitor to closely observe the monuments, revisit and consult the information available, and even interact with each other and locals to make the decision. The more areas they visit and unlock, the more points they earn for themselves. The overall progress in the city can be viewed in the leaderboards page, accessible from the map view.

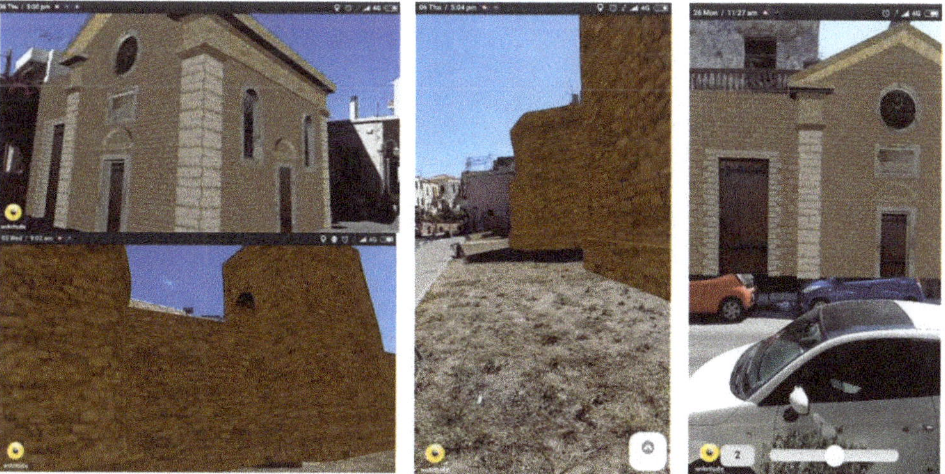

Figure 14. 3D depictions of the demolished towers of the Byzantine wall and the facial restoration of the Rocco temple.

The 3D models of the historical monuments are the main feature of the MAR activity, as shown in Figures 14 and 15, situating 3D information in the context of the respective real-world surroundings. The 3D models are accessed from the navigation views. When the user is in close proximity to the monuments, the location event handler informs the views to update the content and enable the AR experiences. In this screen, a reconstructed 3D model of the monument is overlaid on the camera and the GPS and inertial sensors are exploited to display the monument on its real-world location. The users can freely move around the real-world site to view the monuments from all available angles. They can click on the model to get information or access the slider, available at the bottom right, to change the representation. The user can shift between visualizing the full-scale 3D model or specific reconstructed parts. It is possible (but not in this case) to download multiple reconstructions relevant to one building, potentially with each one representing more than one historical period, signifying the monument's evolution over time.

Figure 15. 3D depiction of the Glass Mosque featuring the now demolished minaret, as seen by the mobile's camera [36].

Figure 16. Map activity user interface.

The library page is where a collection of historical information is displayed. It consists of a view pager containing historical periods in chronological order. Each page has a historical briefing consisting of an image showing the active area for that period and a list including the monuments that have been correctly classified. The locked monuments are contained in a separate list at the end of the page.

The profile and leaderboard page follow the same architecture. In the profile page, the user can swap through pages containing personalized information, as well as observe local progress and lists of saved and visited places. By clicking on the progress plates, the users can be transported to the library for the selected period. By selecting a monument, users are transferred to specified detail pages unique to each monument. The leaderboard page can be used to get information about the city. The user's position is highlighted on a list showing the current standings for the city. User progress can be compared with that of other visitors (Figure 16).

6. Evaluation

6.1. Technical Evaluation

Field tests were conducted by employing two devices, that is, a Samsung Galaxy S3 Neo (Android 4.4, RAM 1.4 GB, 1.2 GHz quad-core, Wi-Fi, GPS, geo-magnetic sensor, accelerometer, gyroscope) and a Xiaomi Redmi note 4xAndroid 6.0, RAM 4GB, 2GHz octa-core, GPS, Wi-Fi, geo-magnetic sensor, accelerometer, gyroscope). A large difference between the two devices concerning the accuracy of their GPS receivers was observed. The S3 Neo presented an average accuracy of eight meters, while the Redmi Note presented the expected accuracy of three meters. When placed at wider areas, both devices performed better, 3m for S3 Neo and 1.4m for Redmi Note. Although geo-location accuracy represents the confidence level of the receiver and is not precise, actual measurements remained consistent with the acquired values. This indicates that depending on the deployed device, the accuracy of the resulting AR registration may vary independent of the underlying implementation. In relation to orientation tracking, both devices performed as expected. Asboth devices are equipped with the same sensors, they performed similarly during the instant tracking testing.

A well-known issue concerning the instant tracking method is the inconsistency between our calculation of the device's orientation and the respective orientation calculated by the API. This measurement has been observed to vary by up to five degrees when the geo-magnetic sensor reports low accuracy values. This error is minimized by recalibrating the sensor while moving the device in an 'eight' figure motion. The user is informed when the sensor accuracy is low.

Another well-known AR issue occurs when the users leave the AR application's screen during tracking; for instance, when the phone rings. The users are required to return to the AR camera view and reinitiate tracking.

6.2. User Evaluation

In order to evaluate the resulting application, continuous informal usability tests were conducted throughout the development of the system, adopting the 'think aloud' method, while users experienced early versions of the app, on-site and in the lab. Quantitative evaluation metrics were not employed because of the complexity of the MAR experience in an outdoor, warm, busy environment. Detailed user comments were recorded.

The more persistent user comments discussed the accuracy of registration of the AR camera view during navigation. Most users found that the use of the camera instead of the map limited their movements and perception of their surroundings and refrained from using it besides when locating specific sites and classifying the monuments. During the classification process, the AR camera was useful as it helped locate the specified monument.

The AR camera in conjunction with the constant usage of the GPS for extended time periods resulted in rapid battery consumption. Addressing such issues, the AR camera view was defined as a standalone activity instead of a replacement to the map in the main activity.

User comments drove the development of the gallery as well as the 'save' option placed on the monument details page. The implemented functionality allowed users to save monuments from either the list in the collections screen or from the info window when pressing on the marker at the map screen. The info window functionality was mostly missed and the users found it useful to be able to save a monument to their visited places when browsing through its details. Therefore, this was transferred to the details page, while the list functionality remained as is. In relation to the gallery option, the initial design contained a link, which when pressed, the users transferred to a full screen view of the images. The users suggested that a gallery region should be included that shows the images in the details page and when an image is pressed, the application will then proceed to the full screen view of the images. The implemented region follows a carousel fashion where the users can swipe right and left in order to browse through the images. When an image is pressed, it is brought forward in full screen.

In relation to users' views regarding the AR experiences, users' reactions were quite positive. Most users had never been acquainted with a similar capability through their smart phones and were excited to view specific 3D reconstructions superimposed on real-world surroundings, visualized on their mobile phone's screen. Although the registration difficulty of accurately aligning 3D content was commented on by most users, the geo-location approach proved easy-to-use and intuitive. The instant tracking method proved challenging for an unaccustomed audience. However, after initial explanation and minimal guidance, users proceeded to experiment with 3D models placed on the annotated area, overlaid over the real-world building they represented, in its past state.

7. Conclusions

In this paper, we presented the design of a MAR application for consumer-grade mobile phones with the ultimate goal of increasing the synergy between visitors and cultural heritage sites. In addition to exploring standard web and application content, the proposed platform offers a novel approach for visualizing historical information on-site. By employing 3D models of the historical monuments through AR, user experience is enhanced, bridging the gap between digital content and the real-world environment. An expandable platform was put forward that can easily envelop additional historical sites requiring little preparation. The proposed system will enable future experts to display their digitized collections using varied forms of data presentation.

Although outdoor mobile augmented reality presents several localization and registration challenges, it provides novel experiences to a wide audience. The availability and technological advances of modern smartphones allow for the development of reliable AR experiences that enhance the understanding of historical datasets.

AR is a field that has just reached the wider public. Mobile AR is based on the limited processing power and tracking characteristics of today's technologies, making robust AR hard to achieve. However, AR is rapidly evolving with the addition of sophisticated tracking and registration algorithms, as well as specialized hardware.

Tracking and registration in AR are far from solved. Plugins that allow the development of AR applications based on game engines such as Unity3D and the Unreal engine, as well as new platforms for developers and novel displays, would allow a new generation of sophisticated interactions and high fidelity graphics, and would greatly improve user experience.

User-friendly authoring tools in order to create and publish content would contribute to AR's wide adoption. Experts could create new AR scenes by uploading media assets and placing them on a map, while users would be able to browse through the geo-located information and even create and modify their own content. By reducing the size of the assets, AR experiences could then be available over the Internet.

In its current state, the application presented in this paper does not support social interactions between the users, apart from the posted overall rankings. Extending the platform to include comments, likes, shares, and so on by adding a communication layer would increase users' interest in heritage sites and enhance motivation to engage with cultural applications.

Gamification around heritage sites and applications is gaining increasing value as it engages visitors and allows for new means of interacting with cultural heritage information. AR and VR are tools that the cultural heritage community could exploit so that interaction with historical datasets can be complemented by new forms of visualization and interaction. Combined with scavenging and treasure hunts, location-aware storytelling could enhance exciting immersive experiences related to cultural heritage awareness and increase visitor involvement and engagement.

Author Contributions: C.P. is responsible for data analysis, carrying out the on-site development, and implementation of the software. L.R. verified the geo-location methods. K.M. supervised the technical development of this work and provided the theoretical formalism. D.D. planned the historical routes and checked the historical validity of the imagery results. All authors discussed the results and contributed to the final manuscript.

Funding: This research received no external funding.

Acknowledgments: In this part, we would like to thank Wikitude for providing us with an educational license for their Augmented Reality SDK.

Conflicts of Interest: The authors declare no conflict of interest.

References

1. Milgram, P.; Kishino, F. A taxonomy of mixed reality visual displays. *IEICE Trans. Inf. Syst.* **1994**, *E77-D*, 1321–1329.
2. Milgram, P.; Takemura, H.; Utsumi, A.; Kishino, F. Augmented reality: A class of displays on thereality-virtuality continuum. In Proceedings of the Telemanipulator and Telepresence Technologies, Boston, MA, USA, 31 October–1 November 1994; Volume 2351, pp. 282–292.
3. Mekni, M.; Lemieux, A. Augmented reality: Applications, challenges and future trends. In Proceedings of the 13th International Conference on Applied Computational Science of Recent Advances in Computer Engineering Series, Kuala Lampur, Malaysia, 23–25 April 2014; Volume 20, pp. 205–214.
4. Kersten, T.P.; Tschirschwitz, F.; Deggim, S. Development of a virtual museum including a 4d presentation of building history in virtual reality. In Proceedings of the International Archives of the Photogrammetry Remote Sensing and Spatial Information Sciences, 2017 3D Virtual Reconstruction and Visualization of Complex Architectures, Nafplio, Greece, 1–3 March 2017; Volume XLII-2/W3, pp. 361–367.
5. Kersten, T.P.; Tschirschwitz, F.; Lindstaedt, M.; Deggim, S. The historic wooden model of Solomon's Temple: 3D recording, modelling and immersive virtual reality visualisation. *J. Cult. Herit. Manag. Sustain. Dev.* **2018**. [CrossRef]
6. Azuma, R.T. A survey of augmented reality. *Presence* **1997**, *6*, 355–385. [CrossRef]
7. Amin, D.; Govilkar, S. Comparative Study of Augmented reality SDKs. *Int. J. Comput. Sci. Appl.* **2015**, *5*, 11–26.
8. Mosque YialiTzami. Available online: http://www.explorecrete.com/chania/EN-chania-07-mosque.html (accessed on 10 July 2018).
9. Church of Saint Rocco. Available online: http://www.chaniahistory.gr/index.php?option=com_k2&view=item&id=208:church-of-st-rocco&Itemid=1474&lang=en (accessed on 10 July 2018).
10. Byzantine Wall of Chania. Available online: https://www.kastra.eu/castleen.php?kastro=kydonia (accessed on 10 July 2018).
11. Hildebrand, A.; Daehne, P.; Christou, I.T.; Demiris, A.; Diorinos, M.; Ioannidis, N.; Almeida, L.; Weidenhausen, J. Archeoguide: An Augmented Reality based System for Personalized Tours in Cultural Heritage Sites. In Proceedings of the International Symposium of Augmented Reality, Munich, Germany, 5–6 October 2000.
12. Stricker, D.; Karigiannis, J.; Christou, I.T.; Gleue, T.; Ioannidis, N. Augmented reality for visitors of cultural heritage sites. In Proceedings of the International Conference on Artistic, Cultural and scientific Aspects of Experimental Media spaces, Bonn, Germany, 21–22 September 2001.

13. Vlahakis, V.; Ioannidis, N.; Karigiannis, J.; Tsotros, M.; Gounaris, M.; Stricker, D.; Gleue, T.; Daehne, P.; Almaida, L. Archeoguide: An augmented reality guide for archaeological sites. *IEEE Comput. Gr. Appl.* **2002**, *22*, 52–59. [CrossRef]

14. Vlahakis, V.; Karigiannis, J.; Tsotros, M.; Gounaris, M.; Almeida, L.; Stricker, D.; Gleue, T.; Christou, I.T.; Carlucci, R.; Ioannidis, N. Archeoguide: First results of an augmented reality, mobile computing system in cultural heritage sites. In Proceedings of the 2001 VAST '01Conference on Virtual Reality, Archeology, and Cultural Heritage, Athens, Greece, 28–30 November 2001; ACM: New York, NY, USA, 2001; pp. 131–140.

15. Haugstvedt, A.C.; Krogstie, J. Mobile Augmented Reality for Cultural Heritage: A Technology Acceptance Study. In Proceedings of the IEEE International Symposium on Mixed and Augmented Reality, Science and Technology, Atlanda, Georgia, 5–8 November 2012; pp. 247–255.

16. Noh, Z.; Sunar, M.S.; Zhigeng, P. A Review on Augmented Reality for Virtual Heritage System. In Proceedings of the Edutainment '09 4th International Conference on E-Learning and Games: Learning by Playing, LNCS 5670, Banff, AB, Canada, 9–11 August 2009; Springer: Berlin/Heidelberg, Germany, 2009; pp. 50–61.

17. Zollner, M.; Keil, J.; Wüst, H.; Pletinckx, D. An augmented reality presentation system for remote cultural heritage site. In Proceedings of the 10th International Symposium on Virtual Reality, Archaeology and Cultural Heritage VAST 2009, Julians, Malta, 22–25 September 2009; Debattista, K., Perlingieri, C., Pitzalis, D., Spina, S., Eds.; University of Malta: Msida, Malta, 2009.

18. Sylaiou, S.; Mania, K.; Karoulis, A.; White, M. Exploring the Relationship between Presence and Enjoyment in a Virtual Museum. *Hum. Comput. Stud.* **2010**, *68*, 243–253. [CrossRef]

19. Wojciechowski, R.; Walczak, K.; White, M.; Cellary, W. Building virtual and augmented reality museum exhibitions. In Proceedings of the 9th International Conference on 3D Web Technology, Monterey, CA, USA, 5–8 April 2004; pp. 135–144.

20. Niedmermair, S.; Ferschin, P. An Augmented Reality Framework for On-Site Visualization of Archaeological Data. In Proceedings of the 16th International Conference on Cultural Heritage and New Technologies, Vienna, Austria, 12–15 November 2011; pp. 636–647.

21. Choudary, O.; Charvillat, V.; Grigoras, R.; Gurdjos, P. MARCH: Mobile augmented reality for cultural heritage. In Proceedings of the 17th MM '09ACM International Conference on Multimedia, Beijing, China, 19–24 October 2009.

22. Papagiannakis, G.; Magnenat-Thalmann, N. Virtual Worlds and Augmented Reality in Cultural Heritage Applications. In *Recording Modeling and Visualization of Cultural Heritage, Proceedings of the International Workshop, Centro Stefano Franscini, Monte Verita, Ascona, Switzerland, 22–27 May 2005*; Baltsavias, M., Gruen, A., van Gool, L., Pateraki, M., Eds.; Taylor and Francis Group: Boca Raton, FL, USA, 2006; pp. 419–430.

23. Papagiannakis, G.; Magnenat-Thalmann, N. Mobile Augmented Heritage: Enabling Human Life in Ancient Pompeii. *Int. J. Arch. Comput.* **2007**, *5*, 395–415. [CrossRef]

24. Pacheco, D.; Wierenga, S.; Omedas, P.; Wilbricht, S.; Knoch, H.; Verschure, P.F.M.J. Spatializing experience: A framework for the geolocalization, visualization and exploration of historical data using VR/AR technologies. In Proceedings of the Virtual Reality International Conference, Laval, France, 9–11 April 2014.

25. Julier, S.J.; Schieck, A.F.; Blume, P.; Moutinho, A.; Koutsolampros, P.; Javornik, A.; Rovira, A.; Kostopoulou, E. VisAge: Augmented reality for heritage. In Proceedings of the 5th PerDis'16ACM International Symposium on Pervasive Displays, Oulu, Finland, 20–22 June 2016.

26. Lee, J.Y.; Lee, S.H.; Park, H.M.; Lee, S.K.; Choi, J.S.; Kwon, J.S. Design and implementation of a wearable AR annotation system using gaze interaction. In Proceedings of the Consumer Electronics (ICCE), Digest of Technical Papers, Las Vegas, NV, USA, 9–13 January 2010; pp. 185–186.

27. Hable, R.; Rößler, T.; Schuller, C. evoGuide: Implementation of a tour guide support solution with multimedia and augmented-reality content. In Proceedings of the 11th International Conference on Mobile and Ubiquitous Multimedia, Ulm, Germany, 4–6 December 2012.

28. Fritz, F.; Susperregui, A.; Linaza, M. Enhancing cultural tourism experiences with augmented reality technologies. In Proceedings of the 6th International Symposium on Virtual Reality, Archaeology and Cultural Heritage (VAST), Pisa, Italy, 8–11 November 2005.

29. Martinez, M.; Munoz, G. Designing augmented interfaces for guided tours using multimedia sketches. In *MIXER*; Citeseer: University Park, PA, USA, 2004.

30. Střelák, D.; Škola, F.; Liarokapis, F. Examining User Experiences in a Mobile Augmented Reality Tourist Guide. In Proceedings of the 9th PETRA '16ACM International Conference on PErvasive Technologies Related to Assistive Environments, Corfu Island, Greece, 29 June–1 July 2016.
31. Ragia, L.; Sarri, F.; Mania, K. Precise photorealistic visualization for restoration of historic buildings based on tacheometry data. *J. Geogr. Syst.* **2018**, *20*, 115–137. [CrossRef]
32. Azuma, R.T.; Baillot, Y.; Behringer, R.; Feiner, S.; Julier, S.; MacIntyre, B. Recent advances in augmented reality. *IEEE Comput. Gr. Appl.* **2001**, *21*, 34–47. [CrossRef]
33. Liarokapis, F.; Brujic-Okretic, V.; Papakonstantinou, S. Exploring Urban Environments using Virtual and Augmented Reality. *J. Virtual Real. Broadcast.* **2006**, *3*, 1–13.
34. Guidi, G.; Russo, M. Diachronic 3D reconstruction for lost Cultural Heritage. International Archives of the Photogrammetry. *Remote Sens. Spat. Inf. Sci.* **2011**, *38*, 371.
35. Rodríguez-Gonzálvez, P.; Muñoz-Nieto, A.L.; del Pozo, S.; Sanchez-Aparicio, L.J.; Gonzalez-Aguilera, D.; Micoli, L.; Haynes, I. 4D Reconstruction and visualization of Cultural Heritage: Analyzing our legacy through time. *Int. Arch. Photogramm. Remote Sens. Spat. Inf. Sci.* **2017**, *42*, 609. [CrossRef]
36. Panou, C.; Ragia, L.; Dimelli, D.; Mania, K. Outdoors Mobile Augmented Reality Application Visualizing 3D reconstructed Historical Monuments. In Proceedings of the 4th International Conference on Geographical Infrormation Systems Theory, Applications and Management, GISTAM, Madeira, Portugal, 17–19 March 2018; pp. 59–67.
37. Guidi, G.; GonizziBarsanti, S.; Micoli, L.L.; Malik, U.S. Accurate Reconstruction of the Roman Circus in Milan by Georeferencing Heterogeneous Data Sources with GIS. *Geosciences* **2017**, *7*, 91. [CrossRef]
38. Fielding, R.T. Architectural Styles and the Design of Network-Based Software Architectures. Ph.D. Thesis, University of California, Berkeley, CA, USA, 2000.

MDPI
St. Alban-Anlage 66
4052 Basel
Switzerland
Tel. +41 61 683 77 34
Fax +41 61 302 89 18
www.mdpi.com

ISPRS International Journal of Geo-Information Editorial Office
E-mail: ijgi@mdpi.com
www.mdpi.com/journal/ijgi